First Edition

Aerial Apparatus Driver/Operator Handbook

Michael A. Wieder, Writer
Carol Smith, Editor

Validated by the International Fire Service Training Association
Published by Fire Protection Publications, Oklahoma State University

RECYCLABLE

Cover photo courtesy of Capt. Chris Mickal, New Orleans Fire Department Photo Unit

The International Fire Service Training Association

The International Fire Service Training Association (IFSTA) was established in 1934 as a "nonprofit educational association of fire fighting personnel who are dedicated to upgrading fire fighting techniques and safety through training." To carry out the mission of IFSTA, Fire Protection Publications was established as an entity of Oklahoma State University. Fire Protection Publications' primary function is to publish and disseminate training texts as proposed and validated by IFSTA. As a secondary function, Fire Protection Publications researches, acquires, produces, and markets high-quality learning and teaching aids as consistent with IFSTA's mission.

The IFSTA Validation Conference is held the second full week in July. Committees of technical experts meet and work at the conference addressing the current standards of the National Fire Protection Association and other standard-making groups as applicable. The Validation Conference brings together individuals from several related and allied fields, such as:

- Key fire department executives and training officers
- Educators from colleges and universities
- Representatives from governmental agencies
- Delegates of firefighter associations and industrial organizations

Committee members are not paid nor are they reimbursed for their expenses by IFSTA or Fire Protection Publications. They participate because of commitment to the fire service and its future through training. Being on a committee is prestigious in the fire service community, and committee members are acknowledged leaders in their fields. This unique feature provides a close relationship between the International Fire Service Training Association and fire protection agencies which helps to correlate the efforts of all concerned.

IFSTA manuals are now the official teaching texts of most of the states and provinces of North America. Additionally, numerous U.S. and Canadian government agencies as well as other English-speaking countries have officially accepted the IFSTA manuals.

ISBN 0-87939-180-4 *Library of Congress LC# 00-103795*

First Edition, First Printing, July 2000 *Printed in the United States of America*
First Edition, Second Printing, April 2001

If you need additional information concerning the International Fire Service Training Association (IFSTA) or Fire Protection Publications, contact:

Customer Service, Fire Protection Publications, Oklahoma State University
930 North Willis, Stillwater, OK 74078-8045
800-654-4055 Fax: 405-744-8204

For assistance with training materials, to recommend material for inclusion in an IFSTA manual, or to ask questions or comment on manual content, contact:

Editorial Department, Fire Protection Publications, Oklahoma State University
930 North Willis, Stillwater, OK 74078-8045
405-707-3020 Fax: 405-707-0024 E-mail: editors@ifstafpp.okstate.edu

Table of Contents

Preface

This is the first edition of the **Aerial Apparatus Driver/Operator Handbook.** This text details the important responsibilities of firefighters who are assigned to drive and operate fire department vehicles that are equipped with an aerial device.

Acknowledgement and special thanks are extended to the IFSTA validating committee members who contributed their time, wisdom, and knowledge to this manual:

Committee Chair
Fire Chief Wayne Sandford
East Haven (CT) Fire Dept.

Fire Chief Bob Anderson
Spokane County (WA) Fire District #9

Firefighter Scott Carrigan
Exeter (NH) Fire Dept.

Boyd Cole
Libertyville, Illinois

Michael Robertson, Fire Instructor
United States Air Force Fire Protection
Career Field
San Angelo, Texas

Fire Marshal Tom Ruane
Peoria (AZ) Fire Department

Chief Training Officer R. Peter Sells
Toronto (ON) Fire Services

Lee Cooper, President
Fire Instructor's Association of Minnesota

Alan Joos, Certification Coordinator
Utah Fire and Rescue Academy

Assistant Chief Steve Kreis
Phoenix (AZ) Fire Department

Battalion Chief Bob Madden
Bend (OR) Fire & Rescue

Gregory Ranard, Fire Instructor
United States Air Force Fire Protection
Career Field
San Angelo, Texas

Lieutenant Michael Wilbur
Fire Department of the City of New York

Captain Michael Witten
Tulsa (OK) Fire Department

Special thanks are extended to Mr. Jim Birmingham and Mr. Mike Couture of the Connecticut State Fire Academy who also provided a thorough review of the rough draft of this manual.

Special appreciation is also extended to the following fire departments that provided their personnel, resources, and time to assist our staff in shooting many of the new photographs that were necessary to complete this manual:

PHOENIX, ARIZONA FIRE DEPARTMENT
Asst. Chief Steve Kries
Deputy Chief David Vera
FF Raymond Martinez
FF/PM Nolberto A. Gem (E705/B)
FF Susie Fawcett
Eng. Kim Kottmer (L4/B)
Eng. Rodney Imhoff (L11/B)
FF Tony Dominquez (Rover/B)
FF Greg Relf (R22/C)
FF Scott Crowley (R7/B)
Eng. Jerry Witt (L33/C)
Eng. Bill Sawyers (L33/A)
FF Mitch Finley (L33/A)

PEORIA, ARIZONA FIRE DEPARTMENT
Chief Mike Witt
Captain John Woodall (L193/B)
Eng. Stacy Irvine (L193/B)
FF Bill Van Gotum (L193/B)
FF Ken Wier (L193/B)
Captain John Davirro (L193/A)

TORONTO, ONTARIO, CANADA FIRE SERVICES
Chief Training Officer Peter Sells
Aerials 5, 8, 22, 26
Training Captain Robert Feeney
FF Shane Elliot
FF Darryl Gage
FF Russell Gurevsky
FF Todd Graves
FF Timothy Green
Nathaniel Green

GOODYEAR, ARIZONA FIRE DEPARTMENT
Chief Mark A. Gaillard
Eng. Robert Marshal

DUNCAN, OKLAHOMA FIRE DEPARTMENT
Chief Perry Brinegar
Asst. Chief John Holden
Major Bobby Biffle
Driver Mike Jennings
FF Jeff Bruehl
FF Rodney Martin
FF Donnie Wainscott

TULSA, OKLAHOMA FIRE DEPARTMENT
Chief Tom Baker
Visual Comm. Coordinator Frank Mason
The crew of Ladder 29-C

A manual of this scope would not be possible without the assistance of many other people and organizations who assisted us by providing pictures and information vital to the project's completion:

Underwriters Laboratories, Inc.
James E. Johannessen, Manager of Aerial Testing
Sun City, Arizona Fire Department
El Reno, Oklahoma Fire Department
Chief Ronnie Martin
Mustang, Oklahoma Fire Department
Chief Alvin McClung
Bob Esposito, Trucksville, Pennsylvania
Ron Jeffers, Union City, New Jersey
Warren Gleitsmann, Timonium, Maryland
Joel Woods, University of Maryland Fire & Rescue Institute, College Park, Maryland
Rocky Hill, Connecticut Fire Department
Pennsburg, Pennsylvania Fire Department
Oklahoma City, Oklahoma Fire Department
Conoco Oil Company, Ponca City, Oklahoma Refinery
Emergency One, Inc.
Harvey Eisner, Tenafly, New Jersey
Capt. Chris Mickal, New Orleans (LA) Fire Department Photo Unit
Tom McCarthy, Chicago (IL) Fire Department
Ed Prendergast, Chicago, Illinois
Index by Kari Kells, Index West

Gratitude is also extended to the following members of the Fire Protection Publications staff, whose contributions made the final publication of the manual possible:

Cynthia Brakhage, Associate Editor
Tara Gladden, Editorial Administrative Assistant
Don Davis, Coordinator, Publications Production
Ann Moffat, Graphic Design Analyst
Desa Porter, Senior Graphic Designer
Ben Brock, Senior Graphic Designer
Tara Carman, Student Technician
Bob Crowe, Research Technician
Lee Noll, Research Technician

Introduction

The need for carrying out fire fighting operations from an elevated position dates back to the construction of the first two-story structure. In these early days of fire fighting history, two basic strategies were available: use rickety, wooden ground ladders to accomplish as much as possible or allow the fire to burn down to a level at which it could be reached. In many cases, even if the first strategy was selected, the second soon became the reality. As time went on, the construction of ground ladders for fire fighting operations improved; however, this resulted in little overall improvement to fire fighting operations. Greater reach and capabilities were needed.

Progress was finally realized in 1870 when San Francisco Firefighter Daniel Hayes developed the first fire department aerial ladder, appropriately called the Hayes Aerial. Mounted on a horse-drawn cart, the manually raised, two-section, wooden ladder had a length of 85 feet (26 m). Though a few manufacturers dabbled with pneumatically raised ladders, the manually raised, wooden, 85-foot (26 m) aerial ladder remained the standard until the introduction of the spring-raised aerial around 1905. This would be the most common type of aerial ladder used for the next 30 years.

The first power-operated, 100-foot (30 m) steel ladder was introduced by the Peter Pirsch Company in 1936. That unit was truly the beginning of the modern aerial ladder. In the years since, advances in aerial ladder technology have resulted in ladders as tall as 144 feet (44 m), and ladders constructed of various metals including steel alloys and aluminum.

In comparison to the aerial ladder, the elevating platform aerial device is a relative newcomer to the fire service. The first applications of this type of aerial device can be traced back only as far as 1950. At that time, a demonstration using an electric utility's "cherry picker" was conducted at the Connecticut State Fire School. However, it was not until 1958 that the Chicago Fire Department placed into service the first elevating platform designed specifically for fire fighting operations. These articulating devices were built by the Pitman Manufacturing Company, which eventually became Snorkel Economy. The articulating elevating platforms quickly became popular with the fire service. It was not until the mid-1960s that the telescoping elevating platform was introduced. The early 1970s saw the introduction of a hybrid of the elevating platform and the aerial ladder — called the aerial ladder platform. This aerial device consists of a heavy-duty aerial ladder with a platform attached to the end of the fly section.

Aerial ladders, elevating platforms, and water towers are all widely used in today's fire service. Progressive technology has given us aerial devices with capabilities that are vastly improved over their predecessors. Piped airways, 1,000 pound (454 kg) platform load ratings, and piped waterway systems capable of flowing 3,000 gpm (12 000 L/min) are now commonplace.

One thing has not changed. The trucks still cannot put out fires by themselves. Competent, well-trained driver/operators, fire officers, and truck company members are essential for the optimum use of the aerial apparatus. The driver/operator must be thoroughly versed in the proper operation of the vehicle and the aerial device, in the proper positioning of the vehicle, and in the care and maintenance of the vehicle and its equipment. The fire officer and the truck company members must be prepared to carry out any of the common truck company functions, including search and rescue, ventilation, salvage, forcible entry, exposure protection, and elevated master stream operations.

Present-Day Requirements for Fire Apparatus

Certain performance functions are required of present-day aerial apparatus, such as the ability to readily deploy the aerial device when required, the ability to carry the large amount of equipment required by the truck company, and to safely transport the crew members to and from the emergency scene. These requirements have resulted in the development of various types of aerial apparatus to meet specific needs. Currently, suggested specifications for aerial apparatus are published in NFPA 1901, *Standard for Automotive Fire Apparatus.*

The NFPA committee memberships are chosen so that representation is given to the general public through consultants, insurance organizations, manufacturers, trade associations, and members of the fire service. These standards are subject to periodic review by the committees and to a complete revision every five years. All proposals for revisions must be approved, after general public comments, by the committees and the membership of NFPA at its semiannual convention.

The specifications in NFPA 1901 are such that a purchaser may take advantage of bids from a number of manufacturers. In this way, restrictive features that tend to limit competitive bidding can be avoided as much as possible. The suggested specifications are merely informative and are not mandatory unless a purchaser so states.

Municipal officials are urged to work toward a long-range plan for systematic replacement of fire apparatus. Such a plan should be based upon the ability of fire apparatus to meet the requirements of the jurisdiction it will be serving. Maintenance costs, parts replacement difficulties, and performance reliability should all be carefully considered before purchasing apparatus.

Third-party testing agencies, such as Underwriters Laboratories®, provide testing and certification of fire apparatus. These tests, although not mandatory, are universally accepted; apparatus specifications should require such testing. Specifications should include the required performance of the pump or aerial device, road acceleration and braking, piping flow requirements, and possibly maximum speed. Several detailed performance requirements are outlined in NFPA 1901. Tests for aerial devices are outlined in NFPA 1914, *Standard for Testing Fire Department Aerial Devices.* Upon delivery, tests should be conducted to determine if all specification requirements have been met. Specific tests for a given locale may need to be added. The vehicle should also be weighed to determine if it meets gross vehicle weight and weight distribution limitations as set forth in the specifications. Once in service, the aerial device should be tested yearly in accordance with the directions found in NFPA 1914.

Purpose and Scope

This first edition of the **Aerial Apparatus Driver/ Operator Handbook** is designed to educate the driver/operators responsible for operating fire apparatus equipped with aerial devices. These include aerial ladders, aerial ladder platforms, articulating elevating platforms, telescoping elevating platforms, and water towers. It includes the information necessary to meet the job performance requirements of NFPA 1002, *Standard for Fire Apparatus Driver/Operator Professional Qualifications,* Chapters 1, 2, 4, and 5. The operation of apparatus equipped with fire pumps is covered in IFSTA's **Pumping Apparatus Driver/Operator Handbook.**

The operation of aerial devices requires specialized training in a variety of areas. These include vehicle operation, vehicle stabilization, aerial device deployment and utilization, maintenance and testing, and specifications of new apparatus. The purpose of this manual is to present general principles of aerial device operation, with application of those principles wherever feasible. It is also intended to guide the driver/operator in the proper operation and care of the apparatus.

This manual covers an overview of the qualities and skills needed by the driver/operator, safe driving techniques, types of aerial apparatus, positioning of the aerial apparatus, vehicle stabilization, the operation of telescoping and articulating aerial devices, fire fighting tactics as they apply to the use of aerial devices, and apparatus maintenance and testing.

Notice on Gender Usage

In order to keep sentences uncluttered and easy to read, this text has been written using the masculine gender rather than both the masculine and female gender pronouns. Years ago it was traditional to use the masculine pronouns to refer to both sexes in a neutral manner. This usage is applied to this manual for the purpose of brevity and is not intended to address only one gender.

Chapter 1

The Driver/Operator

Job Performance Requirements

This chapter provides information that will assist the reader in meeting the following job performance requirements from NFPA 1002, *Standard on Fire Apparatus Driver/Operator Professional Qualifications*, 1998 edition. Particular portions of the job performance requirements (JPRs) that are met in this chapter are noted in bold text.

1-3.1 **The fire department vehicle driver/operators shall be licensed to drive all vehicles they are expected to operate.**

1-3.2 **The fire department driver/operator shall be subject to periodic medical evaluation, as required by NFPA 1500, *Standard on Fire Department Occupational Safety and Health Program,* Section 8-1, Medical Requirements, to determine that the driver/operator is medically fit to perform the duties of a fire department vehicle driver/operator.**

The fire apparatus driver/operator is responsible for safely transporting firefighters, apparatus, and equipment to and from the scene of an emergency or other call. Once on the scene, the driver/operator must be capable of operating the apparatus properly, swiftly, and safely. The driver/operator must also ensure that the apparatus and the equipment it carries are ready at all times.

In general, driver/operators must be mature, responsible, and safety conscious. Because of their wide array of responsibilities, often under emergency conditions, driver/operators must be able to maintain a calm, "can-do" attitude under pressure. Psychological profiles, drug and sobriety testing, and background investigations may be necessary to ensure that the driver/operator is ready to accept the high level of responsibility that comes with the job.

To perform their duties properly, all driver/operators must possess certain mental and physical skills. The required levels of these skills are usually determined by each jurisdiction. In addition, National Fire Protection Association (NFPA) 1002, *Standard for Fire Apparatus Driver/Operator Professional Qualifications (1998),* sets minimum qualifications for driver/operators. It requires any driver/operator who will be responsible for operating an aerial device to also meet the requirements of NFPA 1001, *Standard for Fire Fighter Professional Qualifications (1997),* for Fire Fighter I. The following sections discuss some of the basic mental and physical skills that may be required.

Skills and Physical Abilities Needed by the Driver/Operator

Anyone seeking the responsibility for operating emergency vehicles must possess a number of important cognitive and physical skills. Some of the more important skills are highlighted in the following sections.

Reading Skills

Driver/operators must be able to read. They will be constantly required to understand the written word. Some examples of duties that call for good reading comprehension are as follows:

- Reading maps and flowcharts
- Reviewing manufacturer's operating instructions
- Studying prefire plans
- Reviewing printed computer dispatch instructions
- Reading and working on a mobile data terminal (MDT) (Figure 1.1)

Writing Skills

The driver/operator must also be able to convey information completely and accurately in writing. Some examples of job functions that require writing skills are completing maintenance reports, equipment repair requests, and fire reports. Each driver/operator should be evaluated for the ability to write clearly and concisely.

Mathematical Skills

Basic mathematical skills are important to the driver/operator. Every day the driver/operator uses math in hydraulic calculations and numerous other situations. The driver/operator should be able to add, subtract, multiply, divide fractions and whole numbers, and determine square roots. The driver/operator should also be able to solve simple equations such as those used in friction loss problems. It is not the purpose of this manual to review basic mathematical skills. If a prospective candidate is deficient in these skills, attempts should be made to provide the assistance necessary to correct this situation. Often, a local educational institution will offer such programs. These programs will not only be of benefit to prospective candidates but also to current driver/operators who need to brush up on their math skills.

Physical Fitness

The driver/operator often must perform rigorous physical activities while setting up the aerial apparatus at a fire scene. These activities include tasks such as:

- Connecting a ladder pipe and supply hose to the aerial ladder (Figure 1.2)
- Deploying ground ladders
- Placing heavy stabilizer pads beneath the stabilizers
- Performing ventilation activities
- Operating power tools

The driver/operator must be prepared to perform the lifts, bends, and strenuous actions needed to complete these tasks. The driver/operator must be subjected to a periodic medical evaluation in accordance with NFPA 1500, *Standard on Fire Department Occupational Safety and Health Program (1997)*. NFPA 1582, *Standard on Medical Requirements for Fire Fighters(1997),* may be used as the criteria upon which the medical evaluation may be based.

Figure 1.1 Driver/operators must have good reading ability to work on a mobile data terminal.

Figure 1.2 Many of the tasks a driver/operator performs necessitate strength and flexibility.

Vision Requirements

The safe operation of an apparatus depends greatly upon the driver/operator's ability to see. NFPA 1582 requires that the firefighter have a corrected far visual acuity of 20/30 with contact lenses or spectacles. The standard contains further information on uncorrected vision and diseases of the eye. Consult the standard for specific details.

Hearing Requirements

Emergency vehicles en route to and on the emergency scene generate high levels of noise. Amid the noise of engines, sirens, air horns, and radio traffic, the driver/operator must be able to recognize different sounds and their importance. For example, he must be able to distinguish between the siren on his vehicle and that on another emergency vehicle.

The driver/operator must also be able to focus on particular sounds, such as radio instructions, for placing apparatus. Failure to hear such orders may result in placing the apparatus in a less effective or even unsafe position. NFPA 1582 recommends rejecting the firefighter candidate who has a hearing loss of 25 decibels or more in 3 of 4 frequencies (500-1000-2000-3000 Hz) in the unaided worst ear. It also recommends rejecting a candidate who has a loss greater than 30 decibels in any one of three frequencies (500-1000-2000 Hz) and an average loss greater than 30 decibels for four frequencies (500-1000-2000-3000 Hz). Consult NFPA 1582 for more specific details.

Other Skills

Several other skills, although not required, will help the driver/operator perform well. Mechanical ability aids in understanding the operation and maintenance of the apparatus. Because the driver/operator is often responsible for the apparatus while the officer is absent, basic supervisory skills and an understanding of fireground tactics will help in coordinating activities on the fireground.

Selection of Driver/Operators

In career fire departments, driver/operators are most often selected from the rank of firefighter. Selection is frequently based upon a required time of service with the department, written or performance tests, or a combination of service and tests. Whatever the method used, selection should be based upon skill and ability rather than simply upon seniority or position.

Volunteer fire departments may use the same criteria as the career departments for selecting a driver/operator. More commonly, chief officers select members who they feel are ready for the responsibility of fire apparatus operation and train them toward that goal. The candidate should have to pass some form of examination before being approved as a driver/operator on emergency calls. Some volunteer fire departments may also allow lateral entry into a driver/operator position. This is common for members who have truck driving experience. To meet the intent of NFPA 1002, these candidates will also have to attend and complete a Firefighter I course.

Every fire department, whether career, volunteer, or industrial, must have an established and thorough training program for prospective fire apparatus driver/operators. Simply letting firefighters drive the truck around the block a few times and showing them how to raise the aerial device is not adequate. Those with previous truck driving experience must focus on the significant differences between driving a commercial truck under routine conditions and driving a fire apparatus during an emergency response. They must learn the critical new skills necessary to operate an emergency vehicle before being put in an emergency response situation.

An effective training program consists of appropriate amounts of classroom (theoretical) instruction, practical training in the field (application), and testing to ensure that the person is ready for the responsibility in a real-world setting. For more direction on establishing a driver/operator training program, consult NFPA 1451, *Standard for a Fire Service Vehicle Operations Training Program (1997).*

Driving Regulations

Driver/operators of fire apparatus are regulated by state or provincial laws, city ordinances, and departmental standard operating procedures (SOPs). All driver/operators must be fully cognizant of all pertinent laws and SOPs. It is commonly known that ignorance of laws and ordinances is no defense if they are broken. More importantly, failure to know/follow department SOPs can have deadly consequences during emergency or even routine situations.

In general, a fire apparatus driver/operator is subject to all statutes, laws, and ordinances that govern any vehicle operator. Most laws and statutes concerning motor vehicle operation are maintained at the state or provincial level. Individual states or provinces may define what constitutes an emergency vehicle and exempt it from certain laws or statutes, such as following posted speed limits and parking requirements. Driver/operators must understand these exemptions and their parameters. For example, a particular state or province may allow emergency vehicles to proceed through a red light or stop sign. However, it may also specify that this can only be done when responding to an emergency, under safe road conditions, and after coming to a complete stop. Driver/operators exceeding these parameters are subject to penalty and/or liability.

Most driving regulations pertain to dry, clear roads. Driver/operators must adjust their speed to compensate for wet roads, darkness, fog, or any other condition that makes normal emergency vehicle operation more hazardous. Under all circumstances, the fire apparatus driver/operator must exercise care for the safety of others and must maintain complete control of the vehicle.

In most jurisdictions, emergency vehicles are not exempt from laws that require vehicles to stop for school buses when their flashing red signal lights indicate that children are being loaded or unloaded. Fire apparatus should proceed only after the bus driver or a police officer gives the proper signal. Even then, the driver/operator should proceed with extreme caution as the children may not be aware of the approaching fire apparatus and the bus driver may panic and give a premature signal to proceed. Driver/operators must obey all traffic signals and rules when returning to quarters from an alarm and should not have warning devices operating at this time.

Recent legal decisions have held that a driver/operator who does not obey state, local, or departmental driving regulations can be subject to criminal and civil prosecution if the apparatus is involved in a collision (Figure 1.3). If the driver/operator is negligent in the operation of an emergency vehicle and becomes involved in an accident, both the driver/operator and the fire department may be held criminally and/or civilly responsible.

Licensing Requirements

In the United States, the federal Department of Transportation (DOT) establishes the basic requirements for licensing of drivers. In Canada, Transport Canada (TC) has the same authority. Other nations have similar organizations. Both the DOT and TC establish special requirements for licensing drivers of trucks and other large vehicles. While these are national guidelines, each state or province has limited authority to alter them as it deems necessary for its jurisdiction. Some states and provinces require a fire apparatus driver/operator to obtain a commercial driver's license (CDL) for driving a large truck. Others have exempted fire service personnel from these licensing requirements. However, exempting fire apparatus driver/operators from CDL licensing requirements does not exempt them from the other laws pertaining to driving large trucks. This includes regulations relative to alcohol/drug consumption prior to operating the vehicle. Each fire department must be aware of the requirements within its jurisdiction and make sure that its driver/operators are licensed accordingly.

Figure 1.3 Driver/operators who operate the apparatus in a reckless manner may be subject to criminal or civil penalties. *Courtesy of the Tulsa, Oklahoma Fire Department.*

Chapter 2

Types and Construction of Aerial Fire Apparatus

Job Performance Requirements

This chapter provides information that will assist the reader in meeting the following job performance requirements from NFPA 1002, *Standard on Fire Apparatus Driver/Operator Professional Qualifications*, 1998 edition. Particular portions of the job performance requirements (JPRs) that are met in this chapter are noted in bold text.

2-3.7* Operate all fixed systems and equipment on the vehicle not specifically addressed elsewhere in this standard, given systems and equipment, manufacturer's specifications and instructions, and departmental policies and procedures for the systems and equipment, so that each system or piece of equipment is operated in accordance with the applicable instructions and policies.

(a) *Requisite Knowledge:* Manufacturer specifications and operating procedures, policies, and procedures of the jurisdiction.

(b) *Requisite Skills:* The ability to deploy, energize, and monitor the system or equipment and to recognize and correct system problems.

A fire department has a wide variety of aerial apparatus from which to choose when designing or selecting an apparatus to serve its jurisdiction. The type of aerial apparatus it chooses will often be based on a variety of factors including maneuverability, length of reach needed, most common tasks required of the aerial device, and amount of equipment that the apparatus needs to carry. All aerial apparatus should meet the design requirements contained in NFPA 1901, *Standard for Automotive Fire Apparatus,* that was in effect at the time the apparatus was manufactured. It should be noted that during the period from 1991 to 1996, the requirements for aerial apparatus were in a separate standard titled NFPA 1904, *Standard for Aerial Ladder and Elevating Platform Apparatus*. However, in 1996 the requirements for aerial apparatus were merged back into the NFPA 1901 standard.

NFPA 1901 classifies aerial devices into three basic categories: aerial ladders, elevating platforms, and water towers. These classifications make sense from the organizational standpoint of the NFPA 1901 standard. However, from the educational perspective of this manual, it is more logical to divide the three basic categories of aerial devices into five major types. The five types of aerial devices that IFSTA recognizes are:

1. Aerial ladders
2. Aerial ladder platforms
3. Telescoping aerial platforms
4. Articulating aerial platforms
5. Water towers

This groups the devices based on how they are used, and the tactical advantages and disadvantages more easily fit these types.

This chapter describes each of these types of apparatus and their major features. This chapter also covers the basic features common to all aerial devices and the types of fixed systems and portable equipment that are commonly carried on aerial apparatus.

Aerial Ladder Apparatus

Aerial ladder apparatus is the most common type of aerial fire apparatus operated in North America. An *aerial ladder* is a power-operated ladder that allows firefighters to easily climb and descend between the tip of the ladder and the turntable. The primary function of the *turntable* is to provide continuous rotation of the aerial device on a horizontal plane. The fully extended length, also referred to as the working height, of North American-made aerial ladders range from 50 to 135 feet (15 m to 41 m). The working height for aerial ladders is measured from the ground to the highest ladder rung with the ladder at maximum elevation and extension (Figure 2.1). Models manufactured in other countries may exceed these heights.

The main uses of aerial ladders are rescue, ventilation, elevated master stream application, and gaining access to upper levels. To accomplish these objectives, aerial ladder apparatus carry a complement of ground ladders, tools, and other equipment. The aerial ladder may be mounted on either a two- or three-axle, single-chassis vehicle or on a three-axle tractor-trailer vehicle (Figures 2.2 a through c). The single-chassis vehicle is usually equipped with dual rear wheels and is shorter than the tractor-trailer vehicle.

Tractor-trailer aerial apparatus, also known as tiller trucks, are equipped with steerable rear wheels on the trailer (Figure 2.3). A tiller operator is required to steer the rear wheels of this type of vehicle. In general, tractor-trailer trucks are more maneuverable than single-chassis vehicles. Single-chassis vehicles with rear-wheel steering may rival tillers in on-scene positioning. The tiller's maneuverability is an asset when the apparatus must negotiate narrow streets or heavy traffic and assists with positioning the apparatus on the emergency scene. As we will learn later in this manual, the tiller's ability to be jackknifed may also aid in stabilizing the apparatus when the aerial device is deployed.

The Ladder

A fire department aerial ladder must be constructed to perform safely under a wide variety of potential fire and rescue conditions. The supporting beams of the ladder are of truss construction, which provides needed strength. Strength is needed because aerial ladders are subject to stress whether the tip is supported or unsupported.

Trusses consist of an assembly of bars or rods to form a rigid framework. A solid beam of equal size would provide adequate strength, but its weight would be prohibitive for aerial ladder operation. Trusses may be constructed by several methods, but engineering design has proven that trusses lend more strength if their assembled members form triangles or a combination of triangles (Figure 2.4). A properly designed truss permits tension and com-

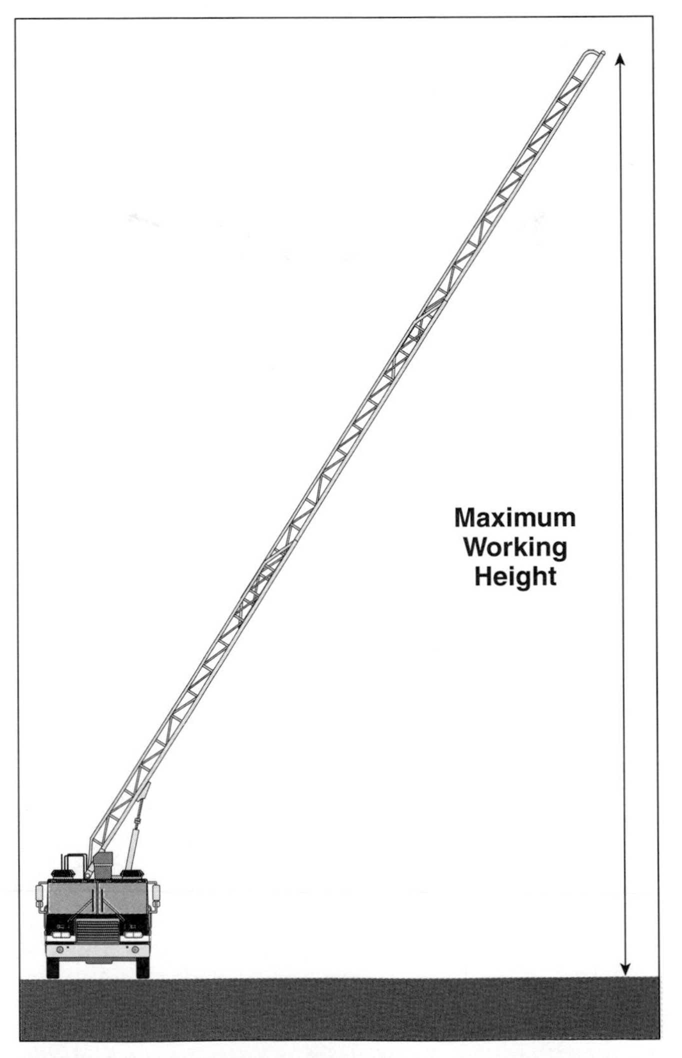

Figure 2.1 The working height is measured from the ground to the highest rung when the ladder is fully elevated and extended.

Figure 2.2a A two-axle aerial ladder apparatus. *Courtesy of Ron Jeffers.*

Figure 2.2b A three-axle aerial ladder apparatus. *Courtesy of Warren Gleitsmann.*

Figure 2.2c A tillered aerial ladder apparatus. *Courtesy of Joel Woods, Maryland Fire and Rescue Institute.*

pression stresses to be distributed over the length of the ladder for maximum strength with minimum weight.

NFPA 1901 requires that aerial ladders have at least two sections and a minimum reach of 50 feet (15 m). The bottom section of the aerial ladder is referred to as the *base* (or bed) section. The second

Figure 2.3 The steerable rear wheels of the tiller apparatus make it highly maneuverable.

Figure 2.4 Aerial ladders receive their strength from the truss work used in their design.

and subsequent sections that extend beyond the base section are called the *fly* sections (Figure 2.5). The manner by which each section slides into the other and certain safety factors are characteristic of each manufacturer's design.

At the time this manual was written, the two most common materials for construction of aerial ladders were heat-treated aluminum alloy and steel. The fabricated structural members may be fastened by rivets, or the entire truss segment may be welded (steel) or helliarched (aluminum) (Figures 2.6 a and b).

There are three main portions of an aerial ladder to which firefighters and driver/operators commonly refer. These are the base rails, the top rails, and the rungs. The *base rails,* also commonly referred to as the beams, are the lower chords of the aerial ladder to which the rungs, trussing, and other portions of the ladder are attached (Figure 2.7). The

Figure 2.5 The fly sections of the aerial ladder extend out from within the base section.

Figure 2.6a Most aerial ladders are constructed by welding pieces of metal together.

Figure 2.6b Some manufacturers choose to construct their aerial ladders with rivets as opposed to welds.

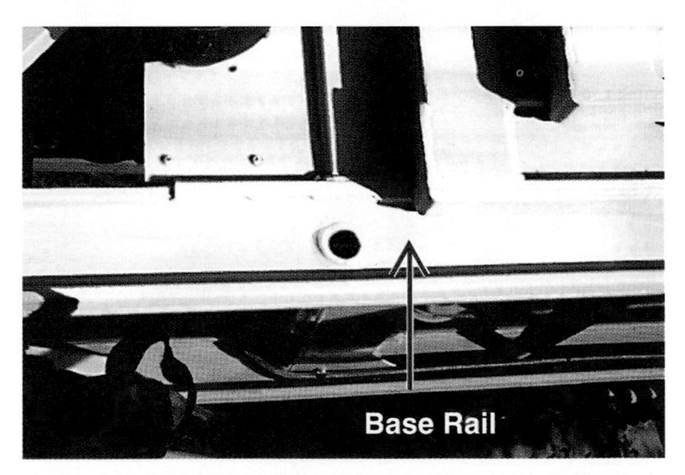

Figure 2.7 The beam of the ladder is the bottom structural member to which the rungs are attached.

top rails, also known as the handrails, are the top chords to which the opposite ends of the trussing are attached (Figure 2.8). These are the parts of the ladder that firefighters hold onto as they ascend and descend. The *rungs* are the portions that are attached between the two base rails and are used as steps for personnel on the aerial ladder (Figure 2.9).

NFPA 1901 requires that no section of the aerial ladder be narrower than 18 inches (457 mm). The rungs should be spaced on 14-inch (356 mm) centers and be covered with a nonskid finish. The top (hand) rail must be at least 1 inch (25 mm) wide and at least 12 inches (305 mm) above the base rail. Folding steps must be provided at the tip of the ladder so that a ladder pipe operator has a place to stand for extended periods of operation (Figure 2.10).

NFPA 1901 also requires that an aerial ladder with a rated vertical height of 110 feet (34 m) or less, with stabilizers set, shall be capable of being raised from the bedded position to its maximum elevation and extension and rotated 90° in 120 seconds or less. Aerial ladders that exceed 110 feet (34 m) are allowed 180 seconds to perform the same evolution. All aerial ladders manufactured since 1991 must have a minimum tip load of 250 pounds (114 kg) when the ladder is fully extended and at any elevation within its normal range of motion.

Figure 2.8 The top rail is also referred to as the handrail.

Figure 2.10 Aerial ladders must be equipped with folding steps at the top of the ladder.

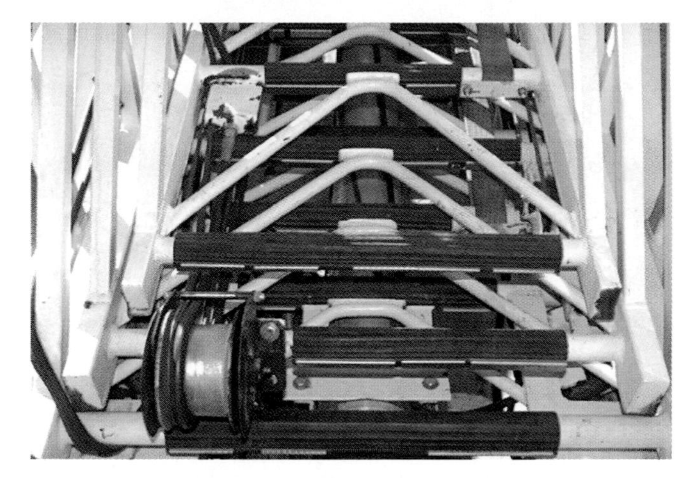

Figure 2.9 The rungs are attached to the beams of the aerial ladder.

Figure 2.11 A typical elevating platform.

Elevating Platform Apparatus

IFSTA divides the NFPA classification of elevating platform apparatus into three distinct types of aerial devices: aerial ladder platforms, telescoping aerial platforms, and articulating aerial platforms. Each device has it own characteristics and, from a training standpoint, warrants its own discussion. However, as noted in NFPA 1901, there are some design requirements that are common to all types of elevating platforms, and we will discuss these first.

A platform leveling system is required so that the platform's position is horizontal to the ground at all times. The required minimum floor area of a platform is 14 square feet (1.3 m^2). Each platform must have a rail completely enclosing it with no opening under the railing greater than 24 inches (610 mm) (Figure 2.11). Two gates below the top railing provide for firefighter/victim access and exit. A kickplate is required at floor level and should be 4 inches (100 mm) high. This kickplate prevents a firefighter's feet from sliding off the platform in case of a slip or fall. Drain openings must be provided to prevent water from building up on the floor of the platform. All elevating platforms require two operating control stations — one at street level and one in the platform (Figures 2.12 a and b). A backup hydraulic system is required.

A heat-protective shield is also required to protect the firefighters in the platform from the effects of radiated heat. An upsurge of tremendous heat occurs at the time of a wall or roof collapse or similar condition. If there are firefighters in the platform at this time and the shield is not in place, the possibility for serious injuries exists. This shield can be dislocated during practice drills, during periods of apparatus maintenance, or through the entrapment of water or ice. The reflective shield can lose its ability to reflect heat through age, dirt,

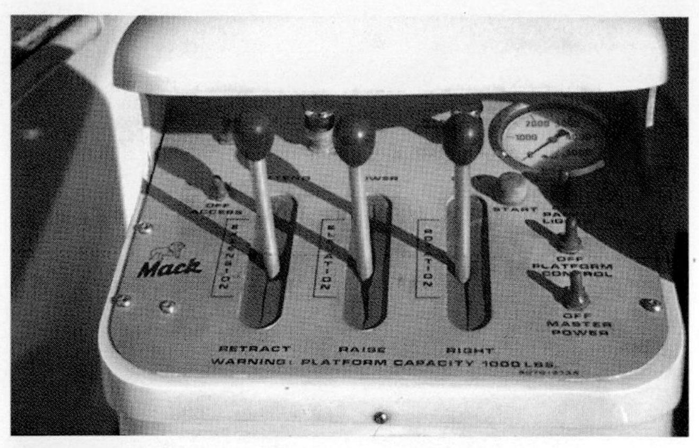

Figure 2.12a Most aerial devices have controls located on the turntable, the rear of the apparatus, or at the pump panel.

Figure 2.12b Elevating platform apparatus have aerial device controls located in the platform.

Figure 2.13 Most elevating platforms have a fog curtain nozzle located on their underside.

Figure 2.14 The fog curtain nozzle is activated by either a foot pedal or a control lever.

or tears. Shields that have been damaged in any way will not provide adequate protection for firefighters in the platform and must be replaced.

A protective water fog curtain nozzle must be located on the bottom of the platform (Figure 2.13). It is operated by a quick-actuating lever (Figure 2.14). It is designed to be operated when the platform is exposed to high levels of heat to provide an additional barrier. NFPA 1901 requires this nozzle to flow at least 75 gpm (284 L/min).

All elevating platforms must have a minimum load capacity of 750 pounds (340 kg) when the aerial device is fully extended, at any elevation within its normal range of motion, and with no water in the piping system. This load requirement drops to 500 pounds (227 kg) when the water delivery system is charged. The water delivery system must be capable of discharging at least 1,000 gpm (3 785 L/min) when the aerial device is in any position. Elevating platforms that are 110 feet (34 m) or shorter should be able to be raised from the bedded position to the maximum elevation and extension and be rotated 90° in 150 seconds or less. There is no time requirement for elevating platforms taller than 110 feet (34 m).

Fire fighting equipment in the platform typically includes a permanently mounted turret nozzle that is supplied by a water system incorporated within the booms or the ladder (Figure 2.15). Electrical, air, and hydraulic outlets are usually provided in

Figure 2.15 All elevating platforms are equipped with a piped waterway and turret nozzle.

Figure 2.16 Some elevating platforms are equipped with a hose connection for a handline to be extended from the aerial device.

the platform. Floodlighting and forcible entry equipment may also be on or attached to the platform. A communication system between the two control stations is also necessary. Many platforms are provided with discharges to provide handlines and a mobile standpipe to upper floors (Figure 2.16).

Aerial Ladder Platforms

Aerial ladder platform apparatus are similar to aerial ladder apparatus except that a work platform is attached to the end of the aerial ladder. The aerial ladder platform combines the safe work area of a platform with a safe, climbable aerial ladder. The working height of all types of elevating platforms is measured — with the aerial device at maximum extension and elevation — from the ground to the top surface of the highest platform handrail. Aerial ladder platforms commonly range in size from 85 to 110 feet (26 m to 34 m).

The most common aerial ladder platform apparatus in service today is the straight chassis, three-axle vehicle with a rear-mounted aerial device (Figure 2.17). However, midship-mounted models have increased in popularity in recent years (Figure 2.18). At least one manufacturer has experimented with an aerial ladder platform mounted on a tractor-tiller apparatus, but this has not gained wide acceptance in the field. Another European manufacturer has also introduced an aerial device to the North American market that features an aerial ladder with a detachable platform. This allows the driver/operator or company officer the option of using the apparatus as an aerial ladder or aerial ladder platform, depending on the situation.

Figure 2.17 Most rear-mounted aerial ladder platforms are large, three-axle apparatus. *Courtesy of Joel Woods, Maryland Fire and Rescue Institute.*

Figure 2.18 Midship-mounted aerial ladder platforms are becoming more popular in the fire service. *Courtesy of Ron Jeffers.*

Telescoping Aerial Platforms

NFPA lists aerial ladder platforms and telescoping aerial platforms under the same definition. However, each type of apparatus has different capabilities on the fireground. The primary difference is that aerial ladder platforms are designed with a large ladder that allows firefighters and victims to routinely climb back and forth from the platform. Telescoping aerial platforms are not intended for this purpose. They are generally equipped with a small ladder attached to the boom (Figure 2.19). This ladder is designed primarily as an emergency escape ladder for firefighters in the platform. Common sizes of telescoping aerial platforms in use in North America range from 75 to 100 feet (23 m to 30 m).

Telescoping aerial platform devices have two or more sections and are made of either box-beam construction or tubular truss-beam construction. *Box-beam* construction consists of four sides welded together to form a box shape with a hollow center (Figure 2.20). Hydraulic lines, air lines, electrical cords, and waterways may be encased within the center or on the outside of the box beam. *Tubular truss-beam* construction is similar in design to the truss construction of aerial ladders (Figure 2.21). Tubular steel is welded to form a box shape using cantilever or triangular truss design.

Figure 2.19 A telescoping aerial platform.

Figure 2.20 Some telescoping aerial platforms have solid, box beam aerial device sections. *Courtesy of Ron Jeffers.*

Articulating Aerial Platforms

Articulating aerial platform apparatus are similar to the telescoping aerial platform apparatus. The primary difference is in the operation of the aerial device. Instead of telescoping into each other, the boom sections are connected by a hinge, and they fold like an elbow (Figures 2.22 a and b). The booms

Figure 2.21 An example of a telescoping aerial platform that is constructed of an open truss design. *Courtesy of Joel Woods, Maryland Fire and Rescue Institute.*

Figure 2.22a A typical articulating platform aerial apparatus.

Figure 2.22b The articulating aerial platform has hinged sections.

are constructed in basically the same manner as telescoping platforms. These units perform many of the same functions as telescoping aerial apparatus. These functions include rescue, ventilation, master stream application, and gaining access to upper floors. Standard articulating aerial platforms range in height from 55 to 85 feet (17 m to 26 m).

Over the years, several manufacturers have offered aerial devices that both telescope and articulate (Figure 2.23). These devices typically consist of two or more telescoping booms on each side of the articulating hinge. Most of these devices resemble articulating aerial platforms that have telescoping sections. However, at least one manufacturer has a model that is an aerial ladder with an articulating platform on the end of the last fly section. These combination aerial devices range in height from 90 to 174 feet (28 m to 53 m).

Figure 2.23 This model of an articulating aerial platform also has telescoping sections within the articulating boom sections. This allows the device to reach heights of 150 feet (45 m).

Pumpers Equipped with Water Towers

Many fire departments choose to outfit their pumpers with water towers. These are telescoping or articulating aerial devices whose primary function is to deploy elevated master streams (Figures 2.24 a and b). The movement of the water tower and control of the fire stream are remotely controlled by the driver/operator from ground level. These controls are located either at the rear of the apparatus or on the midship pump panel. Most water towers are designed so that their fire streams may be deployed at a range of elevations starting from a few degrees below horizontal to nearly 90° from the ground. Common sizes for these devices range from 50 to 130 feet (15 m to 40 m). They are capable of maximum flows ranging from 1,000 to 5,000 gpm (4 000 L/min to 20 000 L/min).

Historically, some manufacturers of telescoping water towers have equipped them with ladders

Figure 2.24a An articulating water tower apparatus. *Courtesy of Ron Jeffers.*

Figure 2.24b A telescoping water tower apparatus. *Courtesy of Joel Woods, Maryland Fire and Rescue Institute.*

(Figure 2.25). In order for them to be truly considered an aerial ladder, the ladder affixed to the water tower must meet the same requirements as those listed for aerial ladder in NFPA 1901. Otherwise, it is recommended that these devices be used only as emergency escape routes.

Figure 2.25 Some telescoping water towers are also equipped with a ladder.

Quints

The term quint is derived from the phrase quintuple fire apparatus, meaning a piece of apparatus that has five major features or functions. A *quint* is a fire apparatus that is equipped with a fire pump, water tank, ground ladders, and fire hose in addition to the aerial device. The reasons that some fire departments prefer to operate quints rather than standard aerial apparatus vary from jurisdiction to jurisdiction. Some departments feel that it is not effective to tie up a separate pumper when it is necessary to deploy an elevated master stream. Other departments have the philosophy that every major piece of apparatus should have the capability of extinguishing a fire should it be the first apparatus to arrive on the scene. In a few cases, fire departments have experimented with quints as replacements for traditional pumper and ladder companies.

Figure 2.26a A pumper equipped with an aerial ladder.

In order to be considered a quint as defined by NFPA 1901, the apparatus must have the following:

- A fire pump with a capacity in excess of 1,000 gpm (3 785 L/min)
- A tank that holds at least 300 gallons (1 136 L) of water
- 800 feet (244 m) of 2½-inch (65 mm) or larger supply hose and 400 feet (122 m) of attack hose
- 85 feet (26 m) of ground ladders, to include at least one attic ladder, one straight ladder with folding roof hooks, and one extension ladder
- An aerial ladder or elevating platform with a permanently piped waterway

Figure 2.26b A pumper equipped with an aerial ladder.

Quints come in various sizes. Some departments use a pumper-sized apparatus with a 50- to 75-foot (15 m to 23 m) aerial device (Figures 2.26 a and b). Other quints are full-sized aerial apparatus (Figures 2.27 a and b). Each jurisdiction must decide which type of apparatus most suits its needs.

Figure 2.27a An aerial ladder platform quint apparatus.

Figure 2.27b A full-sized aerial ladder quint apparatus.

Primary Features of Aerial Devices

There are a number of features and systems that are common to most types of aerial devices. These are referred to throughout the rest of the manual. The intent of the following sections is to familiarize the readers with important information that they will need to know as they work through this book.

Basics of Aerial Device Hydraulic Systems

In the early days of aerial apparatus, aerial devices and stabilizers were operated manually by any one of several designs including spring-operated and gear-driven. Today, all aerial devices are hydraulically powered, as are virtually all stabilizers (Figure 2.28). Hydraulic fluid is the medium used within the system to transmit force. Hydraulic fluid is used because it is practically incompressible, and it allows force to be transmitted over a relatively large distance with little loss of power. Depending on the manufacturer of the aerial apparatus, the fluid may be under pressures of 3,500 psi (24 500 kPa) or more. It is for this reason that personnel searching for leaks in a hydraulic system under pressure should use extreme care and should never attempt to block a leak with any part of the body. Pinhole leaks at high pressure are capable of cutting through human tissue and may also cause severe burns.

Force is created on the fluid by a hydraulic pump. The pump is powered by a power take-off (PTO) arrangement off the vehicle's main engine. Depending on the manufacturer of the aerial device, the pump used for the hydraulic system is either a rotary vane or rotary gear positive-displacement

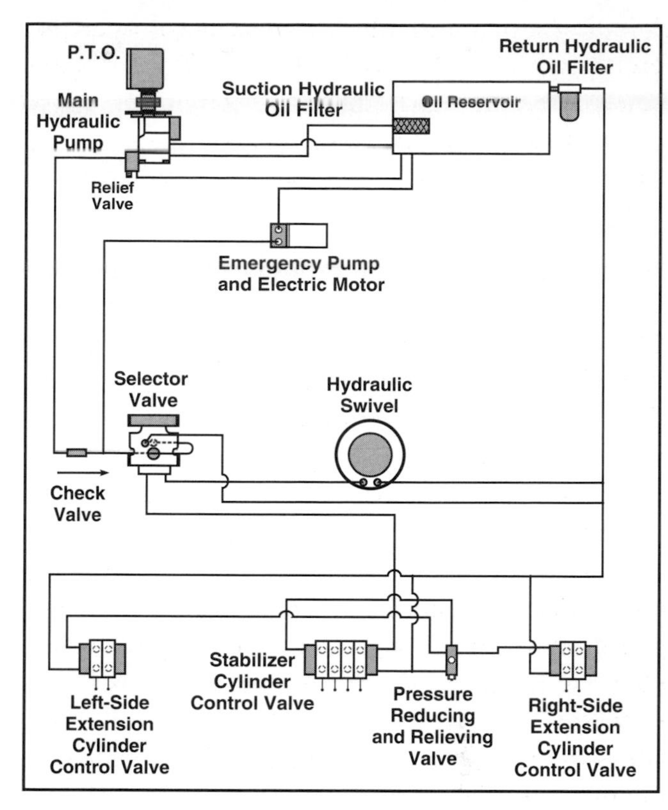

Figure 2.28 This schematic shows the parts of a typical aerial device hydraulic system.

pump. Both pump a fixed amount of fluid (measured in gpm or L/min) at a given speed (measured in rpm). This allows for predictable operation of the hydraulic components.

The hydraulic fluid moved into the hydraulic system is supplied from the hydraulic reservoir. Fluid displaced from the system flows back into the reservoir for storage before being recirculated through the system. The reservoir is designed to supply an adequate amount of fluid to operate the hydraulic system and to condition the fluid while it is stored in the tank. Baffles located in the tank slow the movement of the fluid through the tank. This allows air, heat, and foreign matter to be released from the fluid before it is reintroduced into the hydraulic system. Filters and exchange-type oil coolers are also used to supplement the reservoir's ability to condition the fluid. Coolers, usually located in front of the vehicle's radiator, remove heat from the fluid as it returns to the reservoir from the hydraulic system.

The hydraulic fluid is supplied through the system by a series of tubing and hoses. Most manufacturers use steel tubing and aircraft-type, steel-braided hose. Hose must be rated to burst at a pressure that is at least four times stronger than normal operating pressure. Hydraulic fluid is supplied to components above the rotation point of the turntable through a high-pressure hydraulic swivel assembly to permit continuous rotational capability of the device.

In order to control the flow of hydraulic fluid through the system, a number of different valves are used. Valves start, stop, regulate, and direct the flow of oil to control pressure in the system by allowing the pressure to be built up or released. Valves may be controlled manually, electrically, hydraulically, mechanically, or by a combination of these methods. There are numerous types of valves used in an aerial apparatus hydraulic system. *Check valves* prevent fluid from flowing backwards through a component and act as a safety feature in the event that a leak develops in the system. *Relief valves* limit the pressure built up in the system, thus preventing damage due to overpressurization. *Counterbalance valves* prevent unintentional or undesirable motion of the device from position.

One of the more important valves in the system is the selector valve (also known as the transfer or diverter valve) (Figure 2.29). The *selector valve* is a three-way valve that directs fluid to either the stabilizer control valves or the aerial device control valves. A sliding spool in a housing directs fluid flowing through the valve to one work port or the other, at the same time blocking flow through the opposite work port. By blocking flow to the system not in use, the selector valve acts as an interlock to prevent both the stabilization and aerial device systems from operating at the same time. Thus, it is difficult to raise the aerial if the stabilizers have not already been set. When the aerial is deployed, the selector valve makes inadvertent operation of the stabilizers unlikely.

Once the hydraulic fluid is directed into one system or the other, actuator valves, monitor valves, stack valves, and proportional directional control valves are used to direct and control the power in that system. These valves tend to be four- or five-way valves in order to accomplish two-directional control. This final group of valves supplies hydraulic fluid to the actuators, which are the devices that convert the fluid power developed in the system back into mechanical force. The actuators in these systems are hydraulic cylinders. Hydraulic cylinders convert the energy in the system into linear mechanical force or motion. This is accomplished when pressurized hydraulic oil is directed into a chamber created by fitting a piston into a cylindrical barrel. Hydraulic cylinders used for elevation and telescopic control of the aerial device and for operating the stabilizers are double acting. *Double-acting cylinders* are capable of receiving oil under pressure from both sides of the piston so that force can be created in either direction.

Figure 2.29 The selector valve is used to provide power to the stabilizer or aerial device portions of the hydraulic system.

All manufacturers of aerial devices provide an auxiliary hydraulic pump for use in the event of a failure of the main hydraulic pump. This is a requirement of NFPA 1901. The auxiliary pump is typically a 12-volt DC, electrically operated pump connected directly to the vehicle's battery. However, some manufacturers have an auxiliary pump that runs on 110-volt AC power. This direct connection allows the pump to be operated even if the main vehicle engine goes down. These auxiliary pumps provide all the same motions as the main pump, but at a reduced speed. However, the use of these pumps should be limited to bedding the aerial device after a main system failure occurs. They are not designed nor intended to be used for long periods of time. In fact, they should be operated only for intervals of about one minute, with an equal amount of rest between operations. Auxiliary motors are subject to overheating if they are operated for longer periods of time. These motors should be tested regularly according to manufacturer's instructions. Some aerial devices — depending on the manufacturer of the device — are also equipped with hand cranks that rotate the turntable during hydraulic system failures.

As previously mentioned, the function of the hydraulic system is to provide power for operating the various hydraulic cylinders needed to stabilize the apparatus and deploy the aerial device. The following sections highlight the major hydraulic cylinders associated with an aerial apparatus.

Stabilizer Cylinders

Most modern aerial apparatus are equipped with hydraulically operated stabilizers (Figure 2.30). The stabilizers are deployed anytime the aerial device is raised from its bed. They prevent the apparatus from tipping over when the aerial device is raised and maneuvered. Depending on the design of the apparatus, each stabilizer may be operated by one or two hydraulic cylinders. The cylinders force the stabilizer arms out and down to take much of the weight of the apparatus off the apparatus suspension. This makes the truck more stable as the aerial device is operated. For more information on stabilizing the apparatus, see Chapter 6 of this manual.

Aerial Device Hoisting Cylinders

Hoisting cylinders are sometimes referred to as elevating cylinders. Their function is to elevate the aerial device from its stowed position (Figure 2.31). Hoisting cylinders are heavy, seamless steel outer shells bored to an extremely smooth inside surface. The cylinders are often chrome-plated on the inside to resist friction and wear. The cylinders are enclosed at the top and the bottom with removable caps. A solid piston rod, packed and tightly fitted with rings, is inside each cylinder. The piston rods are also chrome-plated to resist wear.

The end of the piston rod outside the cylinder is fitted with a trunnion that is anchored to either the aerial device or to the turntable, depending on the make of the apparatus (Figure 2.32). The trunnion holds the piston rod in contact with the aerial device or turntable. This allows the piston to transmit the force of the hydraulic fluid in the cylinders to the aerial device. The amount of force exerted by the hydraulic fluid depends on the hydraulic system pressure and the piston surface area. The larger the piston area, the more upward force that is produced. Normal hydraulic pressure is from 850 to 3,000 psi (5 950 kPa to 21 000 kPa).

Figure 2.30 The stabilizers are powered by hydraulic cylinders.

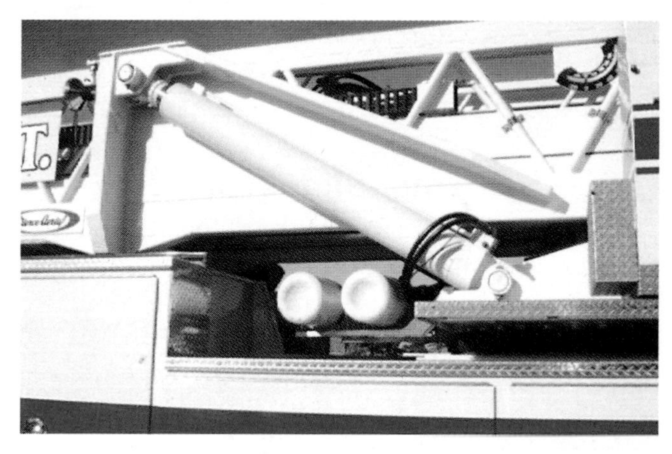

Figure 2.31 Hoisting cylinders are used to elevate the aerial device.

Figure 2.32 The hoisting cylinder is attached to the turntable by a trunnion.

Many new hoisting cylinders are equipped with integral safety valves that lock and hold oil in the cylinder in the event of a leak or a blowout. If this occurs, the ladder can only be lowered by manually bleeding the pressure from the cylinder through a bleed-down valve. In older systems, the hydraulic oil between the top heads and the piston is forced through metering holes. This action retards the downward travel of the aerial device in case of a leak or system blowout and allows the aerial device to settle down slowly. This safety feature prevents damage to the aerial device and injury to any firefighters on the device.

Aerial Device Extension Cylinders

Most newly designed aerial devices use extension cylinders. These are dual cylinders that are fitted on the base section and are used to extend the second section of the device (Figure 2.33). Their design and operating principles are similar to those previously described for hoisting cylinders. In some designs, the extension cylinders simply operate a series of pulleys and cables that in turn extend the fly sections.

Cable Systems, Slides, and Rollers

On most aerial devices, the third, fourth, and fifth (if present) sections of the aerial device are extended by a system of cables and pulleys (Figure 2.34). The extension and retraction cables are usually galvanized aircraft-type cables made from plow steel, and they are resistant to abrasion. The cables are designed and constructed to be much stronger than necessary for normal operation, providing a built-in safety factor. A dual-chain drive may also be used to retract and extend the aerial device (Figure 2.35). Aerials using a main

Figure 2.34 The aerial ladder is raised and lowered using a series of pulleys and cable.

Figure 2.33 Some aerial devices have an extension cylinder between the base section and the first fly section of the aerial ladder.

Figure 2.35 Some manufacturers use a chain drive for extending the aerial ladder.

cable drum can be powered mechanically, hydraulically, or electrically depending on the make of the truck.

The slide and rollers allow the various sections of the aerial device to extend and retract without causing excessive wear on the parts of the device that come in contact with each other during movement. Most devices use a combination of slide pads and rollers to facilitate this motion (Figure 2.36).

The Turntable

The aerial device turntable decking is constructed of rigid metal plates and provides a surface on which the operator may stand (Figure 2.37). On most aerial apparatus, the lower aerial device control pedestal is located on the turntable.

Turntables generally operate on large sheer ball-type bearings. One side of the bearing is fastened to the turntable and the other side to the aerial device structure. Gear teeth are mounted on the periphery of the turntable, and these teeth engage gears that rotate the turntable. The gears are driven by a mechanical drive or an electric or hydraulic motor.

Although aerial turntables may differ in mechanical features, their operating principles are similar. Most aerial manufacturers provide a manually operated crank in addition to power controls in case of system failure (Figure 2.38).

Control Pedestals

The operator's control pedestal usually stands about waist high and is positioned on the turntable to the side of the aerial device (Figure 2.39). For apparatus

Figure 2.36 Slides or pads are used to reduce wear when the aerial device is extended and retracted.

Figure 2.37 The turntable.

Figure 2.38 The manual crank may be used to rotate the aerial device in the event of a loss of hydraulic power.

Figure 2.39 Most aerial devices have a control pedestal on the turntable.

equipped with a platform, NFPA 1901 also requires a control station in the platform (Figure 2.40). This allows remote operation of the aerial device from the platform. If there are two control pedestals on the aerial device, the turntable control station must have the ability to override the platform control station. The top of the pedestal is usually designed with a sloping panel to make it easier to read the instruments mounted on it. Some types of aerial devices, particularly water towers and small quints, have control stations located on the rear of the apparatus or at the pump panel (Figure 2.41).

Figure 2.40 The platform control station allows firefighters in the platform to move the aerial device as they desire.

Most control pedestals use three separate levers to control elevation, extension, and rotation. Other controls and instruments that may be located on the control pedestal include the following:

- Engine speed switch
- Light switch
- Hydraulic oil pressure gauge
- Extension indicators
- Stop and lock controls
- Rung alignment indicator
- Inclinometer
- Engine starter switch
- Communications equipment
- Elevated master stream controls
- Hydraulic lock valve

Some aerial devices, particularly water towers, telescoping platforms, and articulating platforms, feature a single "joystick" type of control capable of making all aerial device movements.

Figure 2.41 These aerial device controls are located on the pump panel of a quint.

Water Delivery Systems

Water delivery systems are used to discharge elevated master streams for fire attack from the aerial device. There are several types of water delivery systems, depending on the type of aerial device. The following sections describe some of the more common ones.

Piped Aerial Ladder Waterways

Many aerial ladders are equipped with piped waterways that eliminate the need for laying hose up the aerial ladder to a master stream nozzle. There are two common types of piped waterway systems: bed ladder systems and telescoping waterway systems.

The oldest type of piped aerial ladder waterway is the bed ladder pipe (Figure 2.42). This is a nontelescoping section of pipe, usually 3 or 3½ inches (77 mm or 90 mm) in diameter, attached to

Figure 2.42 Bed ladder pipes are common on older aerial ladders.

the underside of the bed (base) section of the aerial ladder. The master stream nozzle is attached directly to the tip end of the pipe. The bed ladder pipe is supplied through a connection at the turntable end of the pipe. Frequently, this intake consists of a two- or three-way siamese. Bed ladder pipes are usually equipped with solid stream nozzles because their inability to telescope usually prevents them from being positioned for effective fog stream application. Automatic remote-controlled bed ladder pipe nozzles are rare. Most bed ladder pipes are equipped with manually operated nozzles. These nozzles may be operated from the tip of the retracted aerial device or, preferably, from the ground or turntable through the use of rope(s) attached to the nozzle (Figure 2.43). One piece, or end, of the rope is attached to the nozzle near the outlet tip, and the other is attached to the end of the handle. This allows the nozzle to be moved up and down. Some bed ladder pipes are equipped with a gear-operated mechanism for raising and lowering the fire stream. A hand crank at the base of the ladder is used to maneuver the nozzle.

Figure 2.43 Most bed ladder pipes are operated manually with a rope. *Courtesy of Ron Jeffers.*

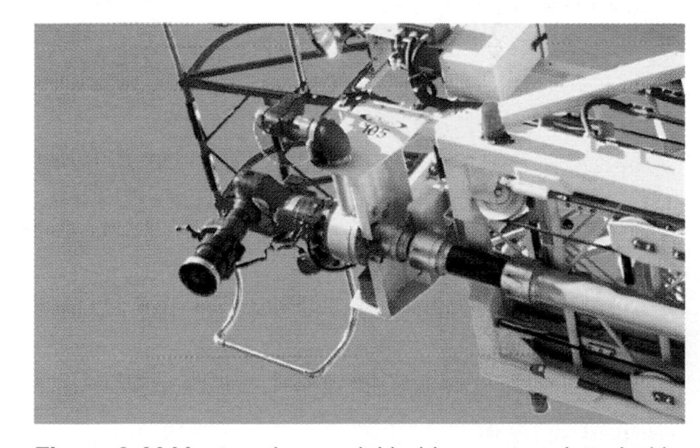

Figure 2.44 Most modern aerial ladders are equipped with telescoping waterways.

Many newer aerial ladders are equipped with a telescoping waterway that extends toward the top of the ladder (Figure 2.44). Most ladders that are 110 feet (34 m) or shorter and equipped with a telescoping water system have piping that extends directly to the tip of the fly section. Taller ladders may have a water system that ends before reaching the tip. The telescoping piping system consists of three or four sections that reduce in size — the largest is attached to the bed section of the ladder and the smallest attached to the fly section. These pipes are made of aluminum or other metal. Generally, the minimum internal diameter in these piping systems is 4 inches (100 mm). The bottom end of the piping is connected to additional piping that runs through the turntable to water inlets, usually found on the rear of the vehicle. The piping that goes through the turntable is equipped with a continuous swivel joint that allows 360° rotation while flowing water. On quints, the telescoping waterway system is also connected directly to the fire pump so that the truck may supply itself if necessary.

These prepiped systems usually have remote-controlled fog nozzles operated by switches located near the tip of the fly section or on the turntable operator's control pedestal (Figures 2.45 a and b). These switches allow the water stream to be opened or closed, moved in any direction, or adjusted from straight to fog stream patterns. Most telescoping water systems are capable of flows up to 1,000 gpm (4 000 L/min).

Detachable Ladder Pipe Systems

Aerial ladders not equipped with piped waterway systems have detachable ladder pipe systems. These systems are generally stored on the vehicle and attached to the ladder only when needed. The main components of these systems are a detachable ladder pipe nozzle, fire hose, hose straps, and a clappered siamese (Figure 2.46). The ladder pipe is designed to be clamped to the top two rungs of the last fly section. For additional security, it may also be lashed to the ladder using a short section of rope or hose straps. The ladder pipe may be equipped with either a solid stream or a fog stream nozzle. It can be controlled from the tip of the aerial ladder or from the ground by ropes similar to those described for bed ladder pipes. Detachable ladder pipes are typically limited to flows of less than 750 gpm (3 000 L/min).

The ladder pipe is usually supplied by a single 3- or 3½-inch (77 mm or 90 mm) hoseline that is run directly up the center of the ladder rungs and attached prior to elevating the ladder. Placing the hose either side of center causes unnecessary torsional stress on the aerial ladder. The hose should be secured by hose straps at a minimum of two or three locations on the ladder. The hose should be long enough that the end opposite the nozzle rests on the ground. A two- or three-inlet clappered siamese should then be attached to the bottom end of the hoseline (Figure 2.47). This siamese can be supplied by two or three 2½- or 3-inch (65 mm or 77 mm) hoselines. A large diameter supply hose fitted with a reducer may be substituted for the siamese and multiple smaller supply lines, if desired.

Figure 2.45a Remote nozzle controls are commonly found at the control pedestal.

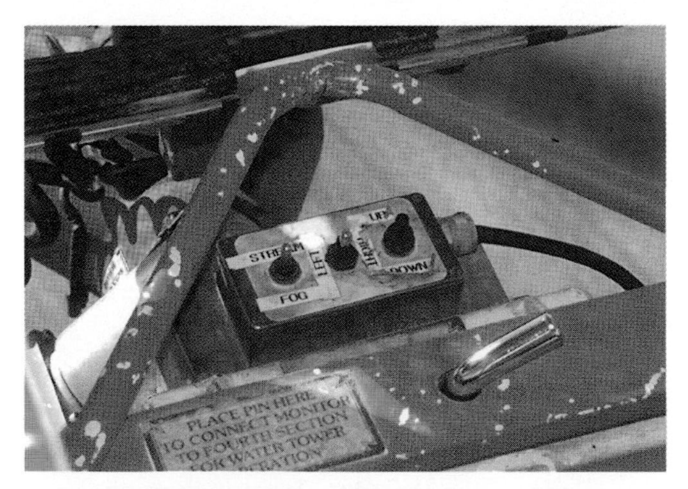

Figure 2.45b Some aerial ladders may have nozzle controls at the tip of the aerial ladder.

The backward thrust of master streams, both solid and fog, may cause an excessive side twist to an extended aerial ladder if the stream is projected toward either side of the ladder. Most ladders with detachable ladder pipe systems restrict the sideways movement of the nozzle to about 15° either

Figure 2.46 A typical detachable ladder pipe.

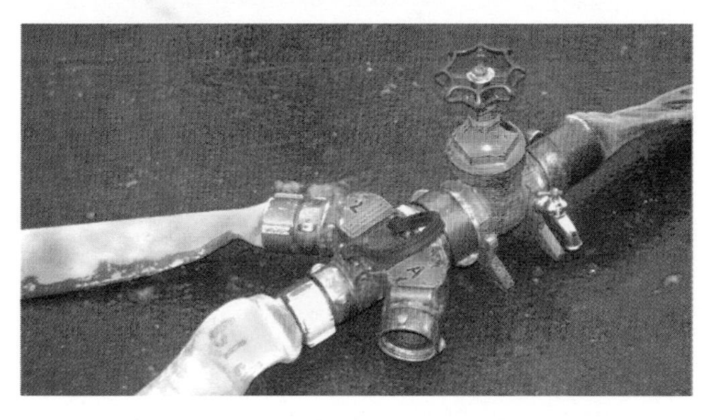

Figure 2.47 The supply hose to a detachable ladder pipe is often equipped with a siamese appliance to increase the flow to the nozzle.

side of center (Figure 2.48). Rotation of the aerial turntable permits a ladder pipe to cover a greater area. Most ladder pipes should not be operated from an aerial positioned at a 90° angle, or vertical position. If at all possible, the ladder should be placed between 70° and 80°. This will provide the safest service and optimal stream penetration.

Elevating Platform Waterway Systems

In general, elevating platform waterway systems are similar to those previously described for piped

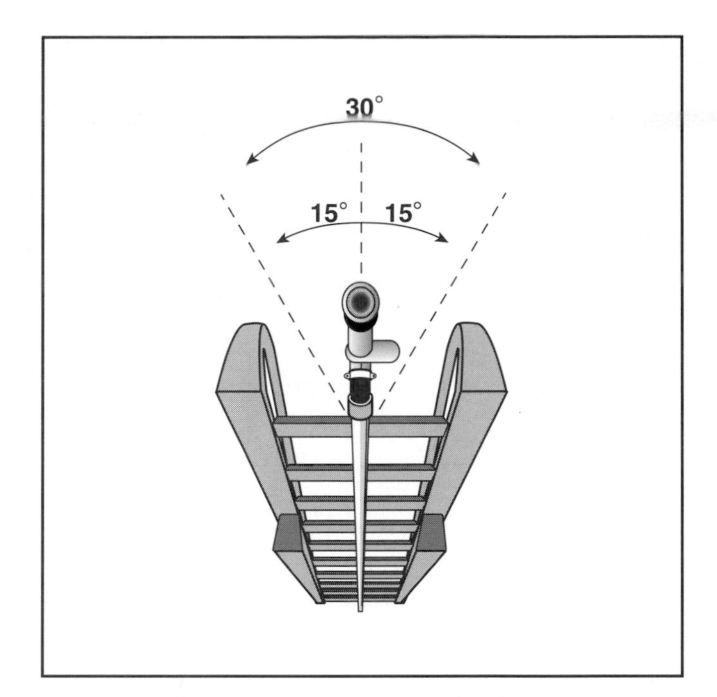

Figure 2.48 Detachable ladder pipes should not have a horizontal range of movement exceeding 15° in either direction.

Figure 2.49 Some elevating platforms are equipped with two turret nozzles.

aerial ladder systems. The primary difference is that the nozzle is located in the platform, rather than being at the end of a ladder. Because elevating platforms typically have greater load capacities than aerial ladders, the piping on these systems may be larger, and flows of up to 2,000 gpm (8 000 L/min) are possible on some models. When it is necessary to place a firefighter at the end of the aerial device during elevated master stream operations, the elevating platform is safer and more comfortable than the aerial ladder. Some elevating platforms are equipped with two nozzles in the platform (Figure 2.49). These nozzles may be operated individually or at the same time to provide maximum stream placement flexibility. For example, one stream may be used to attack the main body of fire, while the second stream is used to protect an exposure. Most elevating platform systems are also equipped with at least one 2½-inch discharge in the platform (Figure 2.50). This allows handlines to be stretched from the platform if they are needed on upper levels of a structure. Some departments have preconnected attack lines attached to this discharge (Figure 2.51).

Water Tower Systems

As described earlier in this chapter, elevating water towers are devices designed specifically for the deployment of elevated master streams. The water delivery system on these devices is similar in design to those described for aerial ladders with piped water systems. Control of both the motion of the water tower and the nozzle is accomplished from a control panel located either on the rear step of the apparatus or on the pump panel (Figure 2.52). Most commonly, the water discharged from a water tower is pumped from the fire pump on the apparatus.

Water towers may be equipped with a number of options that enhance their use under special conditions. Some water towers are equipped with video cameras at the tip of the aerial device. A corresponding video monitor is then placed at the control panel location. This allows the driver/operator to have an aerial view of where the fire stream from the water tower is being directed.

Another option that is included on some water towers is the addition of a piercing nozzle at the end of the aerial device. This hardened steel nozzle may be poked through sheet metal buildings, aircraft skins, and metal roofs. Once the nozzle has pierced the desired structure, a broken stream nozzle at its tip may be turned on at the operator's panel. This stream is capable of flows up to 300 gpm (1 200 L/min).

Figure 2.50 A standpipe connection on an elevating platform.

Figure 2.51 Some jurisdictions choose to place a preconnected handline on the elevating platform.

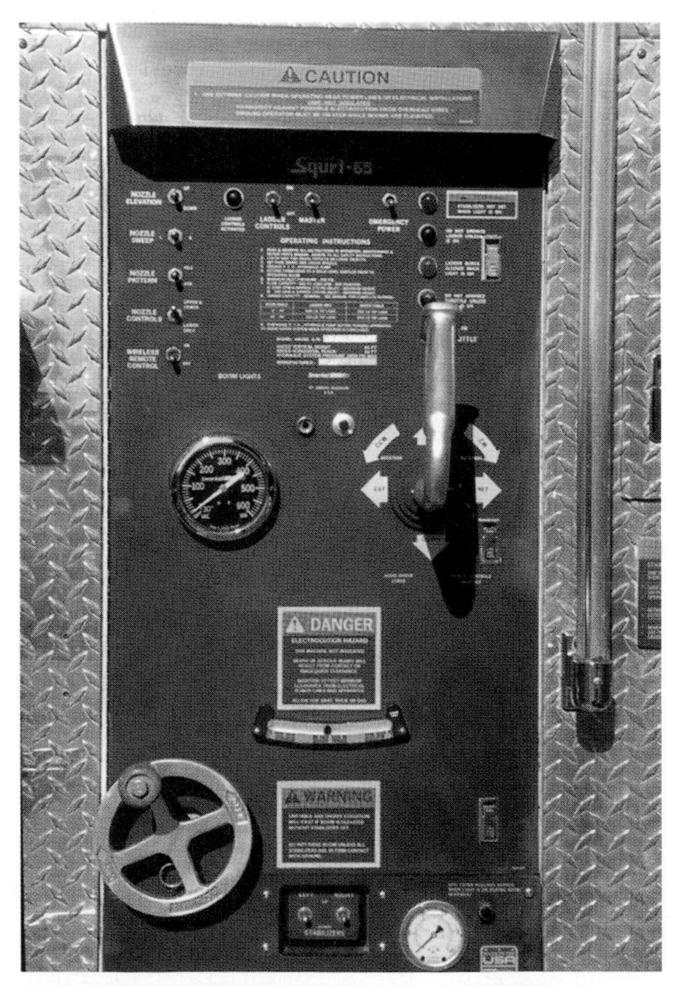

Figure 2.52 A typical water tower control panel.

Communication Systems

Voice communication systems are used to permit firefighters at the tip of the aerial device to communicate with the driver/operator at the turntable control station. The system consists of a "hands-free" transmitter/receiver at the tip of the ladder or in the platform and a second transmitter/receiver at the turntable control station (Figures 2.53 a and b). The transmitter/receiver at the turntable control station does not have to be a hands-free type unit. These receivers may either be connected by hard wire or may be of the radio type.

Prior to the release of the 1996 edition of NFPA 1901, only aerial devices equipped with platforms were required to have communication systems. However, many fire departments chose to install them on aerial ladders as well. The 1996 edition of NFPA 1901 changed to require communication systems installed on all types of aerial devices constructed from that time forward. IFSTA recommends, when possible, that all aerial devices still in service without a voice communications system be retrofitted to include one.

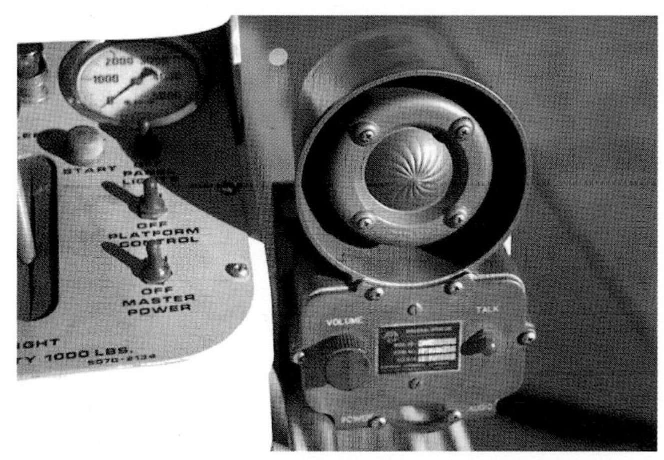

Figure 2.53a Intercom controls and the speaker are located on the control pedestal.

Figure 2.53b A hands-free microphone and speaker are located at the tip of the aerial ladder.

Breathing Air Systems

Some aerial devices are equipped with fixed breathing air systems. These systems allow one or more firefighters operating at the tip of the aerial device to breathe clean air without the need to don a standard fire department self-contained breathing apparatus (SCBA). While at the tip of the ladder or in the platform, the firefighter dons a facepiece that is fitted with a special supply hose. This supply hose is connected to an air outlet fitting on the aerial device (Figure 2.54). The air outlet is connected to compressed breathing air cylinders mounted at the bottom of the aerial device by a system of hoses and/or tubing (Figure 2.55). In some cases, the facepiece air hose may be long enough to allow the

Figure 2.54 Many aerial apparatus are equipped with air systems that allow the firefighter at the tip to breathe off the system as opposed to an SCBA.

Figure 2.55 The aerial device breathing air system is supplied by large capacity cylinders.

firefighter the ability to leave the aerial device and work remotely from it while still connected to the breathing air system.

NFPA 1901 does not require aerial devices to be equipped with breathing air systems. However, the standard does contain some requirements for the design of breathing air systems if they are installed on aerial devices with elevating platforms. Breathing air systems in elevating platforms must be capable of providing air for at least two firefighters. The compressed air cylinders that are connected to the system must have a capacity of at least 400 cubic feet (11.3 m³). All piping system components must be designed for three times the working pressure of the system. A low-air warning should sound at both the upper and lower control stations when the system's capacity drops below 20 percent remaining. All breathing air cylinders and equipment must also conform to appropriate governmental regulations. In the United States, the breathing air cylinder installation must comply with Title 49 *Code of Federal Regulations* (CFR), Part 178, Subpart C, "Specifications for Cylinders."

The quality of the air in the cylinders must meet the requirements set forth in NFPA 1500, *Standard on Fire Department Occupational Safety and Health Program*. This standard specifies that the air quality should meet the requirements for Grade D air quality established by the American National Standards Institute (ANSI) and the Compressed Gas Association. These requirements are found in ANSI/CGA G7.1, *Commodity Specification for Air*. IFSTA recommends that, when practical, existing aerial apparatus be equipped with breathing air systems.

Apparatus-Mounted Special Systems

There are a variety of special systems commonly mounted on aerial apparatus. In most jurisdictions, the driver/operator is the person responsible for operating these systems in addition to the fire pump or aerial device. The following sections provide general information on the types of special equipment that may be found on apparatus. The exact procedures for operating these systems vary depending on the manufacturer and design of the particular equipment found on the apparatus. Consult the owner's manual for exact operating instructions.

Electric Power Generation Equipment

At one time, rescue and specialty vehicles were the only apparatus likely to feature electrical generation equipment, but today's aerial apparatus is often being equipped with it as well. This equipment is used to power floodlights and other electrical tools or equipment that may be required on the emergency scene. Regardless of the type of system that is used, its design and installation should meet pertinent electrical codes and include proper grounding of the system.

Inverters (alternators) are used on aerial apparatus when the local jurisdiction determines that it is not necessary for the apparatus to be able to generate large amounts of power. The *inverter* is a step-up transformer that converts the vehicle's 12- or 24-volt DC current into 110- or 220-volt AC current (Figure 2.56). Advantages of inverters are fuel efficiency and low or nonexistent noise during operation. Disadvantages include small capacities and limited mobility from the vehicle. These units are generally capable of providing approximately 1,500 watts (1.5 kW) of electric power. They are most commonly used to power vehicle-mounted floodlights.

Generators are the most common power source used for emergency services. They can be portable or fixed to the apparatus. Portable generators are powered by small gasoline or diesel engines and generally have 110- and/or 220-volt capacities (Figure 2.57). They can be operated in the compart-

Figure 2.56 Inverters provide a small quantity of AC electrical power.

Figure 2.57 Portable electric generators are typically stored in an apparatus compartment.

ment of the apparatus, or they can be carried to a remote location. Most portable generators are designed to be carried by either one or two people. They are extremely useful when electrical power is needed in an area that is not accessible to the vehicle-mounted system. Portable generators are designed with a variety of power capabilities — 5,000 watts (5 kW) of power is the largest.

Vehicle-mounted generators usually have a larger capacity than portable units (Figure 2.58). In addition to providing power for portable equipment, vehicle-mounted generators are responsible for providing power for the floodlighting system on the

Figure 2.58 Fixed generators provide large quantities of AC electrical power.

vehicle. Vehicle-mounted generators can be powered by gasoline, diesel, or propane engines or by hydraulic or power take-off systems and should be installed according to the pertinent electrical code. Fixed floodlights are usually wired directly to the unit through a switch or circuit breaker, and outlets are also provided for other equipment. These power plants generally have 110- and 220-volt capabilities — capacities up to 12,000 watts (12 kW) are common on aerials. The generators on rescue vehicles are usually larger than the generators on pumpers or aerials. Capacities of up to 50,000 watts (50 kW) are common.

Scene Lighting and Power Distribution Equipment

Most modern fire department aerials have scene lighting and electric power distribution equipment on the apparatus. The driver/operator must be familiar with the location and proper operation of this equipment.

Lighting equipment can be divided into two categories: portable and fixed. *Portable lights* are used where fixed lights are not able to reach or when additional lighting is necessary. Portable lights generally range from 300 to 1,000 watts. They may be supplied by a cord from the power plant or may have a self-contained power unit. The lights usually have handles for safe carrying and large bases for stability and placement. Some portable lights are connected to telescoping stands that eliminate the need for personnel to either hold them or find something to set them on (Figure 2.59).

Figure 2.59 Portable floodlights can be deployed to remote locations.

Figure 2.60 Fixed floodlights can illuminate the area around the apparatus.

Fixed lights are mounted to the vehicle or aerial device, and their main function is to provide overall lighting of the emergency scene. Fixed lights are usually mounted so that they can be raised, lowered, or turned to provide the best possible lighting (Figure 2.60). Often, these lights are mounted on telescoping poles that allow this movement. More elaborate setups include electrically, pneumatically, or hydraulically operated booms with a bank of lights. The bank of lights generally has a capacity of 500 to 1,500 watts per light. Lights that are placed at the end of the aerial device are helpful to firefighters who are performing rescues, ventilation, or other operations near the tip of the device. NFPA 1901 does not contain requirements for lights at the end of the aerial device. However, when possible, suitable floodlights should be placed on all aerial devices.

The amount of lighting should be carefully matched with the amount of power available from the power-generating device carried on that vehicle. Overtaxing the power-generating device will give poor lighting, may damage the power-generating unit or the lights, and will restrict the operation of other electrical tools.

A variety of other electric distribution equipment may be used in conjunction with power-generating and lighting equipment. Electrical cables or extension cords are necessary to conduct electric power to portable equipment. The most common size cable is the 12-gauge, 3-wire type. The cord may be stored in coils, on portable cord reels, or on fixed automatic rewind reels (Figure 2.61). Twist-lock receptacles provide secure, safe connections. Electrical cable must be adequately insulated, contain a ground, be waterproof, and have no exposed wires.

Junction boxes may be used when multiple connections are needed. The junction is supplied by one inlet from the power plant and is fitted with several outlets. Junction boxes are commonly equipped with a small light on top to make them easier to find and plug into and to give visual assurance that they are energized (Figure 2.62).

In situations where mutual aid departments frequently work together and have different sizes or types of receptacles (for example: one has two prongs — the other has three), adapters should be carried so that equipment can be interchanged. Adapters should also be carried so that firefighters can plug their equipment into standard electrical outlets.

Figure 2.61 An example of a fixed electrical cord reel on the apparatus.

Figure 2.62 The junction box allows more than one appliance to be connected to the electric supply.

Figure 2.63 Many aerial apparatus carry a supply of hydraulic rescue tools.

Hydraulic Extrication Tool Systems

It is becoming increasingly common for fire department aerial apparatus to carry a variety of extrication equipment (Figure 2.63). The most commonly used tools are powered hydraulic extrication tools. There are four basic types of powered hydraulic tools used by the rescue service: spreaders, shears, combination spreader/shears, and extension rams. The wide range of uses, the speed, and the superior power of these tools have made them the primary tools used in most extrication situations. These tools receive their power from hydraulic fluid supplied through special hoses from a pump. The pumps may receive their power from compressed air, electric motors, two- or four-cycle gas motors, or apparatus-mounted power take-off systems.

These hydraulic tool pumps may be portable and carried with the tool, or they may be mounted on the vehicle and supply the tool through long coiled hoses or a hose reel line (Figure 2.64). Driver/operators must know how the equipment on their apparatus operates and what limitations it has. Most pumps are not capable of supplying full power to the tool when the hose length between the pump and tool exceeds 100 feet (30 m). Apparatus-mounted power systems generally have a standard method for engaging the device. This may be as simple as flipping an electric switch, or it could entail engaging the apparatus power take-off system.

The driver/operator should also be familiar with the number of tools that may be hooked to the system. Most manufacturers of powered hydraulic extrication tools have manifold blocks that can be connected to the end of a supply hose. Multiple tools may be attached to these manifold blocks (Figure 2.65). The driver/operator should understand the limitations of this equipment and make sure that not too many tools are being attached to the system.

Figure 2.64 The hydraulic rescue tool hose reel allows tools to be deployed from the apparatus without the need to carry the power unit along with them.

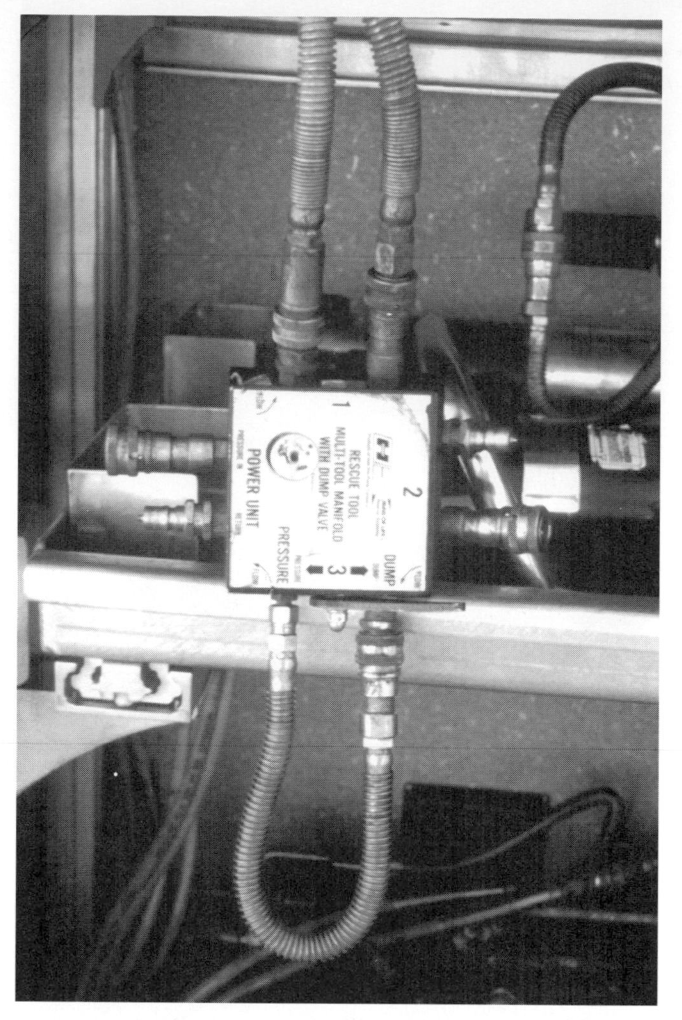

Figure 2.65 The manifold block allows more than one hydraulic rescue tool to be operated off the system at the same time.

For more detailed information on powered hydraulic extrication equipment and its operation, see IFSTA's **Principles of Extrication** manual.

◆ Portable Equipment Commonly Carried on Aerial Apparatus

In addition to the apparatus-mounted systems previously described, aerial apparatus may also carry a large array of portable tools and equipment. There are a number of reasons why this is true. One of the most obvious is that the large size of most aerial apparatus lends itself to more compartment space than is normally found on a fire department pumper. The crews assigned to aerial apparatus historically have been assigned a variety of fireground functions, including forcible entry, ventilation, search and rescue, and salvage and overhaul. In order to perform these functions, special tools and equipment may be required.

NFPA 1901 contains a list of the minimum equipment with which each aerial apparatus should be equipped. However, most fire departments carry equipment in excess of the required minimum. The follow sections highlight the required types of equipment according to the standard as well as other equipment that is commonly carried.

Ground Ladders

One of the basic tasks of a ladder or truck company is to place ground ladders in service at appropriate locations on the emergency scene. Ground ladders are required when the main aerial device cannot reach a portion of the building or when additional ladders are needed. As NFPA 1901 has been revised over the years, the minimum requirements for ground ladders to be carried on aerial apparatus have changed significantly (Figure 2.66). The current edition of NFPA 1901 requires all aerial apparatus constructed after 1999 to carry a minimum of 115 feet (35 m) of ground ladders. Each jurisdiction can decide the particular length of each ladder based on local requirements; however, the ladder complement must include at least the following:

- One attic ladder
- Two straight ladders with folding roof hooks
- Two extension ladders

As mentioned earlier in this chapter, if the aerial apparatus is going to be operated as a quint, it is only required to carry one of each of those three types of ladders totaling at least 85 feet (26 m). Many departments choose to carry a pole ladder that is 40 feet (12 m) or longer to reach fourth- or fifth-story windows that are not accessible with the main aerial device (Figure 2.67).

The driver/operator should be familiar with the ladder load carried on the apparatus he is assigned to drive. The driver/operator must also know the procedures for inspecting, testing, and cleaning ground ladders. For more information on these practices, see the IFSTA **Fire Service Ground Ladders** manual or the **Essentials of Fire Fighting** manual.

Figure 2.66 All aerial apparatus carry a large supply of ground ladders.

Figure 2.67 Pole ladders are used in locations where aerial apparatus cannot reach.

Forcible Entry Equipment

The firefighters assigned to the aerial apparatus are often charged with gaining access to structures, automobiles, or machinery so that other firefighters or medical personnel may perform their duties. NFPA 1901 requires the following forcible entry equipment, as a minimum, for aerial apparatus:

- Two flathead axes
- Three pickhead axes
- Two 3- to 4-foot (1 m to 1.3 m) pike poles with D-handles
- Two 6-foot (2 m) pike poles
- Two 8-foot (2.4 m) pike poles
- Two 12-foot (3.7 m) pike poles
- Two crowbars
- Two claw tools
- Two 12-pound (5.4 kg) sledgehammers
- One pair of 24-inch (610 mm) or longer bolt cutters
- A variety of smaller hand tools in a tool box

Most modern truck companies carry forcible entry equipment well in excess of this minimum list. Rotary, reciprocating, and chain saws are common appliances on aerial apparatus. They are used for both forcible entry and vertical ventilation operations. Hydraulic door openers (also known as rabbet tools) are used to force entry through swinging doors. Some fire departments equip aerial apparatus with cutting torches for heavy-duty entry operations. Other types of special equipment may be carried depending on local conditions or requirements.

The driver/operator should be familiar with the location, operation, inspection, and servicing of each type of forcible entry equipment carried on the apparatus. For more information on this equipment, see the IFSTA **Forcible Entry** manual or the **Essentials of Fire Fighting** manual.

Ventilation Equipment

Most of the forcible entry equipment previously described can also be used to create ventilation openings in a building that is on fire. In addition to the equipment needed to make the opening, most fire departments use some sort of mechanical means to supplement the natural flow of heat, smoke, and gases from the fire building. There are two common types of equipment used for this purpose: negative-pressure fans and positive-pressure blowers. Most departments choose to carry one type or the other, depending on their ventilation SOPs. Regardless of which type they choose to use, this equipment is most commonly carried on aerial apparatus.

Negative-pressure fans, also called smoke ejectors, are most commonly powered by electric motors. They are placed in window, door, or roof openings for the purpose of drawing out heat, smoke, and gas. *Positive-pressure blowers* are used to increase the air pressure in a structure and thus force out the byproducts of combustion (Figure 2.68). They may be powered by electric-, gasoline-, or water-driven motors.

The driver/operator will be expected to inspect and test these devices on a regular basis. The driver/operator may also be responsible for putting them in service at the emergency scene. For more information on these devices, see the IFSTA **Fire Service Ventilation** manual or the **Essentials of Fire Fighting** manual.

Salvage, Overhaul, and Loss Control Equipment

Salvage and overhaul is the process of seeking and extinguishing hidden fires and protecting building contents from further damage caused by the fire or extinguishing efforts. Truck companies are commonly assigned this function on the fireground. Many of the tools described in the forcible entry section are used to open walls and ceilings to expose fire extension. In addition to that equipment, NFPA 1901 requires aerial apparatus to be equipped with at least six salvage covers (12 x 18 feet [3.6 m x 5.5 m] minimum size) and two scoop shovels to assist with salvage and overhaul efforts.

The concept of loss control has emerged in the fire service in recent years. Loss control starts from the minute companies leave the fire station and extends throughout the entire incident. *Loss control* activities include safe driving skills, water application techniques, ventilation techniques, and salvage and overhaul. Because of their role in support activities on the fireground, truck companies must be acquainted with loss control techniques and equipment. For more information consult the IFSTA **Fire Service Loss Control** manual.

Most fire departments choose to carry loss control equipment beyond that specified by NFPA 1901. Some of the additional equipment that may be carried on the aerial apparatus includes:

- Roll of plastic sheeting that may be cut to cover contents
- Water vacuums

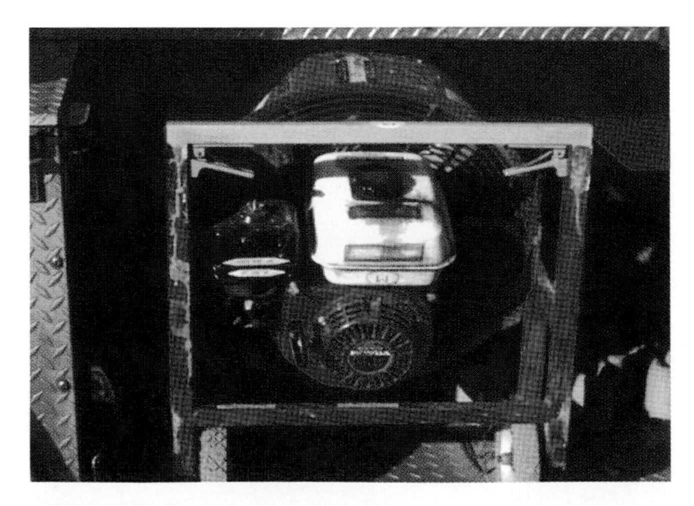

Figure 2.68 Positive-pressure blowers are carried on many aerial apparatus.

- Deodorizing equipment
- Infrared heat detectors
- Thermal imaging cameras
- Portable pumps
- Sprinkler control equipment
- Floor squeegees
- Roof patching materials
- Wood or foam blocks to lift furniture off the floor
- Electric saws and other portable electrical equipment

The driver/operator should inventory this equipment every time the apparatus is inspected. Power equipment should be started and checked to make sure that it is in good running condition. For more information on loss control equipment and practices, see the IFSTA **Fire Service Loss Control** manual or the **Essentials of Fire Fighting** manual.

Other Equipment Required by NFPA 1901

In addition to the various types of equipment previously listed, NFPA 1901 also requires the following equipment be carried on each aerial apparatus:

- One 80 B:C or larger rated portable fire extinguisher
- One 2½-gallon (10 L) or larger water portable fire extinguisher
- One self-contained breathing apparatus and one spare cylinder for each assigned seating position on the apparatus (but not less than four of each)
- One first aid kit
- Four spanner wrenches
- One hose roller/equipment hoist
- Four Class I safety harnesses meeting the requirements contained in NFPA 1983, *Fire Service Life Safety Rope and System Components*
- One 150-foot (46 m) one-person life safety rope as defined in NFPA 1983
- One 150-foot (46 m) two-person life safety rope as defined in NFPA 1983
- Two 150-foot (46 m) lengths of utility rope with a minimum breaking strength of 5,000 pounds (2 268 kg)
- Four wheel chocks that meet the requirements of SAE J348

Chapter 3

Introduction to Apparatus Inspection and Maintenance

Job Performance Requirements

This chapter provides information that will assist the reader in meeting the following job performance requirements from NFPA 1002, *Standard on Fire Apparatus Driver/Operator Professional Qualifications*, 1998 edition. Particular portions of the job performance requirements (JPRs) that are met in this chapter are noted in bold text.

2-2.1* Perform routine tests, inspections, and servicing functions on the systems and components specified in the following list, given a fire department vehicle and its manufacturer's specifications, so that the operational status of the vehicle is verified.

- **Battery(ies)**
- **Braking system**
- **Coolant system**
- **Electrical system**
- **Fuel**
- **Hydraulic fluids**
- **Oil**
- **Tires**
- **Steering system**
- **Belts**
- **Tools, appliances, and equipment**

(a) *Requisite Knowledge:* Manufacturer specifications and requirements, policies, and procedures of the jurisdiction.

(b) *Requisite Skills:* The ability to use hand tools, **recognize system problems, and correct any deficiency noted according to policies and procedures**.

2-2.2 Document the routine tests, inspections, and servicing functions, given maintenance and inspection forms, so that all items are checked for proper operation and deficiencies are reported.

(a) *Requisite Knowledge:* Departmental requirements for documenting maintenance performed, understanding the importance of accurate record keeping.

(b) *Requisite Skills:* The ability to use tools and equipment and complete all related departmental forms.

4-4.1 Perform the routine tests, inspections, and servicing functions specified in the following list in addition to those specified in the list in 2-2.1, given a fire department aerial apparatus, so that the operational readiness of the aerial apparatus is verified.

- **Cable systems (if applicable)**
- **Aerial device hydraulic systems**
- **Slides and rollers**
- **Stabilizing systems**
- **Aerial device safety systems**
- **Breathing air systems**
- **Communication systems**

(a) *Requisite Knowledge*: Manufacturer specifications and requirements, policies, and procedures of the jurisdiction.

(b) *Requisite Skills*: The ability to use hand tools, recognize system problems, and correct any deficiency noted according to policies and procedures.

Fire apparatus must always be ready to respond. Regardless of whether the truck responds to an emergency call once an hour or once a month, it must be ready to perform in the manner for which it was designed. In order to ensure this, certain preventive maintenance functions must be performed on a regular basis. Most apparatus or equipment failures can be prevented by performing routine maintenance checks. Most fire departments require driver/operators to be able to perform these routine maintenance checks and functions. NFPA 1002 also requires the driver/operator to have certain preventive maintenance skills. This chapter covers the basic skills fire apparatus driver/operators should possess.

Before continuing, it is important to differentiate between the terms maintenance and repair. *Maintenance,* as used here, means keeping apparatus in a state of usefulness or readiness. *Repair* means to restore or replace that which has become inoperable. Apparatus or equipment that is said to be in a good state of repair has probably been well maintained. Preventive maintenance ensures apparatus reliability, reduces the frequency and cost of repairs, and lessens out-of-service time. The purpose of preventive maintenance is to try to eliminate unexpected and catastrophic failures that could threaten life and/or property.

Preventive maintenance functions may be carried out by several different people. Fire departments may have a designated apparatus maintenance officer who routinely checks and services the apparatus. Fire departments should have trained mechanics who perform more detailed maintenance procedures. However, the driver/operator should be able to perform basic maintenance functions. In almost all cases, repair functions are carried out by qualified mechanics.

A Systematic Maintenance Program

Every fire department should have standard operating procedures (SOPs) for a systematic apparatus maintenance program. The SOPs should identify who performs certain maintenance functions, when these functions are to be performed, how problems that are detected are corrected or reported, and how the process is documented.

The SOP should clearly dictate those items that driver/operators are responsible for checking and which conditions they are allowed to correct on their own. Most departments allow the driver/operator to correct certain deficiencies such as low fluid levels and burned-out lightbulbs. More detailed repairs need to be made by a certified mechanic. Large fire departments have their own repair shops and mechanics for this purpose (Figure 3.1). These mechanics may have their own vehicles and be able to come to a fire station or incident scene to perform a repair. Smaller fire departments may have a local automotive/truck repair business that assists them with these functions.

The schedule for performing maintenance functions and checks varies from department to department. Typically, career fire departments require driver/operators to perform apparatus inspections and maintenance checks at the beginning of each tour of duty. They may also specify that more detailed work be completed on a weekly or monthly basis. Volunteer fire departments should establish a procedure by which all apparatus are inspected and maintained on at least a weekly or biweekly schedule.

Each fire department apparatus and equipment inspection and maintenance SOP should dictate how maintenance and inspection results should be documented and transmitted to the proper person in the fire department administrative system. Written forms or computer programs may be used to record the information. Appendix A contains several examples of apparatus inspection forms that may be used. Fire departments should maintain an effective filing system that allows the information on these reports to be reviewed, stored, and retrieved when required.

Apparatus maintenance and inspection records serve many functions. In a warranty claim, these records may be needed to document that the necessary maintenance was performed. In the event of an accident, maintenance records are likely to be scrutinized by the accident investigators. Proper documentation of recurrent repairs can also assist

Figure 3.1 Large fire departments typically have their own apparatus maintenance facilities and staff.

in deciding whether to purchase new apparatus in lieu of continued repairs on an older unit. All driver/operators must be trained to use their department's record-keeping system.

 Cleanliness

One often overlooked but very important part of any apparatus inspection and maintenance program is the cleanliness of the vehicle and equipment. Many people look at apparatus cleanliness only from the standpoint of maintaining good public relations. The public sees a piece of fire apparatus as a unit of protection in which it has invested many thousands of dollars. Apparatus marred with dirt, oil, and road grime damages public relations because individuals may feel their investment is not being properly protected.

Although the public relations aspect of cleaning apparatus is important, there are other more important reasons for keeping fire apparatus clean. A clean engine and clean functional parts permit proper inspection, thus helping to ensure efficient operation. Fire apparatus should be kept clean underneath as well as on top. Oil, moisture, dirt, and grime should not be permitted to collect. Some of the more vulnerable areas are the engine, wiring, carburetor or fuel injectors, and controls.

Keeping the apparatus body clean also helps promote a longer vehicle life. This is particularly true in jurisdictions where the use of road salt during inclement winter weather is prevalent (Figure 3.2). These chemicals have a corrosive effect on the steel components of the apparatus body and chassis. Frequent washing reduces the likelihood of damage caused by these chemicals.

Figure 3.2 Departments in areas that receive frequent snow falls must have stringent apparatus cleaning policies to negate the effects of road salt and other similar chemicals that may be detrimental to the apparatus.

On the other hand, overcleaning of the apparatus can also have negative effects. Fire departments whose members are fanatical about apparatus cleanliness often run into problems associated with the removal of lubrication from chassis, engine, pump, and aerial device components. This occurs when any combination of degreasing agents, steam cleaners, and/or pressure washing equipment is used to clean the underside of the apparatus or the aerial device. Care should be taken not to strip all necessary lubrication from the apparatus when washing the apparatus body components. After a particularly heavy cleaning, it may be necessary to perform routine lubricating functions to ensure no unnecessary wear occurs on the apparatus.

Most apparatus manufacturers provide the fire department with specific instructions on how to clean their fire apparatus. If specific instructions are not available, use the following guidelines for apparatus cleaning.

Washing

Washing the exterior of the apparatus, including the aerial device, is the most commonly performed function. Most apparatus manufacturers recommend slightly different procedures for cleaning the apparatus based on its age. Newer apparatus require gentler cleaning procedures than do older apparatus to avoid damage to new paint, detailing, and clear-coat protectants.

During the first six months after an apparatus is received, while the paint and protective coating are new and unseasoned, the vehicle should be washed frequently with cold water to harden the paint and keep it from spotting. To ensure the best appearance of the vehicle in the future and to reduce the chance of damaging new paint and protective coatings, the following washing instructions are recommended:

- Use a garden hose without a nozzle to apply water to the apparatus. The water pressure should be set so that the stream from the end of the hose is no more than 1 foot (0.3 m) in length (Figure 3.3). Higher pressures can drive grit and debris into the finish.

Figure 3.3 When washing new vehicles, the effective range of the water stream should be about 1 foot (0.3 m).

Figure 3.4 Keeping the apparatus glass clean is not only a maintenance issue but also a safety issue.

- Never remove dust or grit by dry rubbing.
- Wash the vehicle with a good automotive shampoo. Follow the shampoo instructions for proper use.
- Do not wash with extremely hot water or while the surface of the vehicle is hot.
- Rinse as much of the loose dirt from the vehicle as possible before applying the shampoo and water. This reduces the chance of scratching the surface when applying shampoo.
- Try to wash mud, dirt, insects, soot, tar, grease, and road salts off the vehicle before they have a chance to dry.
- Never use gasoline or other solvents to remove grease or tar from painted surfaces. Use only approved solvents to remove grease or tar from nonpainted surfaces.
- Dry the vehicle with a clean chamois rinsed frequently with clean water, or use a rubber squeegee. Failure to dry the vehicle or the area around it completely will also encourage corrosion.

Once the new vehicle's finish is properly cured (according to the owner's manual), either a garden hose with a nozzle or a pressure washer may be used to speed cleaning of the apparatus. However, the apparatus still must be regularly hand washed with soapy water to ensure proper cleanliness.

Glass Care

In general, warm soapy water or commercial glass cleaners should be used to clean automotive glass (Figure 3.4). These may be used in conjunction with paper towels or cloth rags. However, dry towels or rags should not be used by themselves because they may allow grit to scratch the surface of the glass. Do not use putty knives, steel wool, or other metal objects to remove deposits from the glass. Avoid attaching decals to the windows that may obstruct the driver/operator's vision.

Interior Cleaning

It is important to keep seat upholstery, dashboard and engine compartment coverings, and floor finishes clean because an accumulation of dirt may cause deterioration of these finishes. Large, loose dirt particles should be swept or vacuumed first (Figure 3.5). Then, warm soapy water or commercial cleaning products may be used to clean the surfaces of these materials. Some manufacturers may specify particular cleaning agents or protective dressings that should be used on their materials.

WARNING!

Many cleaners are toxic, are flammable, or cause damage to interior surfaces. Do not use volatile cleaning solvents, such as acetone, lacquer thinner, enamel reducer, nail polish remover, laundry soap, bleach, gasoline, naphtha, or carbon tetrachloride, to clean interior surfaces. Be sure that the vehicle is well ventilated when using any cleaning products inside the cab or crew-riding area.

Figure 3.5 Depending on the interior finish of the cab, it may be more effective to vacuum rather than sweep out dirt.

Waxing

Fire departments should follow the apparatus manufacturer's instructions regarding the application of wax or similar polishes to the exterior of the apparatus. On many newer apparatus, the application of these products may no longer be necessary and, in fact, may damage some clear-coat protective-seal finishes that are applied over paints. Check with your manufacturer to see which is the case with your apparatus. Wax or polish should not be applied to the apparatus until its paint is at least six months old. In general, before applying polish or wax, the apparatus should be washed and dried. The wax or polish may then be applied with a soft cloth and buffed using a soft cloth or mechanical buffer.

Apparatus Inspection Procedures

The driver/operator should follow a systematic procedure for inspecting his apparatus. Having a sys-

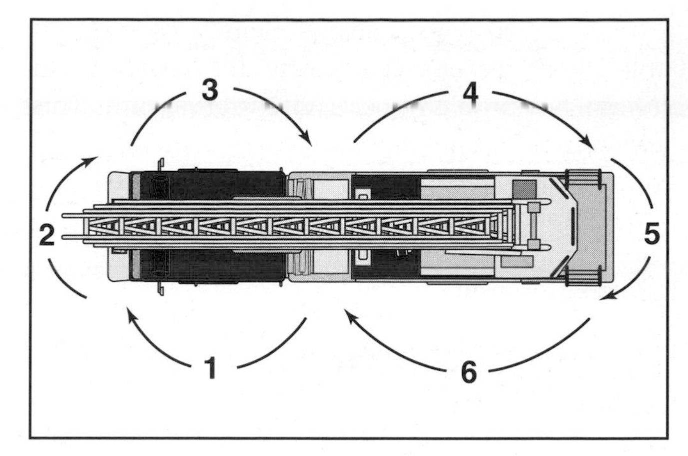

Figure 3.6 This diagram highlights the route for the walk-around inspection procedure.

tematic method helps ensure that all important items are checked every time the inspection is made. The information in the following sections is based on the requirements contained in NFPA 1002 and the government pretrip inspection requirements for obtaining a commercial driver's license (CDL). Even if your jurisdiction does not require that fire apparatus driver/operators obtain a CDL, these pretrip inspection principles provide a sound basis for the type of inspection that all fire apparatus driver/operators should be able to perform. These are the types of checks that career personnel should perform at the beginning of each tour of duty and volunteer personnel should do on a weekly or biweekly basis.

One particular method of performing an apparatus/pretrip inspection is referred to as the *circle* or *walk-around* method. This method involves the driver/operator starting at the driver's door on the cab and working around the apparatus in a clockwise pattern. As the driver/operator circles the apparatus, he checks important areas along the way. In the final step, the driver/operator gets in the cab, starts the apparatus, and performs a functional check on apparatus systems. Figure 3.6 shows one method for performing the walk-around inspection. Exact procedures vary depending on departmental policies and vehicle design. For example, a vehicle with an engine ahead of the cab is checked in a different order than one with an engine behind or underneath the cab. The information contained in this section may be used during the course of the inspection, regardless of the exact order.

If the records from previous inspections are available, the driver/operator may wish to review them to see if any problems were noted at that time. This allows the current inspector to pay extra attention for reoccurrence of those concerns. If possible, the driver/operator may also talk to the last person to inspect or drive the apparatus to get any important information from him.

Approaching the Vehicle

The driver/operator should actually begin the inspection process by being observant when approaching the vehicle. Look for any general problems that are readily apparent by simply looking at the vehicle. Things such as vehicle damage or a severe leaning to one side are examples of things that could be readily apparent. Of course, be sure that a leaning vehicle is not doing so because it is resting on a sloped surface. Some apparatus bay floors are sloped to the extent that the vehicle appears to be leaning.

Look beneath the vehicle for spots that indicate leaking vehicle fluids such as water, coolant, oil, brake fluid, hydraulic fluid, or transmission fluid (Figure 3.7). These could be symptoms of serious problems. If the vehicle is indoors, make sure that proper ventilation equipment is in place or that doors are open to vent vehicle exhaust once the apparatus is started (Figure 3.8). If the weather permits, park the apparatus outside for functional tests. Obviously, it will be necessary to park the apparatus outside the station if the aerial device is going to be deployed. Chock the wheels whenever the apparatus is parked. (**CAUTION:** Do not run gasoline and diesel engines in unvented areas for any period of time. The buildup of carbon monoxide in that area could be harmful to personnel. In particular, diesel exhaust is known to emit benzene derivatives, which laboratory tests have shown to be carcinogenic.)

Figure 3.7 Check under the vehicle for leaking fluids.

Left- and Right-Front Side Inspection

The first portion of the vehicle specifically checked by the driver/operator should be the left (driver's) and right-front sides, or cab, of the vehicle. After the driver/operator inspects the left front of the vehicle, he repeats that procedure on the right (passenger's or officer's) side. The driver/operator should begin this portion of the inspection by making a general observation of the cab on that side of the vehicle for any damage that was not noted in previous inspections.

The driver/operator should then make sure that various aspects of the cab doors are in proper order.

Figure 3.8 Many fire stations are equipped with apparatus exhaust emission control systems.

Each door should close tightly, and the door latch should work as it was designed. There should be little or no play in the operation of the latch. Make sure that all door window glass is intact and clean. Make sure that all steps, platforms, handrails, and ladders are securely mounted and without deformation.

If the vehicle is equipped with saddle fuel tanks beneath the door opening, check to make sure that there are no apparent leaks or other problems with the tank. If the tank contains a fill cap, make sure that it is tightly in place.

The driver/operator should then check the condition of the tire and wheel on that side of the vehicle. A quick visual check of the wheel itself should be made for any signs of missing, bent, or broken studs, lugs, or clamps. Check each lug nut by hand to feel if any are loose. The wheels should not be cracked or damaged because this prevents sealing of the tire to the rim. Look for unusual accumulations of brake dust on the wheel. This could signify a problem with the braking system. Inspect the wheel/tire assembly for other leaks. While seals that retain axle gear oil may show slight seepage and still be serviceable, trails of fluid on the wheel or tire are unacceptable. The driver/operator may also choose to make a quick visual inspection of the suspension components found behind the wheel and tire. Look for defects involving the springs, spring hangers, shackles, U-bolts, and shock absorbers. Springs should not have cracked or otherwise broken leaves. With the vehicle on a level surface, the springs on each side of the vehicle should have approximately the same amount of deflection.

There are a number of things that should be checked relative to the tires themselves. Some of the more important things that driver/operators should check include the following:

- ***Proper tire inflation*** — The tires should be inflated to the pressure noted by the manufacturer on the side of the tire or on a plate inside the door of the apparatus (Figure 3.9). Too much or too little pressure can damage the tire and cause poor road-handling characteristics.
- ***Valve stem condition*** — The valve stem should not be cut, cracked, or loose. The valve stem cap should be securely in place.

Figure 3.9 Check the tires for proper inflation.

- ***Tire condition*** — Check for proper tire type, tread depth (varies according to state or provincial inspection requirements), tread separation, cupping, excessive wear to the sidewalls, cuts, or objects impaled in the tire.

The driver/operator may choose to check equipment in the cab at this time. This includes items such as information resource and map books, portable fire extinguishers, self-contained breathing apparatus (SCBA), and emergency medical equipment. All this equipment must be in proper working order and securely stowed. Pay particular attention to seat-mounted SCBA and their retention systems. Make sure that all the appropriate retention straps are fastened and the SCBA are held securely in place. Consult IFSTA's **Essentials of Fire Fighting** manual for more information on how to properly inspect the fire fighting and rescue equipment carried on the apparatus. Once the left and right sides of the vehicle have been inspected, proceed to the front of the vehicle.

Front Inspection

As with the left- and right-side inspections, the first thing that should be noted when moving to the front of the vehicle is any significant body damage that was not present in previous inspections. The driver/operator may choose to take a quick look beneath the vehicle to note any obvious damage to the front axle, steering system, or pump piping (if present). Be observant for any loose, bent, worn, damaged, or missing parts. Some vehicle air systems have storage tanks that must be drained manually. Refer to the owner's manual for the location and procedures for manual drains.

Check the condition of the windshield and the wiper blades. The windshield should be free of defects and should be clean. The wipers should be held against the windshield with an appropriate amount of tension. The wiper blades should be intact and in good condition.

With the apparatus running or hooked into the electrical charging system, check the condition of all lights, both running and emergency, for proper operation. It may be helpful to have a second firefighter in the cab operate the various light switches so that their operation can be properly checked. The lenses for all lights should be in place and not cracked or broken. Make sure that all bulbs are working. Check the function of high-beam headlights and turn signals. (**NOTE:** It may be desirable to skip this check for now and wait until the apparatus operational checks are made. This is a matter of individual preference.)

Check any audible warning devices on the front of the vehicle for visible damage. This includes electronic siren speakers, mechanical sirens, and air horns. (**CAUTION:** Do not test the operation of audible warning devices with someone standing in front of the vehicle. This can cause hearing damage to that person. The driver/operator can test these devices in the cab when no one else is in a position to be harmed.)

Many modern fire department vehicles have functional emergency equipment on the front bumper area. The driver/operator should check this equipment to make sure that it is in place and operable. The following is a list of some of the more common devices found on the front of fire apparatus and things about them that should be checked.

- ***Pump intakes***—Aerial apparatus equipped with a fire pump may have a pump intake on the front of the apparatus (Figure 3.10). If the intake is normally capped, make sure that the cap is tight enough to prevent air leaks when attempting to draft, but not so tight that it cannot be easily removed. If intake hose is normally preconnected, make sure that the hose is firmly attached. If an intake valve is provided at that location, make sure that it is fully closed. Intake hose should be in good physical condition and properly stowed.
- ***Pump discharges*** — Aerial apparatus equipped with fire pumps may have pump discharges and attack handlines located on the front of the apparatus (Figure 3.11). Check these in the same manner as described for intakes. Hose may be attached to the front bumper in a variety of ways. These include being in recessed wells, as cross

Figure 3.10 If the apparatus has a front fire pump intake, make sure that there are no obstructions and that it is in good working order.

Figure 3.11 Make sure that any handlines on the front of the apparatus are properly packed and secured.

lays, and on booster reels. Make sure that the hose is loaded properly and is secure for road travel. Nozzles should be clean and in place.

- *Hydraulic rescue tool systems* — It is becoming increasingly popular to mount hydraulic hose reels and powered hydraulic rescue tools on the front bumper of the vehicle. These should be checked for proper operation, damage, and cleanliness.
- *Electric cord reels* — Some apparatus have power-rewind, electric cord reels mounted on the front bumper. These allow electric cords to be quickly extended from the apparatus in order to provide power for portable floodlights, smoke ejectors, or other electrical equipment. Make sure that the electric cord and connections are in good physical condition. Also ensure that the reel operates in the manner for which it is intended.

If the apparatus is equipped with a rear-mounted telescoping aerial device, the driver/operator may choose to inspect the platform or tip of the ladder at this time. However, most driver/operators choose to pass on inspecting the aerial device at this point in favor of doing it all at once later in the process. (**NOTE:** Information on inspecting the aerial device is found later in this chapter.)

Some models of telescoping aerial platform apparatus have one or two stabilizers located on the front of the apparatus (Figure 3.12). These stabilizer(s) may be visually inspected at this time for any signs of physical damage. Also, make sure that any stabilizer pads that are supposed to be stowed near the stabilizers are present and in good condition. The operation of the stabilizer(s) will be checked later in the inspection.

Figure 3.12 If the apparatus has stabilizers located on the front bumper area, visually inspect them at this time for any noticeable defects.

Once all front-end equipment has been checked, the driver/operator may proceed to the right-front side of the vehicle and follow the same procedures as listed for the left-front side. Once the right-front side inspection is complete, the driver/operator may continue down the right side of the vehicle using the same procedures as described for the left-front side.

Left- and Right-Rear Side Inspections

This part of the inspection should cover everything from the rear of the cab to the tailboard on each side of the apparatus (Figure 3.13). Again, note any obvious body damage that has occurred since the previous inspection. The same principles listed in the front-side inspection for checking the tires and suspension components may be followed here. Keep in mind that most fire apparatus have dual wheels on the rear axle(s). In addition to checking tire conditions as previously listed, the driver/operator should also make sure that dual tires do not come in contact with each other or other parts of the vehicle. Rear wheels should be equipped with splash guards (mud flaps). Splash guards should be properly attached to the vehicle, be in good condition, and not be dragging on the ground.

Apparatus in cold weather climates may be equipped with automatic snow chains. These are mechanical devices that are activated by a switch in the cab. When activated, a rotating hub with numerous lengths of tire chain swings into place just ahead of the rear tires (Figure 3.14). The chains

Figure 3.13 The side inspection focuses on the portion of the apparatus from the rear of the cab to the rear step.

swing around in a rotating motion so that they fall beneath the rear tires as they move forward. This provides better traction on snow- or ice-covered roads. The driver/operator should inspect the automatic snow chains to make sure that all chains are present and in good condition. During periods of inclement weather, it may be desirable to activate the chains and make sure that they are operating properly.

The driver/operator should open all compartment doors and inspect the equipment inside them. Make sure that all equipment that is supposed to be in each compartment is actually there and properly stowed. The driver/operator may choose to perform equipment inspections at this time. As previously stated, information on inspecting equipment is contained in the IFSTA **Essentials of Fire Fighting** manual. The compartment and the equipment it contains should be neat and clean. When closing each door, make sure that the door closes tightly and latches properly.

Aerial apparatus equipped with a fire pump may have hose stored midship or on the side of the vehicle. This hose should be examined for proper stowing and security. This includes preconnected attack lines that traverse the midship area of the apparatus or are on top of the fender compartments (Figure 3.15). Top-mounted booster reels may also be checked at this time. The water level in the booster tank may also be visually checked through the top vent opening or sight glass. Some aerial apparatus, including those that are not equipped with a fire pump, may have hose bundles for high-rise fire fighting stored on the side of the apparatus (Figure 3.16). Make sure that the hose and other associated equipment in the bundle are in place and in good condition.

Check equipment stored on the exterior of the vehicle to make sure that it is in good physical condition and properly stowed. This includes ladders, intake hose, forcible entry tools, SCBA and/or spare cylinders, handlights, floodlights, cord reels, portable fire extinguishers, and other portable equipment. Equipment that is stored above the pump panel area may also be checked at this time.

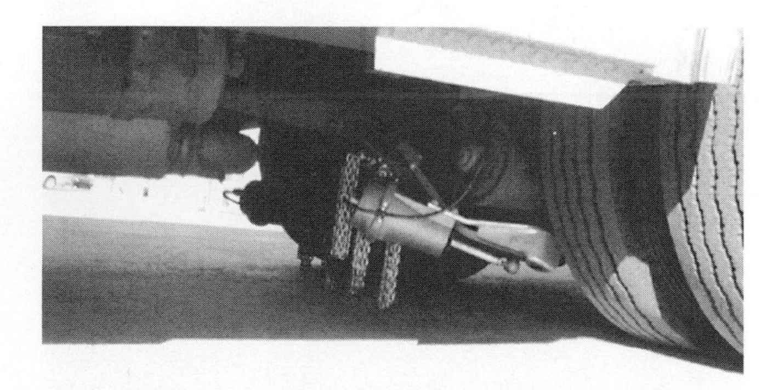

Figure 3.14 Automatic tire chain systems are becoming increasingly popular in jurisdictions that experience freezing precipitation.

Figure 3.15 Inspect any handlines that are found on the midship portion of the apparatus.

Figure 3.16 Make sure that hose bundles are properly constructed and appropriately secured to the apparatus.

Note the condition of reflective striping on the side of the apparatus. Major losses of this striping affect the vehicle's visibility, particularly at night. Also check the operation of side-mounted warning lights. Most aerial apparatus will have at least one stabilizer on each side of the apparatus. Visually inspect the stabilizer for any signs of physical damage. Also make sure that any stabilizer pads that are supposed to be stowed near the stabilizer are present and in good condition. The operation of the stabilizers will be checked later in the inspection. Once everything on the right side of the vehicle has been checked, proceed to the rear of the vehicle.

Figure 3.17 Ground ladders and pike poles are commonly stored on the rear of the aerial apparatus. They may be inspected during this portion of the process.

Rear Inspection

When inspecting the rear of the vehicle, check the rear bumper or tailboard for new damage. Using the same procedure as that used on the front of the vehicle, make sure that all running and warning lights on the rear of the vehicle are in proper working order. Check equipment stored in the rear compartment to make sure that it is present, clean, operable, and properly stowed. Make sure that the rear compartment door opens and closes properly. Any equipment stored on the outside of the rear of the apparatus should be in proper working order and securely stowed. This includes portable fire extinguishers, positive-pressure blowers, spanner wrenches, hydrant wrenches, hydrant valves, portable master stream devices, etc. Check any towing attachments for defects.

If the aerial apparatus is equipped with fire hose, the driver/operator should also inspect the hose loads in the main hose bed. Make sure that the appropriate amount of hose is loaded on the apparatus and that it is loaded and finished correctly. The hose bed cover should be in good condition and in place. If solid hose bed doors are used to protect the hose, check to make sure that they stay open when necessary. In some cases it may be necessary to lift the aerial device in order to fully inspect the hose load. If this is the case, the driver/operator may choose to skip the hose inspection at this time and perform it later when the operational check of the aerial device is conducted. Check any other equipment stored on the rear of the apparatus for proper condition and stowing. This includes ground ladders and pike poles (Figure 3.17).

Located on the rear of the apparatus will be a variety of items directly related to the aerial device. These include the aerial device control station, the stabilizer controls, and waterway inlets. The driver/operator may inspect these items at this time or may wait until the operational check of the aerial device is made. Information on inspecting these items is contained later in this chapter.

Once the rear of the vehicle has been inspected, the driver/operator should proceed to the left-rear side of the vehicle. Check this side using the procedure covered in the previous section.

In-Cab Inspection

When the driver/operator has completed inspecting the outside of the apparatus, he should enter the cab and begin the mechanical check of the apparatus. Before starting the apparatus, if it is not already running, make sure that the seat and mirrors are adjusted in a suitable manner. If the vehicle is not already running, it should be started at this time. Follow the manufacturer's instructions for starting the apparatus. All electrical switches should be in the off position when the apparatus is started to avoid an excessive load on the battery. Sudden temperature changes are not good for any engine. When an engine is started for nonemergency response, do not run it under full load until the engine has had time to warm up to its operating temperature. Likewise, anytime an engine has been running at full load, follow the engine manufacturer's recommendation for shutdown procedures. For more information on proper starting techniques, see Chapter 4 of this manual.

The seatbelt/restraint should be securely mounted and should operate freely without binding. Webbing must not be cut or frayed. Buckles must open and close freely. Mirrors should not be missing or broken. Check the tilt/telescopic steering wheel for proper position (Figure 3.18).

Once the vehicle is running, the driver/operator should ensure that all gauges on the dashboard show the apparatus to be functioning in the normal operating range. Depending on the design of the apparatus, some or all of the following gauges may be found on the dashboard:

- Speedometer/odometer — Indicates the speed of the vehicle in miles or kilometers per hour and the number of miles traveled.
- Tachometer — Measures the speed of the engine in revolutions per minute (rpm).
- Oil pressure — Shows the pressure of the lubricating oil in the engine.

Figure 3.18 The driver/operator should adjust the steering wheel to a comfortable position, if possible.

- Fuel gauge — Indicates the amount of fuel in the vehicle fuel tank(s).
- Ammeter/voltmeter — Gives an indication of the battery voltage and acknowledges the fact that the vehicle's alternator is charging the electrical system.
- Air pressure — Indicates the amount of air pressure available for operating the vehicle's braking system and other pneumatic equipment.
- Coolant temperature — Indicates the temperature of the fluid in the vehicle's engine coolant system.
- Hydraulic pressure gauge — Provides a measure of the pressure on the fluid in the vehicle's hydraulic system.

The speedometer should be at or very near zero with the truck parked (Figure 3.19). If it is showing anything above that, one of the two following possibilities exists:

1. The gauge is defective.
2. The truck is in pump gear or the aerial device power take-off (PTO) has been activated. In either case, deactivate the control and return the apparatus to road gear at this time. Information on how to do this is contained in Chapter 6 of this manual.

The fuel gauge should be checked to make sure that an adequate amount of fuel is in the vehicle's fuel tank. Each department usually has its own policy on the minimum level of fuel that is permissible in the vehicle. As a rule of thumb, it is generally best to keep the fuel tank at least three-quarters full

Figure 3.19 The speedometer should show 0 mph (km/h) when the apparatus is parked.

at all times. This ensures that the apparatus can be operated for an extended period of time without the need to be refueled. The driver/operator should know the size of the fuel tank and how long it will last during prolonged operations so that fuel can be requested in time when needed.

All other gauges should be checked to make sure that they are operating within the intended limits specified by the apparatus manufacturer. Typically, these limits are graphically noted on the face of the gauge. The driver/operator should also ensure that all controls located in the cab are in proper operating condition. These would include the following:

- Electrical equipment switches
- Turn signal switches
- High beam headlight switches
- Heating and air-conditioning controls
- Radio controls
- Audible warning device controls (sirens, auto warning horns, air horns, back-up alarms, etc.) (Figure 3.20) (**NOTE:** Appropriate hearing protection should be worn if any personnel will be exposed to noise levels in excess of 90 decibels.)
- Controls for any computer equipment in the cab (mobile data terminal [MDT], mobile computer terminal [MCT], etc.)
- Windshield wiper controls
- Window defroster controls
- Automatic snow chain control (if applicable)

Figure 3.20 Check the audible warning device controls when the rest of the items in the cab are being checked.

- Interior cab lights
- Auxiliary braking device controls

The driver/operator should keep in mind that newer apparatus may be equipped with electrical load management systems. These devices are intended to prevent an overload of the vehicle's electrical generation system. Overload can be a problem because of the large amount of electrical equipment that tends to be added to modern apparatus. In general, these devices incorporate a load sequencer and load monitor into the same device. The *load sequencer* turns on various lights at specified intervals so that the start-up electrical load for all the devices does not occur at the same time. The *load monitor* "watches" the system for added electrical loads that threaten to overload the system. When an overload condition occurs, the load monitor is designed to shut down less important electrical equipment to prevent the overload. This is referred to as *load shedding.* For example, when activating the electric inverter to turn on two 500-watt floodlights, the load monitor may shut down the cab air-conditioning system or compartment lights. Driver/operators must understand the design of the electrical load management system on their apparatus so that they can determine if it is operating properly. They must also know the difference between load shedding and an electrical system malfunction.

If the apparatus is equipped with a manual shift transmission, the driver/operator should check the adjustment of the clutch pedal. The pedal should not have insufficient or excessive free play (also called *free travel*). *Free play* is the distance that the pedal must be pushed before the throw-out bearing actually contacts the clutch release fingers. Insufficient free play will cause the clutch to slip, overheat, and wear out sooner than necessary. The throw-out bearing will also have a shorter life. Excessive free play may result in the clutch not releasing completely. This can cause harsh shifting, gear clash, and damage to gear teeth. Driver/operators should be familiar with the normal amount of free play in vehicles they are assigned to drive. Any clutch that does not have the appropriate amount of free play should be checked as soon as possible by a certified mechanic.

The steering system should be checked for proper adjustment and reaction. The driver/operator can most effectively accomplish this by checking the steering wheel for excess play that does not result in the actual movement of the vehicle's front tires. In general, steering wheel play should be no more than about 10° in either direction (Figure 3.21). On a steering wheel that has a 20-inch (500 mm) diameter, this will mean a play of about 2 inches (50 mm) in either direction. Play that exceeds these parameters could indicate a serious steering problem that might result in the driver/operator losing control of the apparatus under otherwise reasonable driving conditions.

Making sure that the vehicle's brakes are in proper operating order is an extremely important part of the vehicle inspection process. Many serious fire apparatus collisions have been caused by faulty brakes. There are a number of tests that can be used to check the function of the brakes. Federal, state, or provincial laws may dictate how and when brakes are tested. This section highlights the more common ones.

Most large, modern fire apparatus are equipped with air-operated braking systems. Smaller late-model apparatus and some older large apparatus are equipped with hydraulic braking systems. Most newer apparatus, regardless of whether they have air-operated or hydraulic brakes, are equipped with antilock braking systems (ABS) that reduce the possibility of the apparatus being thrown into a skid when the brakes are fully applied. Driver/operators must know the exact type of braking system on their apparatus so that the apparatus may be tested and driven in an appropriate manner. Procedures for using the different types of braking systems during road travel are covered in Chapter 4 of this manual.

An NFPA 1901 braking test requires that new apparatus be brought to a complete stop from a speed of 20 mph (32 km/h) in a distance not to exceed 35 feet (10.7 m). This test must be performed on a dry, paved surface. It is not recommended to conduct this test on a regular basis because it will cause excessive wear on the braking system components. The standard also requires the parking brake to hold the apparatus in place on a grade of 20 percent (Figure 3.22). In many jurisdictions, this type of grade may not be available, so testing by this method may not be practical.

There are several other NFPA requirements that driver/operators should be aware of and watch for on a regular basis. On apparatus equipped with air brakes, the standard requires the air pressure to build to a sufficient level to allow vehicle operations within 60 seconds of starting. If the driver/operator has to run the vehicle for more than 60 seconds to build sufficient air pressure, the apparatus should be checked by a certified mechanic. In order to maintain adequate air pressures when the apparatus is not in use, it may be necessary to add

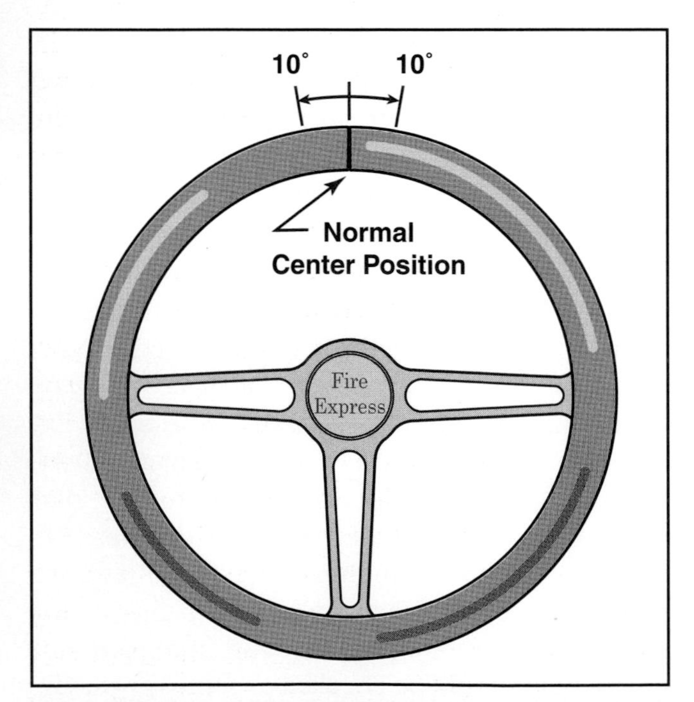

Figure 3.21 The steering wheel should have no more than 10° of free play in either direction from center.

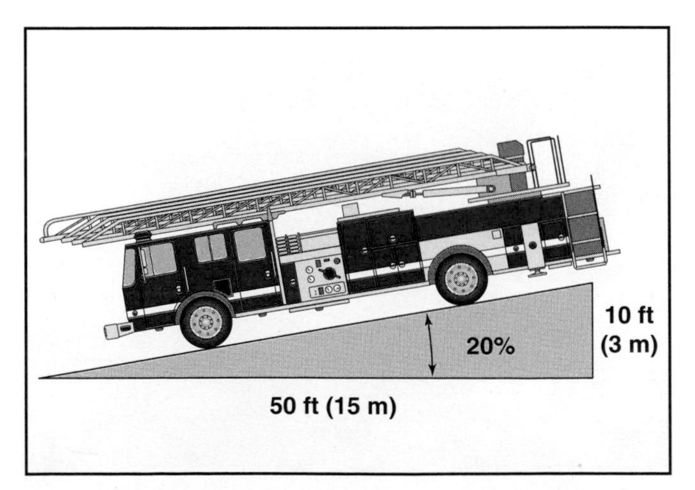

Figure 3.22 The parking brake must hold the apparatus in place on a grade of 20 percent.

an electric air compressor or to hook the apparatus into a fire station compressed air system. This prevents delays in emergency responses.

Apparatus with air brakes are to be equipped with an air pressure protection valve that prevents the air horns from being operated when the pressure in the air reservoir drops below 80 psi (552 kPa). If the driver/operator notices that this is a problem, it must be reported to the appropriate person and corrected.

To test the road brakes, allow the apparatus to move forward at about 5 mph (10 km/h). Then, push down on the brake pedal firmly. The apparatus should come to a complete stop within about 20 feet (6 m). Any of the following conditions may signify a brake problem and will require a mechanic to check the truck more thoroughly:

- The vehicle pulls to one side when the brakes are applied.
- The brake pedal has a mushy or otherwise unusual "feel" when the brakes are applied.
- The vehicle does not stop within about 20 feet (6 m).

The parking brake may also be tested in a similar manner. Allow the apparatus to move forward at about 5 mph (10 km/h). Then, activate the parking brake control (Figure 3.23). The apparatus should come to a complete stop within about 20 feet (6 m).

One of the most common causes of brake malfunction is the buildup of condensation in the air reservoirs supplying the braking system. This condensation forms from humidity in the air that is compressed into the system. The most common

Figure 3.23 A typical push-pull parking brake control.

problems associated with excessive condensation include:

- Diminished braking capacity during road travel
- Moisture freezes in the line to the air system governor causing air pressure to rise above normal in the braking system
- Moisture freezes while the parking brake is set preventing the brake from being released

Newer apparatus may be equipped with automatic bleeding devices that discharge condensation when it reaches an unacceptable level. If the apparatus does not have this type of equipment, it will be necessary to manually bleed the condensation from the braking system. The chassis manufacturer will provide information on when and how this should be performed on its apparatus. It will also provide information on thawing procedures should frozen condensation become a problem. Consult the apparatus manufacturer's operating manual or laws affecting the local jurisdiction for more detailed information on the braking system and brake tests.

Engine Compartment Inspection

Once the entire exterior of the apparatus has been inspected and the in-cab checks have been completed, the driver/operator should shut down the vehicle and prepare to perform some routine checks and preventive maintenance procedures in the engine compartment. Depending on personal preference and departmental SOPs, the driver/operator may prefer to perform these checks before the apparatus has been started. Keep in mind that if this is the case, the readings of some fluid levels (crankcase oil, transmission fluid) have to be adjusted for the cold engine. Just remember that while it is acceptable to perform these checks either before or after the engine has been run, most — with the exception of automatic transmission fluid level — should not be done *while* the engine is running. Also keep in mind that it is best to have the vehicle parked on a flat surface when checking fluid levels. This ensures a more accurate reading.

For vehicles with a tilt cab, check to ensure that the level and/or control mechanism operates freely without binding. Check for the proper operation of cab lift motors and pumps. If the stowed aerial device extends over the cab, it will be necessary to move the aerial device before the cab can be tilted (Figure 3.24). (**CAUTION:** Make sure that there is adequate vertical clearance and that all loose equipment in the cab is removed or secured before the cab is tilted. This will prevent damage to the vehicle.)

There is no established order in which to check the necessary items in the engine compartment. Each driver/operator will probably have his own preferred way of doing it. It is often advisable to follow the order of the required tasks listed on a department apparatus inspection form. As a minimum, the following items should be checked in the engine compartment:

- ***Engine (crankcase) oil level*** — Use the dipstick that is provided to ensure that the oil level falls within the proper parameters (Figure 3.25). Keep in mind that a vehicle whose engine has just been shut off will show a lower oil level than if the same engine had been sitting without running for an extended period of time. It is best to wait a few minutes after the engine is shut off before checking the oil level. If the oil level is low, add a sufficient amount of the approved type of oil through the fill opening on the engine block. Keep in mind that on large trucks the dipstick may be in gallons (instead of quarts as it is on cars and light-duty trucks). Consult the manufacturer's operating manual for direction on the proper type of oil to use. The manufacturer's manual will "recommend" the Society of Automotive Engineers (SAE) numbers for the engine oil. The SAE number indicates only the viscosity. Some other essential characteristics of oil are corrosion protection, foaming, sludging, and carbon accumulation, which may be controlled by the refiner. Do not mix different types of engine oil.
- ***Engine air filter*** — Inspect the air intake system for signs of damage (Figure 3.26). Some air intake systems include an air filter restriction gauge to show when it is time to change the filter. Refer to the owner's manual for details.

Figure 3.24 It may be necessary to raise the aerial device in order to tilt the cab.

Figure 3.25 Check the oil level using the dipstick.

Figure 3.26 The engine air filter should be relatively clean.

- *Emergency shutdown* — Test according to manufacturer's instructions to ensure that the emergency shutdown system operates properly and resets to the proper position. Some diesel engines incorporate an emergency shutdown switch and a manual reset.
- *Exhaust system* — Visually inspect the exhaust system for damage. Ensure that the rain cap on the exhaust system operates freely.
- *Radiator coolant (antifreeze) level* — Check the level by removing the cap on the antifreeze fill opening, commonly located on the coolant system overflow reservoir, or by viewing through the sight glass if one is provided (Figure 3.27). There will generally be at least one mark on the inside of the reservoir indicating the proper level for the antifreeze. Some vehicles' coolant reservoirs have two marks: one for when the engine is hot and the other for when it is cool. Add approved antifreeze when the level is low. The driver/operator should also check the condition of radiator and heater hoses and plumbing to make sure that no leaks are evident. Make sure that no debris, such as leaves or trash, is resting against the radiator intake. This reduces the cooling efficiency of the unit.

WARNING!

Use caution when removing the radiator fill cap on a vehicle that is currently running or has been running recently. Boiling antifreeze and/or steam may be emitted, causing severe injury to the person removing the cap. It is most desirable to check the antifreeze level when the engine and radiator system are cool.

- *Cooling fan* — Inspect the cooling fan for cracks or missing blades.

WARNING!

Some engine cooling fans are activated automatically without warning. Use caution when working around the fan.

- *Windshield washer fluid level* — Check the fluid level, which is usually contained in an appropriately labeled semitransparent tank or pouch. There is no specified level for the fluid in this reservoir. The driver/operator will simply add fluid as it becomes depleted. It is recommended that the reservoir be refilled any time it is less than one-half full. These fluids are commercially available, and compatibility from one brand to another is usually not a concern. Plain water will work; however, it has a tendency to freeze in cold weather conditions.
- *Battery condition* — Check various battery components. Depending on the design of the vehicle, the batteries may be located in the engine compartment or in a separate compartment elsewhere on the vehicle (Figure 3.28). Unsealed

Figure 3.27 The coolant level is usually checked by visually inspecting the level in the overfill reservoir.

Figure 3.28 A typical battery compartment and multiple battery arrangement.

batteries have caps that must be carefully removed so that the electrolyte (water) level in the batteries may be checked. Add distilled water if the internal level is low. Also check the battery connections for tightness and excessive corrosion (Figure 3.29). Corrosion around the battery terminal connections may be cleaned by mixing baking soda (which is chemically a *base*) with water and pouring it on the connections, making sure not to get any of this mixture inside the battery. This will cause a neutralization reaction with the corrosive (*acidic*) buildup. The terminal connection may then be cleaned with a wire brush and rinsed with clear water. Rinse the entire battery compartment thoroughly to ensure that all corrosive residues are flushed. Information on charging or jump-starting dead batteries is presented later in this chapter. If the apparatus is equipped with a built-in battery charger, make sure that this unit is operating properly. (**CAUTION:** Wear appropriate personal protective equipment, including eye protection, when working with batteries. Contact with battery acid may damage skin, eyes, or clothing. Also work in a well-vented area so that any fumes that are developed will dissipate. Make sure that no ignition sources (open flames, cigarettes, etc.) are present during this process.)

- ***Automatic transmission fluid level*** — This fluid level is checked in a manner similar to that for crankcase oil in many vehicles. A special dipstick is provided for this purpose. The correct level of the transmission fluid is noted on the dipstick. Some newer vehicles may be equipped with transmissions that provide an electronic readout of the transmission fluid level. Be familiar with the proper method for checking the fluid level in your vehicle. Add fluid if the reading on the dipstick indicates it is necessary. Depending on the manufacturer's recommendations, it may be necessary to check the transmission fluid level while the vehicle is *running*. If this is the case, the driver/operator may wish to perform this check first, before shutting off the vehicle.

Figure 3.29 Check the actual battery connections to make sure that they are clean.

- ***Power steering fluid level*** — Check the level of this fluid using the indicator marks provided by the manufacturer. Add approved power steering fluid if the level is low. Some power steering systems require the fluid level to be checked while the engine is running at normal operating temperature. Be careful not to overfill the reservoir because damage can occur to the system.
- ***Brake fluid (hydraulic brake systems)*** — Check the level of the brake fluid in the master brake cylinder using the procedure specified by the manufacturer. Add fluid if needed.
- ***Air system*** — Check for leaks in the system. With the air system at normal operating pressure and the engine shut off, walk around the apparatus and listen for leaks. Some apparatus may have an air dryer filter that must be inspected.
- ***Belts*** — Check all belts (water pump, air compressor, fan, alternator, etc.) in the engine compartment for tightness and excessive wear. Most engines have multiple drive pulleys. The driver/operator must be familiar with the proper feel for tightness of each belt when it is properly adjusted.

WARNING!

Never attempt to check the belts while the vehicle's engine is running. This could result in severe injuries or entrapment in the belt and pulley(s).

- ***Leaks*** — Look for leaks of any of the fluids used in the vehicle's engine, including antifreeze, water, windshield wiper fluid, oil, transmission fluid, hydraulic fluid, power steering fluid, and/or battery fluid. Inspect the condition of all hoses and hydraulic lines for the presence of fluids.

- ***Electrical wiring***— Check the general condition of all electrical wiring in the engine compartment. Look for wires that are frayed, cracked, bare, loose, or otherwise worn. Have a mechanic correct any wiring problems that are detected.

Some departments require driver/operators to check chassis lubrication. Proper lubrication saves maintenance and repair dollars and reduces out-of-service time. Effective lubrication depends upon the use of a proper grade of recommended lubricant, the amount used, and the frequency and method of lubrication. To select the proper lubricant, consideration must be given to the requirements of the unit to be lubricated, the characteristics of lubricants, and the manufacturer's recommendations. Always consult the owner's manual for the proper type of oil and the location of fill ports and grease fittings.

If the driver/operator is expected to lubricate the chassis components, he must be familiar with all lubrication fill connections. These connections are similar in appearance to the valve stem on a tire (Figure 3.30). When located, the driver/operator should clean the end of the fitting and then press the end of the lubrication gun fill hose onto the inlet. Operate the pump handle on the gun until no more lubricant enters the inlet (lubricant squeezes out between the hose outlet and the inlet). If the inlet supplies a sealed joint, follow the manufacturer's directions for the appropriate number of pumps. Continue this process around the apparatus until all inlets have been filled. Newer apparatus may be equipped with automatic vehicle lubrication systems. If your apparatus is equipped with such a system, follow the manufacturer's directions for operating the system.

Figure 3.30 Regularly add approved lubrication through all of the fill connections.

In addition to those listed, each fire department may specify other parts of the engine or apparatus for which the driver/operator is responsible for checking. The driver/operator should ensure that all necessary items are checked and documented according to departmental policies.

Posttrip Inspections

All functions described to this point in the chapter are intended to be performed before the operation of the vehicle. It is also a good idea to perform this type of an inspection after the vehicle has been operated for a prolonged period. This could include after operating at a fire for a long period or after an unusually long road trip. Long road trips may occur in remote areas that have large response districts, on mutual aid responses, following parades, or following trips to distant service centers. Each department should have a policy for performing posttrip inspections. IFSTA recommends that the same procedures described for pretrip inspections be used for posttrip inspections.

Charging Batteries

Historically, the charging of vehicle batteries is a function that is commonly performed by fire apparatus driver/operators. Apparatus batteries often require charging or jump-starting due to long periods of inactivity or improper drains on the electrical system. However, apparatus that regularly require charging or jump-starting should be reported to the department mechanic so that corrective measures may be taken.

In the station, these charging functions are generally performed with a battery charger. Because batteries produce explosive hydrogen gas when being charged, chargers must be used correctly to prevent needless accidents.

Driver/operators should be proficient in the use of battery chargers because they often need to be used in a rapid manner to start an apparatus before an emergency response. Use the following procedure for attaching battery charger cables to any vehicle battery. The driver/operator should be wearing safety eye protection whenever performing this evolution.

Step 1: Make sure that the battery and ignition switch(es) are in their off positions.

Step 2: Identify the polarity of the battery to be charged (positive or negative ground).

Step 3: Attach the red (positive or "+") charger cable to the red (positive or "+") battery post (Figure 3.31).

Step 4: Attach the black (negative or "-") charger cable to the black (negative or "-") battery post.

Figure 3.31 Attaching the red (+) cable to the red (+) battery post.

Step 5: Connect the battery charger to a reliable power source (away from gasoline and other flammable vapors).

Step 6: Set the desired battery charging voltage and charging rate if so equipped. (**NOTE:** Switches on battery charger should be in the ***OFF*** position when not in use.)

Step 7: Reverse the procedure to disconnect the battery charger.

If the apparatus needs to be jump-started when it is away from the station, it may be necessary to connect it to another vehicle with jumper cables. Before attempting to jump-start a fire apparatus, the driver/operator must be aware of several important considerations. First, make sure that the vehicle being used as the power source has the same voltage electrical system as the apparatus being jump-started. This will prevent damage from occurring to either or both systems.

Secondly, most fire apparatus have more than one battery on the apparatus. The manufacturer's operations manual should specify which battery should be connected to when jumping or charging the system. Make sure that the jumper cables are connected to the specified battery. As with hooking batteries to a charger, whenever jump-starting vehicle batteries, all people in the area should be wearing appropriate protective clothing and equipment, including eye protection.

The following procedure may be used for jump-starting a fire apparatus:

Step 1: Make sure that the battery and ignition switch(es) on both vehicles are in their off positions.

Step 2: Identify the polarity of the battery to be jump-started (positive or negative ground).

Step 3: Park the vehicles close enough together so that the jumper cables will reach each of the respective batteries, but not so close that the vehicles are actually touching each other. If the vehicles are touching, a ground connection may be formed that will prevent the dead vehicle from being started and could cause damage to both electrical systems. Make sure that the parking brakes are properly set when the vehicles are parked.

Step 4: Open the battery compartments on each vehicle and locate the red (+) and black (-) terminal posts.

Step 5: Connect one of the red (+) jumper cable clamps to the red (+) battery terminal post on the vehicle with the dead battery. Do not let any of the other three clamps touch each other or metal objects.

Step 6: Connect the other red (+) jumper cable clamp to the red (+) battery terminal post on the vehicle with the good battery.

Step 7: Connect the closest black (-) jumper cable clamp to the black (-) battery terminal post on the vehicle with the good battery. Again, do not allow the remaining black (-) clamp to touch anything before the next step is complete.

Step 8: Attach the remaining black (-) jumper cable clamp to some solid metal portion of the apparatus with the dead battery. The connecting point should be at least 18 inches (0.5 m)away from the dead battery and should be away from any moving engine parts (Figure 3.32).

Step 9: Start the vehicle with the good battery and allow the engine to idle for a little while.

Step 10: Try to start the vehicle with the dead battery. If it does not start after a few attempts, discontinue the jumping process and follow your department's standard operating procedures for dealing with a disabled apparatus.

Step 11: When the jumping operation is complete, disconnect the cables in the ***exact reverse order*** as they were connected. Make sure not to touch any other metal objects with the clamps as they are removed or electrical shorting may occur.

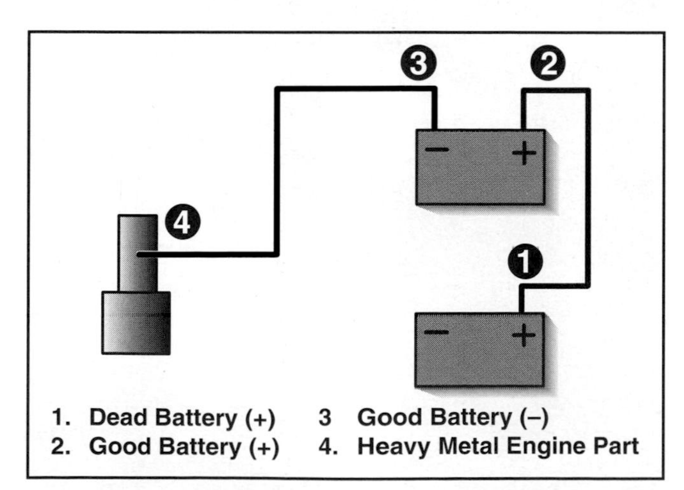

Figure 3.32 When completed, the battery charging/jumping connections should conform to this diagram.

Aerial Device Inspection

Once the general inspection of the apparatus has been completed, the driver/operator should make a visual and operational inspection of the aerial device. If the apparatus is equipped with a fire pump, the pump should also be checked for proper operation. The pump may be checked either before or after the aerial device is checked. The procedures for fire pump maintenance and inspection can be found in the IFSTA **Pumping Apparatus Driver/Operator Handbook**.

The aerial device, like other mechanical devices, is subject to failure when not properly maintained. The driver/operator must closely inspect the operating parts of the aerial device to ensure that the parts are functional. However, regular visual and operational inspections of an aerial device are not adequate to ensure the integrity of the aerial device. Regular structural testing is also important. Testing aerial devices should be done with proper equipment and supervision and is not considered to be routine maintenance. Most fire departments hire outside testing firms to perform this service. More information on apparatus testing can be found in Chapter 10 of this manual.

Visual Inspection

Each fire department should have its own policy that dictates the items that the driver/operator is responsible for checking on a daily or otherwise regular basis. Most commonly these items will be included on an official form that will serve as a checklist for the aerial device inspection. However, as a minimum, it is recommended that the following items be addressed as part of the daily visual inspection of the aerial device:

- *Check the level of the hydraulic fluid in the aerial device hydraulic system.* This can be done by checking the fluid level in the hydraulic reservoir when the system is cold, unless otherwise specified by the manufacturer. Most apparatus have either a dipstick or a site gauge to check the fluid level (Figure 3.33). Add fluid to fill the system to the appropriate level, if required. (**NOTE:** It is important that the fluid level be checked and additional fluid be added only when the stabilizers and the aerial device are all in their stowed positions. The fluid level in the reservoir will necessarily be lowered if any of the hydraulic cylinders are deployed. Adding fluid to the system at this point will result in overfilling the system. This could lead to a spill of hydraulic fluid or damage to the system components when the deployed hydraulic cylinders are retracted.) The manufacturer's operating manual should provide specific direction on how to perform these tasks for its apparatus.

- *Inspect the stabilizers.* The stabilizers should be checked for any signs of visible damage, evidence of hydraulic fluid leakage, damaged hoses, or scoring on the sliding beams or hydraulic pistons. Make sure that the stabilizer warning lights are clean and not visibly damaged. If the device is equipped with manual lockpins, make sure that they are all present and sufficiently secured. Note any elongation or cracks to the lockpin holes on the stabilizers (Figure 3.34). Make sure the stabilizer pads are in place and in good condition.
- *Inspect the turntable assembly.* There are a number of items that should be checked on the turntable assembly. Check the drive pinion and turntable gear teeth for visible damage, proper meshing and alignment, evidence of wear, and adequate lubrication. Also check the turntable bolts to ensure that they are all present and sufficiently tight (Figure 3.35). If department policy allows, the driver/operator should use a properly adjusted torque wrench to ensure that the bolts are tightened to manufacturer's specifications.
- *Inspect the lower control pedestal(s).* Inspect all components of the lower control pedestal for signs of wear and damage (Figure 3.36). All controls should move freely. Electrical connections should be tight and free of wear. An operational

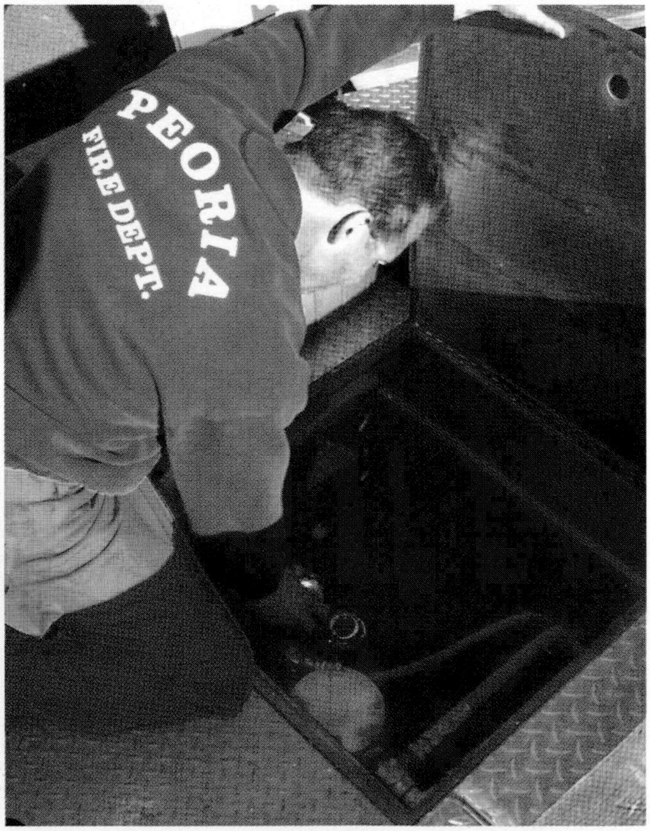

Figure 3.33 Use the means recommended by the apparatus manufacturer to check the level of hydraulic fluid in the aerial device and stabilizing hydraulic system.

Figure 3.34 Check the stabilizer lock pin holes for signs of elongation or other defects.

check of these controls will be conducted when the driver/operator conducts the operational test portion of the inspection. This is detailed later in this section.

- *Inspect the platform control console.* If the aerial device is an elevating platform, make a visual inspection of the aerial device controls on the platform console before any operational testing is conducted (Figure 3.37).
- *Inspect the aerial device communications system.* This test may be conducted either now or when the operational test is conducted. Check all system components for visible damage and proper operation. It may be necessary to position a second firefighter at the tip of the device in order to completely check the operation of the system.
- *Check the status/operation of the breathing air supply system.* If the aerial device is equipped with a breathing air system, check to make sure that there is adequate air in the storage cylinders and that all components are operating correctly. Inspect the cylinders, gauges, regulator, hoses and tubing, and air connections for damage or excessive wear. If possible, check the operation of the low-air warning devices to make sure that they work correctly.
- *Inspect the aerial device extension/retraction system.* Before operating the aerial device, check the extension and retraction system for signs of visible damage and wear. Depending on the manufacturer of the aerial device, this system

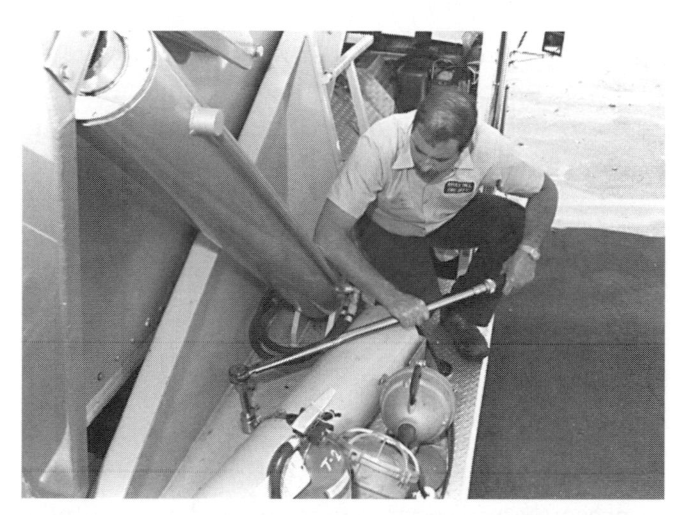

Figure 3.35 A properly calibrated torque wrench is needed to accurately check the tightness of the turntable bolts.

Figure 3.36 Inspect the lower control pedestal for obvious problems.

Figure 3.37 If the apparatus has a platform control pedestal, the controls must also be checked for proper operation.

can include any combinations of hydraulic cylinders, cable systems, and chain-drive systems. If hydraulic cylinders are present, check them for signs of visible damage, fluid leakage, and insecurity. Inspect cables for signs of corrosion, broken strands, excessive wear, stretching (reduction in diameter), or damaged end connections. Sheaves, guards, guides, and any other surfaces that come in contact with the cable should be checked to make sure that they are in good condition and do not have any rough edges that might damage the cable. If the device has a chain drive, check for proper lubrication and signs of damaged chain links or gear drives.

- *Inspect the elevation/lifting cylinders.* All modern aerial devices have these cylinders between the turntable and the lower section of the aerial device. Check for signs of visible damage, fluid leakage, and insecurity (Figure 3.38). Also check the end caps to make sure that they are secure and that no other hardware is missing.
- *Inspect the various sections of the aerial device.* Check the aerial device for signs of wear, cracks in the welds, loose or missing parts, and improper alignment. This includes all beams, rails, locks, alignment systems, and truss work. A quick check of this may be done while the device is stowed, but this will need to be done again once the aerial device is raised.
- *Inspect the elevating platform assembly (if present).* If the aerial device is equipped with an elevating platform, if should be visually inspected for signs of damage. Check the railings, kickplates, deck, heat shields, control platform, nozzle, standpipe connections, floodlights, and other components for obvious signs of damage or missing parts. Inspect the platform leveling system components for signs of damage. This system usually consists of hydraulic cylinders that should be checked for signs of defect in the same manner as the cylinders for extending, elevating, and stabilizing are checked.
- *Inspect the ladder rungs.* Check the ladder rungs for signs of visible damage, looseness, loose rung tread covers, weld cracks, or other potential problems.
- *Inspect the aerial waterway system.* If the aerial device has a piped waterway system, make sure that there is no visible damage to the piping, connections, seals, or other system components. All sections of the piping should be properly aligned and lubricated to manufacturer's recommendations. Again, a preliminary check may be made during the visual inspection but may be done in more detail when the aerial device is deployed.
- *Inspect any equipment that is attached to the end of the aerial device fly section.* Check for the presence and security of items such as axes, pike poles, handlights, roof ladders, and air masks that may be located on the aerial device (Figure 3.39).

Operational Inspection

Once the visual inspection has been completed, most fire departments require the driver/operator

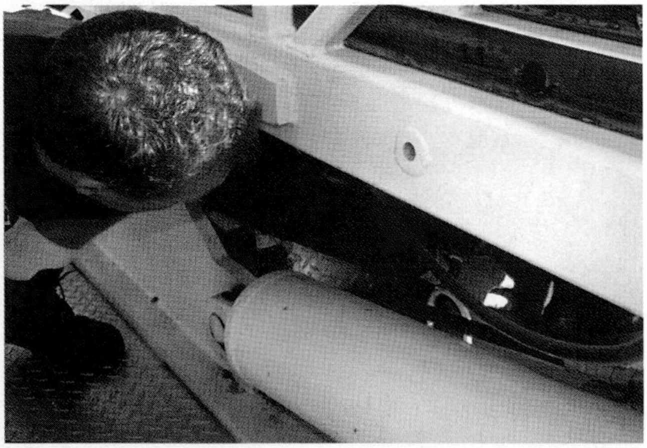

Figure 3.38 Note any fluid leaks from the aerial device lifting (hoisting) cylinders.

Figure 3.39 Check any equipment that is stowed at the end of the aerial device.

to perform an operational inspection of the aerial device. In actuality, the operational inspection serves two useful purposes. The first purpose is obviously to ensure that the aerial device is in proper working condition. The second benefit is that it serves as a review of the aerial device operation for the driver/operator. Even though the driver/operator may drive the apparatus on a regular basis, he may go days, weeks, or months without operating the aerial device on a call. By operating it during the inspection, the driver/operator will be more prepared to operate effectively under emergency conditions.

The exact procedures for stabilizing the apparatus and deploying the aerial device are covered in Chapters 6 through 8 of this manual. However, the following section details the basic actions that should be undertaken when performing an operational inspection of the aerial device.

- *Park the apparatus in a suitable location for operating the aerial device.* In many cases, the driver/operator will be able to perform the operational inspection on the front apron of the fire station (Figure 3.40). If this is not possible, the apparatus should be driven to a suitable location. The chosen location should have a stable parking area that is strong enough to support the weight of the apparatus. There should be no overhead obstructions that might come in contact with the raised aerial device.
- *Transfer power from the drive train to the aerial device hydraulic system.* The procedures for doing this are covered in Chapter 5 of this manual. Make sure that the transfer occurs when the controls are operated. Check to make sure that the transfer indicator light illuminates after the shift is made.
- *Check the operation of the stabilizers.* Lower the apparatus stabilizers according to the manufacturer's instructions and the information contained in Chapter 5 of this manual. Once lowered, check the stabilizers for any signs of physical damage or leaking hydraulic fluid (Figure 3.41). Note if the truck begins to sag toward any particular side after the stabilizers have been deployed for awhile. If the apparatus is equipped with an indicator light, make sure that it illuminates when the apparatus is sufficiently stabilized to allow the aerial device to be lifted from the bed. Once the stabilizers are set, move the hydraulic system selector valve control to the aerial device position.
- *Raise and extend the aerial device.* Operate the appropriate control(s) to fully elevate and extend the aerial device (Figure 3.42). Make sure that operating the control(s) results in the appropriate movement of the aerial device. Look for signs of problems such as jerky motions of the

Figure 3.40 In many cases, the front of the station will be a suitable location for operational testing of the aerial device.

Figure 3.41 Check the deployed stabilizers for signs of leaking hydraulic fluid.

Figure 3.42 Raise and extend the aerial device during the operational test.

aerial device, unusual noises, or unusual bending or twisting of the aerial device. Once the device is raised, further inspect the waterway system, rungs, and extension system for signs of damage or defect. If the apparatus is equipped with an elevating platform, the operation of the aerial device should be conducted from both the lower control station and the platform control console.

- *Rotate the aerial device.* Depending on departmental standard operating procedures and manufacturer's recommendations, rotate the aerial device while extended or while fully retracted. The purpose of this test is to make sure that all turntable rotational equipment is in proper working condition. The aerial device should move in a complete circle without any jerking action as it spins. Listen for unusual sounds and watch for leaking fluids while the rotation is in process.
- *Test the operation of auxiliary equipment on the aerial device.* If the aerial device is equipped with a remote control nozzle, video camera, floodlights, or other equipment, make sure that it is in good working condition.
- *Stow the aerial device and stabilizers.* Once the operational test is complete, all components should be stowed and the apparatus readied for road travel. Make sure that all equipment stows properly, no hydraulic fluid is leaking anywhere, and all indicator devices go out when the appropriate systems are de-energized.

Once the inspection is complete, the driver/operator should make sure that all departmental forms are filled out correctly and routed according to standard operating procedures.

Chapter 4

Driving Aerial Apparatus

Job Performance Requirements

This chapter provides information that will assist the reader in meeting the following job performance requirements from NFPA 1002, *Standard on Fire Apparatus Driver/Operator Professional Qualifications*, 1998 edition. Particular portions of the job performance requirements (JPRs) that are met in this chapter are noted in bold text.

2-3.1* **Operate a fire department vehicle, given a vehicle and a predetermined route on a public way that incorporates the maneuvers and features specified in the following list that the driver/operator is expected to encounter during normal operations, so that the vehicle is safely operated in compliance with all applicable state and local laws, departmental rules and regulations, and the requirements of NFPA 1500, *Standard on Fire Department Occupational Safety and Health Program,* Section 4-2.**

- **Four left and four right turns**
- **A straight section of urban business street or a two-lane rural road at least 1 mile (1.6 km) in length**
- **One through-intersection and two intersections where a stop has to be made**
- **One railroad crossing**
- **One curve, either left or right**
- **A section of limited-access highway that includes a conventional ramp entrance and exit and a section of road long enough to allow two lane changes**
- **A downgrade steep enough and long enough to require down-shifting and braking**
- **An upgrade steep enough and long enough to require gear changing to maintain speed**
- **One underpass or a low clearance or bridge**

(a) ***Requisite Knowledge:*** **The effects on vehicle control of liquid surge, braking reaction time, load factors, general steering reactions, speed, and centrifugal force; applicable laws and regulations; principles of skid avoidance, night driving, shifting, and gear patterns; negotiating intersections, railroad crossings, and bridges; weight and height limitations for both roads and bridges; identification and operation of automotive gauges; and proper operation limits.**

(b) ***Requisite Skills:*** **The ability to operate passenger restraint devices; maintain safe following distances; maintain control of the vehicle while accelerating, decelerating, and turning; maintain reasonable speed for road, weather, and traffic conditions; operate safely during nonemergency conditions; operate under adverse environmental or driving surface conditions; and use automotive gauges and controls.**

2-3.2* **Back a vehicle from a roadway into restricted spaces on both the right and left sides of the vehicle, given a fire department vehicle, a spotter, and restricted spaces 12 ft (3.66 m) in width, requiring 90-degree right-hand and left-hand turns from the roadway, so that the vehicle is parked within the restricted areas without having to stop and pull forward and without striking obstructions.**

(a) *Requisite Knowledge:* Vehicle dimensions, turning characteristics, spotter signaling, and principles of safe vehicle operation.

(b) *Requisite Skills:* The ability to use mirrors, judge vehicle clearance, and operate the vehicle safely.

2-3.3* **Maneuver a vehicle around obstructions on a roadway while moving forward and in reverse, given a fire department vehicle, a spotter for backing, and a roadway with obstructions, so that the vehicle is maneuvered through the obstructions without stopping to change the direction of travel and without striking the obstructions.**

(a) *Requisite Knowledge:* Vehicle dimensions, turning characteristics, the effects of liquid surge, spotter signaling, and principles of safe vehicle operation.

(b) *Requisite Skills:* The ability to use mirrors, judge vehicle clearance, and operate the vehicle safely.

2-3.4* **Turn a fire department vehicle 180 degrees within a confined space, given a fire department vehicle, a spotter for backing, and an area in which the vehicle cannot perform a U-turn without stopping and backing up, so that the vehicle is turned 180 degrees without striking obstructions within the given space.**

(a) *Requisite Knowledge:* Vehicle dimensions, turning characteristics, the effects of liquid surge, spotter signaling, and principles of safe vehicle operation.

(b) *Requisite Skills:* The ability to use mirrors, judge vehicle clearance, and operate the vehicle safely.

2-3.5* Maneuver a fire department vehicle in areas with restricted horizontal and vertical clearances, given a fire department vehicle and a course that requires the operator to move through areas of restricted horizontal and vertical clearances, so that the operator accurately judges the ability of the vehicle to pass through the openings and so that no obstructions are struck.

(a) *Requisite Knowledge:* Vehicle dimensions, turning characteristics, the effects of liquid surge, spotter signaling, and principles of safe vehicle operation.

(b) *Requisite Skills:* The ability to use mirrors, judge vehicle clearance, and operate the vehicle safely.

2-3.6* Operate a vehicle using defensive driving techniques under emergency conditions, given a fire department vehicle and emergency conditions, so that control of the vehicle is maintained.

(a) *Requisite Knowledge:* The effects on vehicle control of liquid surge, braking reaction time, load factors, general steering reactions, speed, and centrifugal force; applicable laws and regulations; principles of skid avoidance, night driving, shifting, and gear patterns; negotiating intersections, railroad crossings, and bridges; weight and height limitations for both roads and bridges; identification and operation of automotive gauges; and proper operation limits.

(b) *Requisite Skills:* The ability to operate passenger restraint devices; maintain safe following distances; maintain control of the vehicle while accelerating, decelerating, and turning; maintain reasonable speed for road, weather, and traffic conditions; operate safely during nonemergency conditions; operate under adverse environmental or driving surface conditions; and use automotive gauges and controls.

4-1.2 Perform the practical driving exercises specified in 2-3.2 through 2-3.5, given a fire department aerial apparatus and a spotter for backing, so that each exercise is performed safely without striking the vehicle or obstructions.

(a) *Requisite Knowledge:* Vehicle dimensions, turning characteristics, spotter signaling, and principles of safe vehicle operation.

(b) *Requisite Skills:* The ability to use mirrors, judge vehicle clearance, and operate the vehicle safely.

4-1.3 Operate a fire department aerial apparatus over a predetermined route on a public way, given the maneuvers specified in 2-3.1, so that the vehicle is safely operated in compliance with all applicable state and local laws, departmental rules and regulations, and the requirements of NFPA 1500, *Standard on Fire Department Occupational Safety and Health Program*, Section 4-2.

(a) *Requisite Knowledge:* The effects on vehicle control on braking reaction time, load factors, general steering reactions, speed, and centrifugal force; applicable laws and regulations; principles of skid avoidance, night driving, shifting, and gear patterns; negotiating intersections, railroad crossings, and bridges; weight and height limitations for both roads and bridges; identification and operation of automotive gauges; and proper operation limits.

(b) *Requisite Skills:* The ability to operate passenger restraint devices; maintain safe following distances; maintain control of the vehicle while accelerating, decelerating, and turning; maintain reasonable speed for road, weather, and traffic conditions; operate safely during nonemergency conditions; operate under adverse environmental or driving surface conditions; and use automotive gauges and controls.

5-2.1 Perform the practical driving exercises specified in 2-3.2 through 2-3.5 from the tiller position, given a qualified driver, a fire department aerial apparatus equipped with a tiller, and a spotter for backing, so that each exercise is performed safely without striking the vehicle or obstructions.

(a) *Requisite Knowledge*: Capabilities and limitations of tiller aerial devices related to reach, tip load, angle of inclination, and angle from chassis axis; effects of topography, ground, and weather on safe aerial device deployment; and use of a tiller aerial device.

(b) *Requisite Skills:* The ability to determine the appropriate position for the tiller, maneuver the tiller into proper position, and avoid obstacles to operations.

5-2.2 Operate a fire department aerial apparatus equipped with a tiller from the tiller position over a predetermined route on a public way, using the maneuvers specified in 2-3.1, given a qualified driver, a fire department aerial apparatus equipped with a tiller, and a spotter for backing, so that the vehicle is safely operated in compliance with all applicable state and local laws, departmental rules and regulations, and the requirements of NFPA 1500, *Standard on Fire Department Occupational Safety and Health Program,* Section 4-2.

(a) *Requisite Knowledge*: Principles of tiller operation, methods of communication with the driver, the effects on vehicle control of general steering reactions, night driving, negotiating intersections, and manufacturer operation limitations.

(b) *Requisite Skills*: The ability to operate the communication system between the tiller operator's position and the driver's compartment; operate passenger restraint devices; maintain control of the tiller while accelerating, decelerating, and turning; operate the vehicle safely during nonemergency conditions; and operate under adverse environmental or driving surface conditions.

The ability to safely control and maneuver the aerial apparatus is one of the most critical aspects of a driver/operator's responsibilities. Simply stated, the first goal of the driver/operator is to get the apparatus and its crew to the scene of an emergency in an expedient, yet safe and efficient manner. You cannot perform the necessary emergency operations until you arrive on the scene, and you cannot perform them at all if you are involved in a collision.

Consider the impact on a response system when an emergency vehicle is involved in a collision while responding to an emergency. For example, suppose that a quint is dispatched to a car fire, drives through a red light, and strikes a passenger car in an intersection while en route. At least one more company is going to be dispatched to respond to the collision (and likely even more resources than that). Also, someone still has to be dispatched to extinguish the car fire. What started as a single-company response now involves at least three companies, chief officers, ambulances, police, and other vital resources. If the driver/operator had only come to a complete stop at the intersection, all of these other resources would still be available for other potential emergencies. The impact on people's lives — those who were hit by the fire truck and those who awaited someone to come put out their car fire — goes without saying. As the old adage goes, be part of the solution not part of the problem.

Statistics compiled annually by the National Fire Protection Association (NFPA) historically show that 15 to 20 percent of all firefighter injuries and deaths are caused by vehicle collisions while responding to or returning from emergency calls (Figure 4.1). This equates to about 25 firefighter deaths per year. United States Department of Transportation studies indicate that about an equal number of civilians are killed annually in apparatus-related collisions. As we will learn in this chapter, most of these collisions are avoidable if safe driving principles are used. Sound driving principles also reduce wear and tear and extend the life cycle of the apparatus.

This chapter discusses the many elements of safe fire apparatus operation. First, it is important to understand the common causes of collisions and how to avoid them. The driver/operator must also understand the proper techniques for starting and driving the vehicle, driving in adverse conditions, and using the warning and traffic control devices. The latter portion of the chapter provides information on performing practical driving exercises required of the driver/operator to meet NFPA 1002.

Figure 4.1 Caution must be exercised to avoid involving the apparatus in a collision.

It is also important to note that while this chapter primarily addresses the operation of aerial fire apparatus, all safe driving practices contained herein also apply to the operation of privately owned vehicles (POVs) by volunteer firefighters. A significant number of collisions involving volunteer firefighters and POVs occur each year. Adherence to the information in this chapter can help reduce the incidence and severity of POV collisions.

Collision Statistics and Causes

It is a commonly held belief that those who fail to recognize history are doomed to repeat it. With that in mind, it is important that we review emergency vehicle collision statistics and causes. Reviewing this information helps us realize where our problems lie and what we can do to correct them.

While the NFPA keeps detailed information on firefighter injuries and deaths, its studies do not provide extensive background on collision causes or conditions. The studies do typically note that things such as failure to wear seat belts, poor road conditions, poor vehicle conditions, driving too fast for conditions, and failure to obey traffic rules are often factors in apparatus-related collisions.

Outside of the work done by individual fire departments relative to their own situations, extensive research information on fire apparatus collision causes is not available. However, two statewide studies on collisions involving emergency medical service (EMS) vehicles do provide keen insight into the problem. Because EMS and fire vehicles operate under similar conditions, the information from these studies is relevant to fire apparatus.

The first study was conducted by Indiana University of Pennsylvania (IUP). It was centered on the 1,079 known providers of EMS in the state of Pennsylvania. During a one-year period, 212 collisions involving EMS vehicles were documented. As indicated by Tables 4.1 and 4.2, most collisions occurred in broad daylight on dry roads.

The second study was conducted by the New York State Department of Health EMS Program. This study looked at 1,412 EMS vehicle collisions that occurred over a four-year period. The results were very similar to the IUP study (Tables 4.3 and 4.4).

What these two studies indicate is that although we must be trained to drive in adverse weather conditions, collisions are most likely to occur during ideal vision and road conditions. Thus, we must look to other reasons as the major causes of fire apparatus collisions. In general, fire apparatus collisions can be grouped into the following five basic causes:

- Improper backing of the apparatus
- Reckless driving by the public
- Excessive speed by the fire apparatus driver/operator
- Lack of driving skill and experience by the fire apparatus driver/operator
- Poor apparatus design or maintenance

A large percentage of collisions occur while backing the vehicle. While they are seldom serious in terms of injury or death, they do account for a significant portion of overall damage costs. Backing collisions occur in a variety of locations (Figures 4.2 a and b). On the emergency scene, it is often necessary to back the apparatus into position for use or to back it out when the assignment is complete. Backing collisions also occur in parking lots while the apparatus is performing either routine or

Table 4.1
Times that Collisions Occur (IUP Study)

Time of Day	Number of Collisions
Daylight	108 (51%)
Dawn/Dusk	23 (11%)
Night	58 (27%)
Unknown	23 (11%)

Table 4.2
Road Conditions when Collisions Occur (IUP Study)

Road Conditions	Number of Collisions
Dry Road	130 (61%)
Wet Road	22 (10.5%)
Snow/Ice	28 (13%)
Muddy Road	1 (0.5%)
Unknown	32 (15%)

Table 4.3
Times that Collisions Occur (NY Study)

Time of Day	Number of Collisions
Daylight	825 (70%)
Dawn/Dusk	52 (5%)
Night	283 (24%)
Unknown	12 (1%)

Table 4.4
Road Conditions when Collisions Occur (NY Study)

Road Conditions	Number of Collisions
Dry Road	891 (63%)
Wet Road	352 (25%)
Snow/Ice	90 (6%)
Muddy Road	4 (1%)
Unknown	77 (5%)

emergency duties. Backing collisions also commonly occur when backing the apparatus into the fire station.

Reckless driving by the public occurs in many forms. Some of the more common problems include:

- Failure to obey posted traffic regulations or directions
- Failure to yield to emergency vehicles
- Excessive speed
- Unpredictable behavior created by a panic reaction to an approaching emergency vehicle
- Inattentiveness

Fire apparatus driver/operators must always be cognizant of the fact that they have little control over the way members of the public react toward them. With this in mind, driver/operators must never put themselves, or the public, in a situation where there is no alternative (other than crashing into each other).

Figure 4.2a Backing accidents are among the most common collisions involving emergency vehicles. *Courtesy of Chris Mickal.*

Figure 4.2b Typical damage from a backing collision. *Courtesy of Ron Jeffers.*

The urgency of the emergency often leads to the driver/operator driving the fire apparatus at speeds faster than should reasonably be used. Excessive speed may lead to one of the two following types of collisions:

- Control of the apparatus is lost on a curve or adverse road surface, which may cause the vehicle to leave the road surface, roll over, or strike another vehicle or object.
- The driver/operator is unable to stop the apparatus in time to avoid a collision with another vehicle or object.

Driver/operators must remember that the fire apparatus they are driving do not handle the same, or stop as fast, as their privately owned vehicles. It takes a much greater distance for a fire apparatus to stop than it does a smaller passenger vehicle because fire apparatus weigh substantially more than standard passenger vehicles. As well, the air brake systems commonly used on fire apparatus take a little longer to activate and stop a vehicle than do the hydraulic/mechanical brake systems on smaller passenger vehicles. More detailed information on stopping distances and braking times is presented later in this chapter. Aerial apparatus also tend to be more top-heavy and severely impacted by quick turns and maneuvers than smaller vehicles (Figure 4.3).

Lack of driving skill by the driver/operator may be attributed to a number of factors, including

Figure 4.3 Apparatus rollover accidents can have very serious consequences. *Courtesy of Ron Jeffers.*

insufficient training and unfamiliarity with the vehicle. Fire departments must ensure that all driver/operator candidates complete a thorough training program before they are allowed to drive aerial apparatus under emergency conditions. ***Never assign someone who has not been trained on a particular vehicle to drive it.*** This unfamiliarity with the driving characteristics of the vehicle can lead to a serious collision.

A number of other factors may contribute to collisions that involve driver/operator error. These factors include the following:

- Overconfidence in one's driving ability — The driver/operator may have an inflated opinion of his capabilities during the adrenaline rush of an emergency response. This may cause him to operate the vehicle in a manner in which neither he nor the vehicle is capable of performing safely.
- Inability to recognize a dangerous situation — Poorly trained driver/operators do not have the ability to sense when they are approaching a hazardous situation. In a study of commercial truck drivers by the Society of Automotive Engineers (SAE), it was determined that in 42 percent of all collisions, the driver/operator was not aware of a problem until it was too late to correct it.
- False sense of security because of a good driving record — This manifests itself in the attitude "I've never had a collision before, why should I worry about one now."
- Misunderstanding of apparatus capabilities — Many driver/operators are ignorant of the potential for disaster that the emergency vehicle presents. They do not know the capabilities of the apparatus that can be used to avoid collisions. They are also unaware of simple vehicle characteristics, such as a fire apparatus is lighter and will travel faster when the water tank is empty. This makes the vehicle more likely to skid under certain conditions.
- Lack of knowledge about how to operate the controls of the apparatus in an emergency — Again, this is a problem associated with insufficient training.
- Adrenaline rush during the emergency response — As professionals, we should not become so excited that it impairs our judgment or actions. Adrenaline rush often leads to the overconfidence in one's driving ability as previously described.

Poor vehicle design and maintenance have been attributed to many serious fire apparatus collisions. However, poor vehicle design problems are usually not as serious with aerial apparatus because they are all typically designed and built by reputable fire apparatus manufacturers. The poor vehicle design problems are more common with "homebuilt" vehicles, such as wildland fire apparatus or water tenders/tankers, which were constructed by members of the fire department or by local mechanics.

Poor maintenance of apparatus can also result in vehicle system failures that lead to collisions. This is particularly true of braking systems. Several fatal fire apparatus collisions have been traced back to improperly maintained apparatus braking systems. By following an effective apparatus maintenance program, the likelihood of mechanical failure leading to collision can be reduced.

Driving Regulations

Driver/operators of fire apparatus are regulated by federal laws, state or provincial motor vehicle codes, city ordinances, NFPA standards, and departmental policies. Copies of these laws and rules should be studied by all members of the fire department. Because the regulations vary from jurisdiction to jurisdiction, those discussed here are general in nature.

Unless specifically exempt, fire apparatus driver/operators are subject to any statute, rule, regulation, or ordinance that governs any vehicle operator. Statutes usually describe those vehicles that are in the emergency category; this classification usually covers all fire department vehicles when they are responding to an emergency. In some jurisdictions, the statutes may exempt emergency vehicles from driving regulations that apply to the general public concerning speed, direction of travel, direction of turns, and parking if they are responding to a reported emergency. Under these circumstances, the driver/operator must exercise care (due regard) for the safety of others and must maintain complete control of the vehicle. All traffic signals and rules must be obeyed when returning to quarters from an alarm or during any other nonemergency driving.

Keep in mind that most driving regulations pertain to dry, clear roads during daylight conditions. Driver/operators should adjust their speeds to compensate for conditions such as wet roads, darkness, fog, or any other condition that makes normal emergency vehicle operation more hazardous (Figure 4.4).

Emergency vehicles are generally not exempt from laws that require vehicles to stop for school buses that are flashing signal lights to indicate that children are boarding or disembarking. Fire apparatus should proceed only after a proper signal is given by the bus driver or police officer. The driver/operator should proceed slowly, watching for any children that may not be aware of the approaching apparatus.

A driver/operator who does not obey state, provincial, local, or departmental driving regulations can be subject to criminal and civil prosecution if the apparatus is involved in a collision (Figure 4.5).

Figure 4.4 Driver/operators must take extra care when driving during adverse weather conditions.

Figure 4.5 It is not safe to pass other vehicles on the right side of the road.

If the driver/operator is negligent in the operation of an emergency vehicle and becomes involved in a collision, both the driver/operator and the fire department may be held responsible.

Starting and Driving the Vehicle

Obviously, before the driver/operator can perform virtually any other part of his duties, he must be able to start the vehicle and drive it in a safe and efficient manner. The following sections detail the standard procedures that should be used for starting and driving fire apparatus. Keep in mind that each apparatus manufacturer has specific directions that apply to its vehicles. Consult the manufacturer's operations manual, supplied with each vehicle, for detailed instructions specific to the vehicle.

The driver/operator should start the vehicle as soon as possible so that it is warmed up when the rest of the crew is assembled and ready to respond. Let it idle as long as possible before putting it into road gear — for nonemergency response this could be 3 to 5 minutes; for an emergency response it may be only a few seconds.

When starting the apparatus under any conditions, but especially emergency response conditions, the first thing the driver/operator needs to know is where the apparatus is going. By taking the time to review the incident location, the driver/operator can consider important factors that may affect the response such as road closings, school crossings, weather conditions, railroad crossings, underpasses, and traffic conditions. ***The vehicle should not be moved until all occupants are in the cab, in a seated position, and wearing seat belts*** (Figure 4.6).

Starting the Vehicle

The following procedure may be used to start the vehicle and begin the response:

Step 1: Disconnect all ground shore lines. Generally, the first thing the driver/operator will do when preparing to board and start the apparatus is disconnect from the apparatus any external electrical cords, air hoses, or exhaust system hoses. Electrical cords

Figure 4.6 Firefighters should be seated and belted before the apparatus moves out.

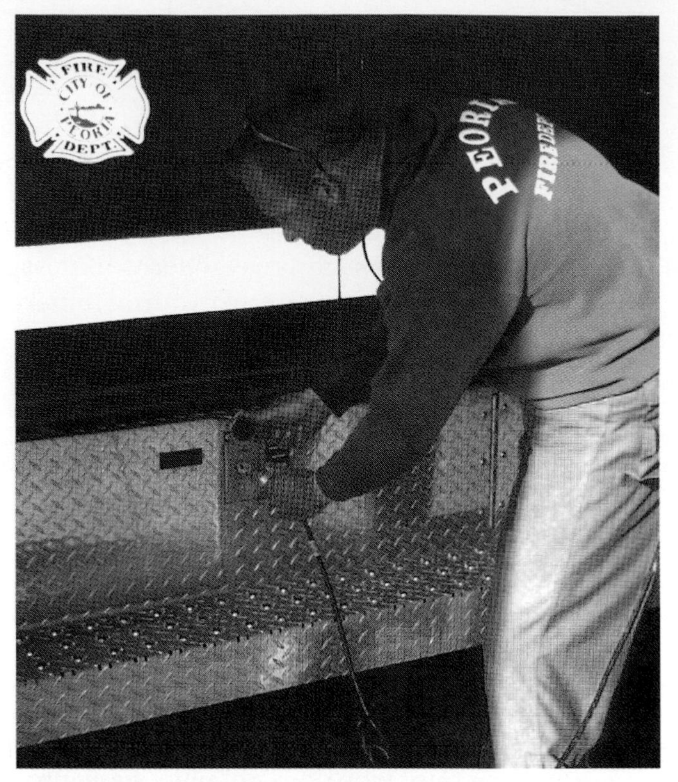

Figure 4.7 The driver/operator should disconnect any shore lines attached to the apparatus.

are used to keep the vehicle's batteries charged at all times (Figure 4.7). Air hoses may be connected to the apparatus to keep an adequate amount of air in the vehicle's air brake system at all times. Exhaust hoses are used to vent diesel fumes to the exterior of the station. The benzene derivatives contained in diesel emissions have been found to be carcinogenic in laboratory studies. Some apparatus are equipped with variations of ground shore lines that are designed to pop off automatically when the apparatus is started or moved. Make sure that they release before the apparatus is driven from the station (Figure 4.8).

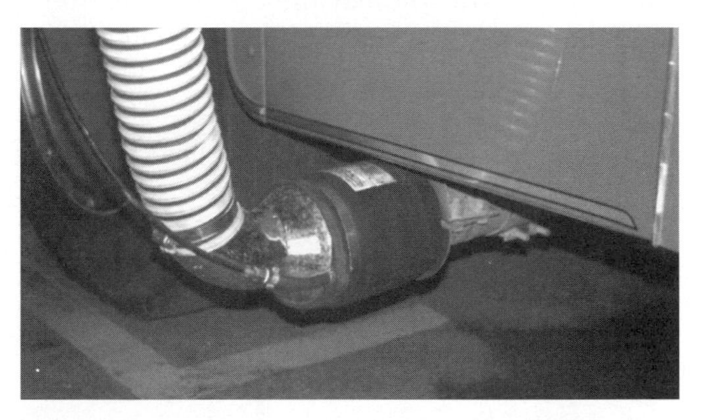

Figure 4.8 Some station exhaust systems are designed to automatically disconnect the hose when the apparatus pulls out of the station.

Step 2: Turn on the vehicle battery(ies). Most fire apparatus are equipped with a battery switch that is intended to turn off all vehicle electrical systems when the apparatus is parked and shutdown. The purpose of this switch is to prevent unwanted elec-

Figure 4.9 Most emergency vehicles have at least two batteries.

trical drains on the battery that might result in a dead battery when the driver/operator attempts to start the vehicle.

Most fire apparatus are equipped with two or four vehicle batteries (Figure 4.9). This is a back-up measure in the event that one of the batteries is dead. It is also necessary because of the larger-than-normal amount of electrical equipment attached to the fire apparatus. On apparatus equipped with two batteries, the battery switch may have four settings: Off, Battery 1, Battery 2, and Both (Figure 4.10). Obviously, this gives the driver/operator the ability to use either or both of the batteries when starting and operating the vehicle. Newer apparatus may have a simple on/off switch (Figure 4.11). Regardless of which type of switch the apparatus is equipped with, the battery switch should never be operated while the engine is running. Follow the apparatus manufacturer's directions and departmental SOPs as to whether one or both batteries should be used. Depending on the location of this switch, the driver/operator may choose to operate this switch prior to entering the cab or immediately upon sitting in the driver's seat.

Step 3a: Start the engine (Manual Transmission). On an apparatus equipped with a manual shift transmission, the driver/operator should start the engine with the drive transmission in Neutral (N) and the vehicle's parking brake set. The driver/operator begins the process by turning on the ignition switches (Figure 4.12). These are usually located on the dashboard. Once these switches are on, press down on the clutch pedal to disengage the clutch. Once the pedal is depressed, you may operate the starter control. There are several designs for fire apparatus starter controls. Some use a single key, similar to a standard passenger vehicle key (Figure 4.13). Others

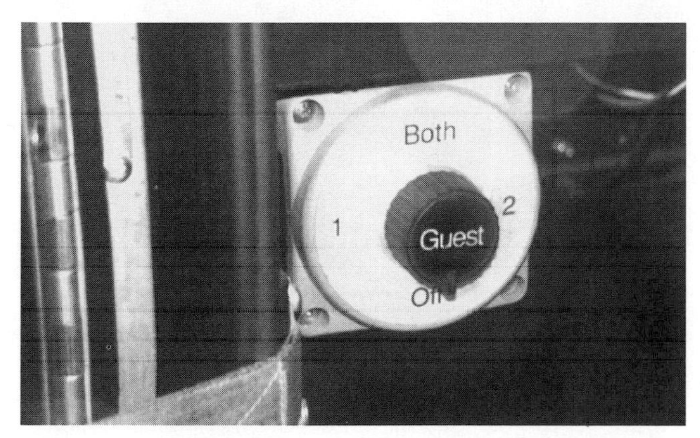

Figure 4.10 A typical four-position battery switch.

Figure 4.12 Ignition switches.

Figure 4.11 A simple on/off battery switch.

Figure 4.13 Some apparatus, particularly those on commercial chassis, have key-operated starters.

use single or dual toggle switches or push buttons (Figures 4.14 a and b). The dual starter controls are used on apparatus equipped with two batteries. Each control is connected to a particular battery. If your SOP is to turn on only one battery, make sure that you use the appropriate starter control to start the apparatus. If you use both batteries, operate both starter controls at the same time. The starter control should be operated in intervals of no more than 30 seconds, with a rest of 60 seconds between each try if the vehicle does not start sooner. The starter may overheat if the controls are operated for longer periods. (**NOTE:** On gasoline-powered apparatus, it may be necessary to operate a manual choke control before operating the starter control [Figure 4.15]. Use the manual choke sparingly in warm weather or after the apparatus is already warm.)

Step 3b: Start the engine (Automatic Transmission). On an apparatus equipped with an automatic transmission, the driver/operator should start the engine with the drive transmission in Neutral (N) or Park (P). The vehicle's parking brake should be set. The driver/operator begins the process by turning on the ignition switches (see Figure 4.12). These are usually located on the dashboard. You may then start the apparatus using the starter control(s). There are several designs for fire apparatus starter controls. Some utilize a single key, similar to a standard passenger vehicle key (see Figure 4.13). Others utilize single or dual toggle switches or push buttons (see Figure 4.14). The dual starter controls are used on apparatus equipped with two batteries. Each control is connected to a particular battery. If your SOP is to turn on only one battery, make sure that you use the appropriate starter control to start the apparatus. If you use both batteries, operate both starter controls at the same time. The starter controls should be operated in intervals of no more than 30 seconds, with a rest of 60 seconds between each try if the vehicle does not start sooner. The starter

Figure 4.14a Toggle switches are used for starters on some apparatus.

Figure 4.14b Push button starters are the most common type found on custom fire apparatus.

Figure 4.15 Apparatus with gasoline engines may have a manual choke control.

may overheat if the controls are operated for longer periods. (**NOTE:** On gasoline-powered apparatus, it may be necessary to operate a manual choke control before operating the starter control (see Figure 4.15). Use the manual choke sparingly in warm weather or after the apparatus is already warm.)

Step 4: Observe the apparatus gauges. Make sure that all gauges on the dashboard move into their normal operative ranges. In particular, pay attention to the oil pressure and air pressure gauges (Figure 4.16). If the oil pressure gauge does not indicate any reasonable amount of oil pressure within 5 to 10 seconds of starting the apparatus, stop the engine immediately and have the lubricating system checked by a trained mechanic. The air pressure gauge should be checked to make sure that adequate pressure is built up to release the parking brake. An interlock should prohibit the parking brake from being disengaged before there is enough air pressure in the system to operate the service brakes. If there is not enough air pressure in the system to release the parking brake (usually at least 60 psi [420 kPa] is required), allow the engine to idle until air pressure is built up to an appropriate level. It may speed the buildup of air pressure if the throttle is increased to a fast idle. This problem should be noted so that the braking system can be checked by a mechanic as soon as possible. The driver/operator should also check the ammeter to make sure that the electrical system is operating/charging properly.

Step 5: Adjust the seat, mirrors, and steering wheel. If the driver/operator was not the last person to drive this vehicle, he should take a moment while the engine is idling/warming to properly adjust the seat and mirrors. It is generally best to adjust the seat first. The seat may be adjusted for height as well as distance from the steering wheel, foot pedals, and other controls. Once the seat is in the desired location, adjust the mirrors so that you can clearly see to the rear of the apparatus. Newer apparatus may have adjustable or telescoping steering wheels that may be adjusted to fit the driver/operator.

The apparatus is now ready to be placed into a road gear and driven from its present location.

Driving the Vehicle

Once the apparatus is running and the air system (if so equipped) has sufficient air pressure, the driver/operator is ready to release the parking brake and place the apparatus into a road gear (Figure 4.17). The following sections discuss the procedures for driving, stopping, engine idling, and shutting down the apparatus.

Figure 4.16 The oil and air pressure gauges are plainly visible on the dashboard.

Figure 4.17 A parking brake control.

Driving Manual Transmission Apparatus

NOTE: The information in this section pertains to the typical 4- or 5-speed manual shift transmission with a single-speed rear axle. For information on driving apparatus with 2-speed rear axles or more than a 5-speed transmission (10, 12, or more gears are common in some jurisdictions), consult the apparatus/transmission manufacturer's operational manual for specific information.

When you are ready to move the vehicle, depress the clutch pedal with your left foot, depress the service brake pedal with your right foot, and release the parking brake (Figure 4.18). Place the gear shifter into a low gear that will allow the vehicle to move without an inordinate amount of wear on the engine. Never attempt to start the apparatus moving while it is in a high or drive gear or when the apparatus is rolling backwards down an incline. This action causes the clutch (on both manual and automatic transmissions) to slip, which may damage the clutch facing. Release the clutch slowly when starting from a standstill because a sudden, rapid release of the clutch throws a heavy load upon the engine, clutch, transmission, and drive components. Take care to avoid vehicle rollback before engaging the clutch.

Keep the apparatus in low gear until the proper speed or revolutions per minute (rpm) is reached for shifting to a higher gear. An engine develops its maximum power up to a certain speed, and excessive engine speed may result in decreased power and excessive use of fuel. It is good practice to keep the transmission in low gear until the apparatus is clear of the station and the driver/operator has an unobstructed view of the street and traffic conditions. When the driver/operator shifts gears, he should disengage the clutch entirely (pedal pushed in all the way). The gear shift lever should be carefully moved, not jammed, into proper position.

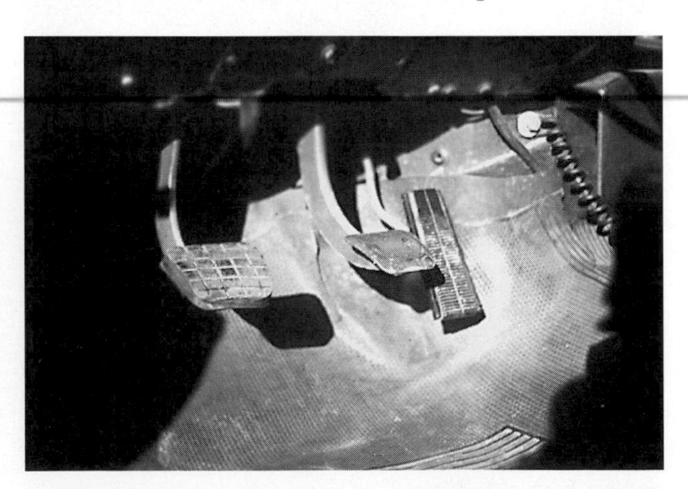

Figure 4.18 This view shows the clutch, brake, and throttle pedals from left to right.

When climbing a hill, shift the transmission to a lower gear. This practice provides adequate driving power and enables the driver/operator to keep the apparatus under control. On sharp curves or when turning corners, shift standard transmissions into a lower gear before entering the curve or intersection. This action maintains peak engine power and apparatus control. When fire apparatus must be driven over rough or rugged terrain, use the lower gears. If the apparatus becomes stuck, such as in mud or snow, do not race the engine or jump the clutch. These actions only cause the apparatus to become further stuck and can lead to mechanical failure. Always maintain front wheels in line with the chassis of the vehicle.

When driving downhill, select a lower gear. A good rule of thumb is to use the same gear whether going up or down the same grade. Remain in gear at all times. The engine provides braking power when the vehicle is in gear. To prevent engine damage, limit downhill speed to lower than maximum governed rpm. The engine governor cannot control engine speed downhill; the wheels turn the driveshaft and engine. Engine rotation faster than rated rpm can result in valves hitting pistons, increased oil consumption, damage requiring overhaul, and injector plug seizures.

Table 4.5 contains information for the appropriate gear selections for both manual shift and automatic transmissions. These gear selections are based on the listed conditions.

Driving Automatic Transmission Apparatus

Apparatus equipped with automatic transmissions eliminate a lot of the decision making by the driver/operator regarding when to shift gears. When the apparatus is ready to move, the driver/operator depresses the interlock on the shifter and moves it to the appropriate gear selection (Figure 4.19). This selection varies depending on the manufacturer of the apparatus or transmission. Some shifters have a normal driving (D) designation on the shifter

Table 4.5
Transmission Gear Selection Criteria

Automatic Shift Selector Position	Manual Gear Shift Position	Operating Conditions
Drive (D), 5, or 1-5	5 or Overdrive	Normal load, grade, and traffic conditions with open road ahead
4 or 1-4	4	Moderate grades and over-the-road operation with moderate speeds
3 or 1-3	3	Operating in heavy traffic
2 or 1-2	2	Need for speed control requiring a hold condition such as descending a steep grade or operation on rough terrain
1	1	Starting gear for initial forward movement

similar to that found on smaller passenger vehicles. Others have a series of numbers on the shifter. They may be single numbers (1, 2, 3, etc.) or range numbers (1-3, 1-4, 1-5, etc.). Newer apparatus may have push-button transmission selectors (Figure 4.20). In this case, the appropriate gear is chosen simply by pushing the appropriate button. Consult the manufacturer's operations manual for recommendations on which gear to select for normal operation.

Driver/operators of apparatus equipped with automatic transmissions should be aware that the pressure placed upon the accelerator influences automatic shifting. When the pedal is fully depressed, the transmission automatically upshifts near the governed speed of the engine. This may result in reduced torque (power) and excessive fuel consumption. If the accelerator is partially depressed, the upshift occurs at a lower engine speed.

Driver/operators have the option of manually selecting a particular gear for operation on apparatus equipped with an automatic transmission. This may be desirable when operating the apparatus at

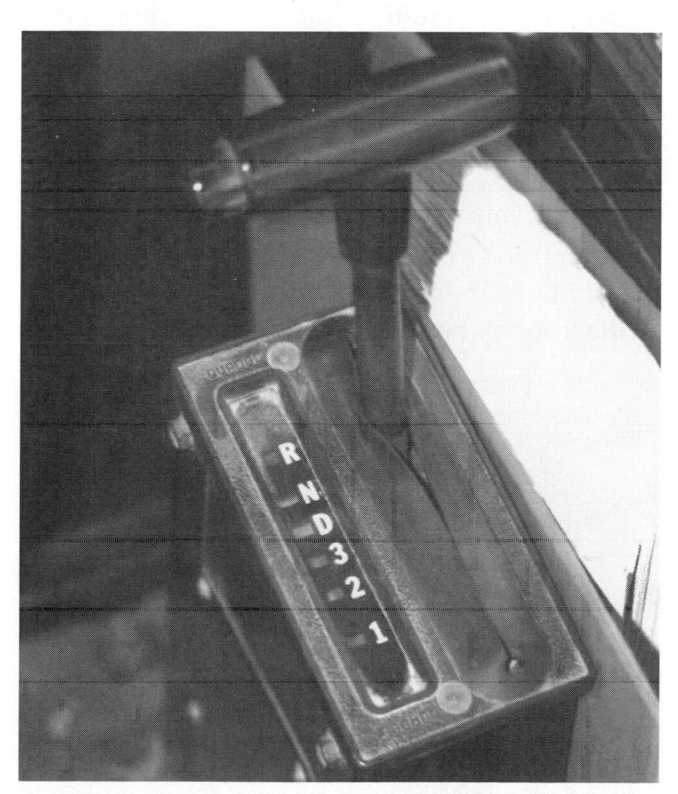

Figure 4.19 Some automatic transmissions have shifting levers.

Figure 4.20 Newer automatic transmissions may be controlled by a series of push buttons.

a slow speed for a long period of time or when driving up a steep hill. Simply move the shifter to the lower gear when this change is desired. Do not attempt to jump more than one gear at a time and never go directly from a forward gear to reverse. Consult Table 4.5 for information on the appropriate gears to select for various conditions.

Cruising

Once the apparatus is moving, accelerate the vehicle gradually. Do not try to reach rated speed in the low gears. Going to rated speed in the low gears can be very noisy and increases engine wear. Stay in the highest gear that allows the apparatus to keep up with traffic and still have some power in reserve for acceleration.

Operators of fire service vehicles can significantly reduce drive train damage and extend apparatus service life by adopting proper operating habits. When driving a fire service vehicle, attempt to maintain engine rpm control through correct throttling. Keeping the engine operating within its power curve ensures adequate power and optimum fuel economy for a given set of conditions.

Whenever possible, avoid overthrottling, which results in lugging. *Lugging* occurs when the throttle application is greater than necessary for a given set of conditions. When this happens, the engine cannot respond to the amount of work being asked for by the operator. Do not allow the engine rpm to drop below peak torque speed if lugging does occur. Automatic transmissions downshift automatically to prevent lugging. Standard transmissions must be downshifted to avoid stalls and prevent lugging. If the apparatus is equipped with a tachometer, it is easier to control the transmission shift points.

When ascending a steep grade, momentary unavoidable lugging takes place. Although this brief lugging is unavoidable, minimize time spent in the lugging state to avoid possible damage to the power plant. Engine rpm drops even with heavy throttle settings. Select progressively lower gears until the combination of available power and the torque multiplication of the transmission gears allow the hill to be climbed easily.

When overthrottling occurs with a diesel engine, more fuel is being injected than can be burned. This results in an excessive amount of carbon particles issuing from the exhaust (black smoke), oil dilution, and additional fuel consumption.

Another point that requires consideration is maximum engine rpm. Fire service apparatus are maintained for a much longer period of time than are commercial vehicles. Over this period of time, the valve springs can become weakened. For this reason, allowing an engine to overspeed as the result of improper downshifting or hill descent should be avoided in an effort to prolong engine life. Choose a gear that cruises the engine at 200 or 300 rpm lower than recommended rpm. This reduces engine wear; power losses caused by the fan, driveline, and accessories; noise; and fuel consumption.

Driver/operators of aerial apparatus must always be conscious of the fact that the aerial device may be hanging several feet (meters) over the front or rear of the apparatus. This must be taken into consideration whenever turning or parking the vehicle. Remember that the bumpers on the cab or rear of the vehicle may not be the only issues of concern for coming into contact with other vehicles or objects.

Because of the large size of many aerial apparatus, there are several points of potential contact under the front, middle, and rear of the apparatus. The driver/operator must keep this in mind when traversing steep ramps, curbs, speed bumps, and similar obstacles in the road. The classic definitions of these areas are as follows (Figure 4.21):

- *Angle of approach* — The angle formed by level ground and a line from the point where the front tires touch the ground to the lowest projection at the front of the apparatus.

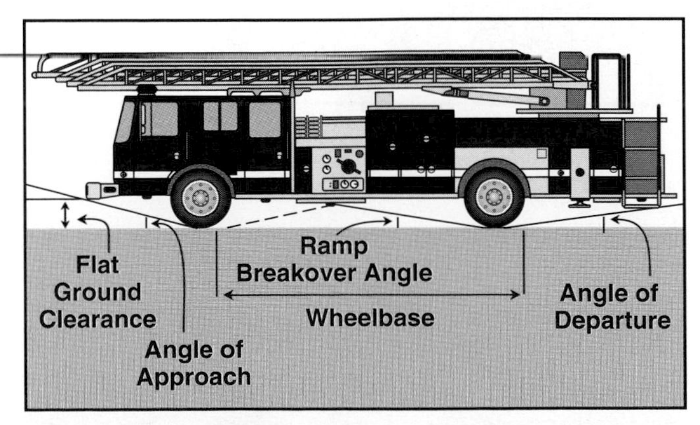

Figure 4.21 This diagram highlights the various portions of the undercarriage that the driver/operator must be concerned with.

- *Angle of departure* — The angle formed by level ground and a line from the point where the rear tires touch the ground to the lowest projection at the rear of the apparatus.
- *Breakover angle* — The angle formed by level ground and a line from the point where the rear tires touch the ground to the bottom of the frame at the wheelbase midpoint.

The driver/operator must have a keen awareness of these three angles for the apparatus he is driving. This will give the driver/operator a sense of which objects or terrain he can safely traverse and which he cannot. Failure to realize this information could result in damage to the apparatus if it "bottoms out" while crossing such an obstacle.

Stopping the Apparatus

The process of braking fire apparatus to a standstill should be performed smoothly so that the apparatus will come to an even stop. Before braking, the driver/operator should consider the weight of the apparatus and the condition of the brakes, tires, and road surface. An abrupt halt can cause a skid, injury to firefighters, and mechanical failure. Some apparatus employ an engine brake or other type of retarding device that assists in braking. The engine brake and retarder are activated when pressure is released from the accelerator. Because they provide most of the necessary slowing action, these devices allow the driver/operator to limit the use of service brakes to emergency stops and final stops. Both devices save wear on the service brakes and make the apparatus easier to manage on hills and slippery roads. Driver/operators of units having retarders should become thoroughly familiar with the manufacturer's recommendations regarding their operation prior to use.

Some air brake systems have limiting systems for varying road conditions. The clutch should not be disengaged while braking until the last few feet (meters) of travel. This practice is particularly important on slippery surfaces because an engaged engine allows more control of the apparatus. More information on proper braking techniques is discussed later in this chapter.

Engine Idling

Shut down the engine rather than leave it idling for long periods of time. This applies to apparatus at an emergency scene that are not being used (Figure 4.22). Long idling periods can result in the use of ½-gallon (2 L) of fuel per hour; the buildup of carbon in injectors, valves, pistons, and valve seats; misfiring because of injector carboning; and damage to the turbocharger shaft seals. When the engine must be left to idle for an extended period of time because of extremely cold weather or during floodlight operations, set it to idle at 900 to 1,100 rpm rather than at lower speeds. Most departments have standard operating procedures for times when the apparatus may be forced to idle for an extended period of time. The driver/operator should be familiar with the SOP and follow it accordingly. As well, some newer apparatus may have computer-controlled throttles that automatically increase the throttle periodically when the apparatus is parked and running. This type of equipment may affect the department's SOP for idling procedures.

Figure 4.22 Apparatus in staging may idle for long periods of time.

Engine Shutdown

Never attempt to shut down the engine while the apparatus is in motion because this action cuts off fuel flow from the injectors. Fuel flow through the injectors is required for lubrication anytime the injector plunger is moving. Fuel pressure can build up behind the shutoff valve and prevent the valve from opening.

Never shut down immediately after full-load operation. Shutting down the engine without a cooling-off period results in immediate increase of engine temperature from lack of coolant circulation, oil film "burning" on hot surfaces, possible damage to heads and exhaust manifolds, and possible damage to the turbocharger that may result in turbo seizure. Allow the engine temperature to stabilize before shutdown. A hot engine should be idled until it has cooled. Generally, an idle period of 3 to 5 minutes is recommended.

The procedure for shutting down the apparatus follows:

Step 1: Place the transmission in the Park (P) or Neutral (N) position once the apparatus is parked in or near the desired shutdown location.

Step 2: Set the parking brake.

Step 3: Allow the engine to idle and cool down for 3 to 5 minutes. If the truck is being prepared to park in a fire station that is not equipped with an exhaust vent system, this cool down idle should be performed on the front apron. The apparatus may then be backed into the station before proceeding.

Step 4: Shut off the engine by moving the ignition key or switch(es) to the off position.

Step 5: Turn the battery switch to the off position.

Step 6: Reconnect all ground shore lines (electric, air, exhaust) if parking in the station.

Safe Driving Techniques

The driver/operator's job is to keep the fire apparatus under control at all times. In the following sections, we review some of the more important issues involving the safe operation of fire apparatus during emergency and nonemergency operations.

Attitude

The first element in learning to drive safely is to develop a safety-conscious attitude. It is critically important that the driver/operator remain calm and drive in a safe manner. Reckless driving, even in response to an emergency, is never acceptable. The driver/operator who drives aggressively, failing to observe safety precautions, is a menace to other vehicles, pedestrians, and other firefighters on the apparatus.

Driver/operators must realize that they cannot demand the right-of-way, although they may legally have it. Actions such as ignoring approaching apparatus, refusing to yield, and driving erratically due to panic may be expected from the public. At all times, the driver/operator must be prepared to yield the right-of-way in the interest of safety. One manner of looking at it is to drive as you would during nonemergency situations and take advantage of the room that clears for you on the road.

In addition to the safety aspects of having a proper attitude, the driver/operator should also consider fire department public image aspects. All fire department members should strive to present a positive fire department image at all times. Reckless operation of the vehicle, degrading gestures, and verbal assaults toward members of the public will not assist in maintaining the positive image you seek. Other actions, such as blaring sirens and air horns on deserted roads at 3 a.m. (unless required by law or department SOP), show the fire department in a negative light. Remember that these are the same people who vote on bond issues and tax referendums and, in the case of volunteer departments, who donate money to your department. They are less apt to do this if they have negative impressions of your organization.

Apparatus Rider Safety

The driver/operator must always ensure the safety of all personnel riding on the apparatus. Riders of emergency vehicles should don their protective gear before getting in the apparatus (Figure 4.23). The one possible exception to this is the driver/operator himself. Some driver/operators are not comfortable driving the apparatus wearing fire boots or bulky protective coats. In this case, the driver/operator should don his protective clothing at the scene.

Figure 4.23 Firefighters should don their gear before boarding the apparatus.

All riders on the apparatus must be seated within the cab and wearing their seat belts before the apparatus is put into motion. Currently, NFPA 1901, *Standard for Automotive Fire Apparatus*, requires that a seat and seat belt be provided within the cab of the apparatus for every firefighter who is expected to ride the truck. NFPA 1500, *Standard on Fire Department Occupational Safety and Health Program*, also specifically states that all riders must be seated and belted. Some states and provinces also have laws that require drivers and passengers in fire apparatus to wear seat belts. NFPA 1500 only provides three exceptions to the seated and belted requirement:

- When providing patient care in the back of an ambulance that makes it impractical to be seated and belted. In these cases, some other type of harness is required.
- When loading hose back onto a fire apparatus.
- When performing training for personnel learning to drive the tiller portion of a tractor-drawn aerial apparatus.

Providing emergency care is not within the scope of this manual. However, driver/operators who drive aerial apparatus equipped with fire pumps are likely to find themselves in the situation of having to load fire hose back onto the vehicle. Loading fire hose while driving the apparatus is particularly common when loading large diameter (4-inch [100 mm] or larger) supply hose. NFPA 1500 provides the following specific directions on how these operations should be performed to maximize safety:

- The procedure must be contained in the department's written standard operating procedures (SOPs), and all members must be trained specifically on how to perform the moving hose-load operation.
- At least one member, other than the one actually loading the hose, must be assigned as a safety observer to the operation. This individual must have complete visual contact with the hose-loading operation, as well as visual and voice communications (usually via a portable radio) with the driver/operator.
- The area in which the hose loading is being performed must be closed to other vehicular traffic.
- The apparatus must be driven in a forward direction (straddling or to one side of the hose) at a speed no greater than 5 mph (8 km/h) (Figure 4.24).
- No members are allowed to stand on any portion of the apparatus.
- Members in the hose bed must sit or kneel while the apparatus is moving.

In many cases, this is a moot point for loading hose on aerial apparatus because it is often necessary to lift the aerial device out of its stored position in order to access the hose bed. It is not possible to move the apparatus when the aerial device is raised.

Tiller training can be especially problematic in that most tiller apparatus only have a single seat within the confines of the tiller operator's enclosure (Figure 4.25). This leaves no fixed riding position for an instructor who wishes to be in close

Figure 4.24 A safety observer equipped with a portable radio should monitor moving hose-loading operations.

Figure 4.25 Most tiller compartments have seating for only one person.

contact with the tiller operator during training. NFPA 1500 allows for a detachable seat to be placed next to the tiller operator's position in which the instructor may sit. This seat must be firmly attached to the apparatus and allow the instructor to be belted into position. The tiller operator and/or the instructor must wear a helmet and eye protection if they are not seated in an enclosed area. Newer apparatus designs may include a tiller operator enclosure that is capable of holding both the tiller operator and an instructor (Figure 4.26). If during the training exercise the apparatus is needed to make an actual emergency response, the training session should be ended. A qualified tiller operator should take the controls of the tiller, and the student and instructor should take seated and belted positions inside the cab of the apparatus.

Although a common practice in the past, firefighters should never be allowed to ride the tailboard or running boards of any moving apparatus. This is specifically prohibited by NFPA 1500. These riding positions offer little protection if the vehicle is involved in a collision or rollover.

WARNING!

Firefighters should never ride on the outside of a moving fire apparatus for any reason, other than those exceptions noted in NFPA 1500 and those previously listed. Serious injury or death could occur if the apparatus is involved in a collision or rollover or if the rider falls from the moving apparatus.

Figure 4.26 Newer tiller apparatus designs may allow for a second person in the tiller enclosure. *Courtesy of Emergency One, Inc.*

Although all newer apparatus are designed with fully enclosed cabs, many older apparatus with jump seat riding areas not totally enclosed are still in service. Some of these are equipped with safety bars or gates that are intended to prevent a firefighter from falling out of a jump seat (Figures 4.27 a and b). These devices are not substitutes for safety procedures that require firefighters to ride in safe, enclosed positions wearing their seat belts. Safety devices that are held in an upright or open position by straps or ropes provide no extra security for the firefighter riding in the jump seat.

Backing the Vehicle

As stated earlier in this chapter, a significant portion of fire apparatus collisions occur while the apparatus is being driven in reverse. This is particularly a problem with aerial apparatus because of its length. All fire departments should have firmly established procedures for backing the vehicle, and these procedures must always be followed by fire apparatus driver/operators.

Figure 4.27a Jump seat safety bars provide minimal protection to the firefighter riding in that location.

Figure 4.27b Safety gates provide a slightly higher level of fall protection than do safety bars.

Figure 4.28 Drive-through bays eliminate the need to back the apparatus into the station.

Figure 4.29 It is preferable that two firefighters, including one equipped with a portable radio, assist the driver/operator in backing the apparatus.

Whenever possible, the driver/operator should avoid backing the fire apparatus. It is normally safer and sometimes quicker to drive around the block and start again. It is most desirable that new fire stations be designed with drive-through apparatus bays that negate the necessity to back the apparatus into them (Figure 4.28). However, there are situations when it is necessary to back fire apparatus. This operation should be performed very carefully. Before backing the apparatus, assign at least one (and preferably two) firefighter(s) to clear the way and to warn the driver/operator of any obstacles obscured by blind spots (Figure 4.29). The firefighter should be equipped with a portable radio. If two spotters are used, only one should communicate with the driver/operator. The second spotter should assist the first one. This is a very simple procedure that can prevent a large percentage of the collisions that occur during backing operations. **Very simply, if you are the driver/operator and you do not have or cannot see the spotters behind you, *do not back the apparatus*!** All fire apparatus should be equipped with an alarm system that warns others when the apparatus is backing up.

Some apparatus are equipped with cameras located on the rear of the apparatus (Figure 4.30). These devices send a visual image of the area be-

Figure 4.30 A video camera may be located on the rear of the apparatus.

Figure 4.31 The monitor in the cab allows the driver/operator to observe the area behind the apparatus.

hind the apparatus to a small monitor screen at the driver/operator's position in the cab (Figure 4.31). This allows the driver/operator a view of the area behind the apparatus when backing. Consult local standard operating procedures to determine whether the availability of this equipment may negate the need to use spotters.

Defensive Driving Techniques

Sound defensive driving skills are one of the most important aspects of safe driving. Every driver/operator should be familiar with the basic concepts of defensive driving. They include anticipating other drivers' actions, estimating visual lead time, knowing braking and reaction times, combating skids, knowing evasive tactics, and having knowledge of weight transfer.

The driver/operator should know the rules that govern the general public when emergency vehicles are on the road. Most laws or ordinances provide that other vehicles must pull toward the right and remain at a standstill until the emergency vehicle has passed, except on freeways. On freeways, vehicles traveling in the inside (left) lanes may be required to pull over to the left (in the median) to make way for emergency vehicles. These laws, however, do not mean that the apparatus driver/operator can ignore stopped vehicles. People may panic at the sound of an approaching siren and accidentally move into the lane of traffic or stop suddenly. Others may not hear warning signals because of car radios, closed windows, or air-conditioning. It must also be remembered that some individuals simply ignore emergency warning signals.

Intersections are the most likely place for a collision involving an emergency vehicle (Figure 4.32). When approaching an intersection, the driver/operator should slow the apparatus to a speed that allows a stop at the intersection if necessary. Even if faced with a green signal light, or no signal at all, the apparatus should be brought to a complete stop if there are any obstructions, such as buildings or trucks, that block the driver/operator's view of the intersection.

Depending on the motor vehicle statutes and departmental SOPs within a particular jurisdiction, fire apparatus on an emergency response may proceed through a red traffic signal or stop sign after coming to a complete stop and ensuring that all cars in each lane of traffic are accounted for and yielding to the apparatus. The determination of what is an emergency response is covered later in this chapter (Warning Devices and Clearing Traffic). Do not proceed into the intersection until you are certain that every other driver sees you and is allowing you to proceed. Simply slowing when approaching the intersection and then coasting through is not an acceptable substitute for coming to a complete stop. When proceeding through the

Figure 4.32 Intersections are the most likely location for a collision. *Courtesy of Ron Jeffers.*

intersection, attempt to make eye contact with each of the other drivers to ensure that they know you are there and about to proceed.

Traffic waiting to make a left-hand turn may pull to the right or left, depending upon the driver. In situations where all lanes of traffic in the same direction as the responding apparatus are blocked, the apparatus driver/operator should move the apparatus into the opposing lane of traffic and proceed through the intersection at an extremely reduced speed (Figure 4.33). Oncoming traffic must be able to see the approaching apparatus. Full use

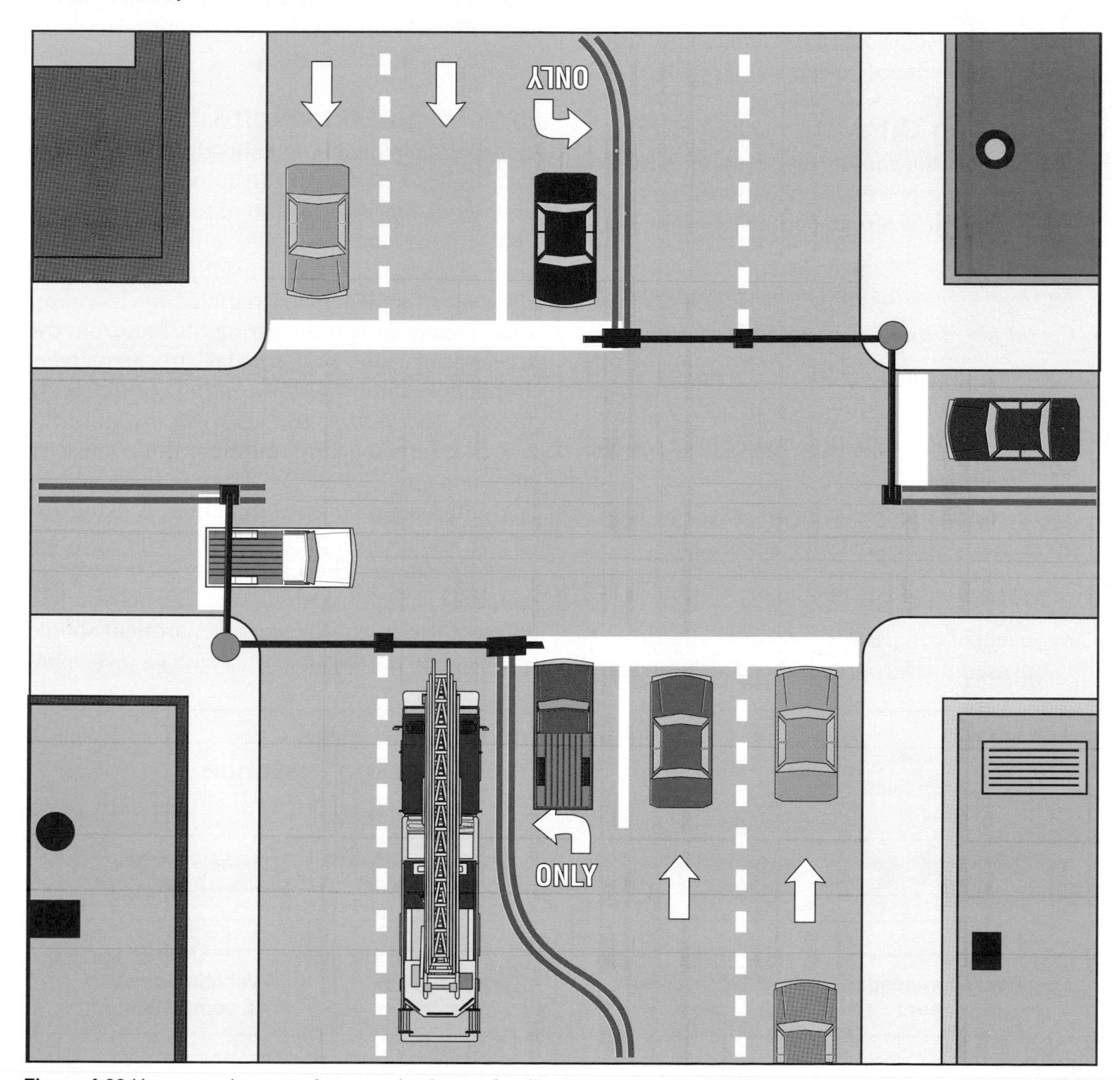

Figure 4.33 Use care when entering opposing lanes of traffic to traverse intersections.

of warning devices is essential. Driving in the oncoming lane is not recommended in situations where oncoming traffic is unable to see the apparatus such as on a freeway underpass or when the intersection is on the crest of a hill. Be alert for traffic that may enter from access roads and driveways. Approaching traffic on the crest of a hill, slow-moving traffic, and other emergency apparatus must be closely monitored.

Even though use of warning sirens, lights, and signals is essential, fire apparatus driver/operators should realize that these signals may be blanketed by other warning devices, street noises, or loud car stereos. Serious collisions and fatalities have been caused by overreliance on warning signals.

Anticipating Other Drivers' Actions

Never assume what another driver's actions will be — expect the unexpected. Anticipation is the key to safe driving. Always remember the following control factors:

- *Aim high in steering*: Find a safe path well ahead.
- *Get the big picture*: Stay back and see it all.
- *Keep your eyes moving*: Scan — do not stare.
- *Leave yourself an "out"*: Do not expect other drivers to leave you an out. Be prepared by expecting the unexpected.
- *Make sure that others can see and hear you*: Use lights, horn, and signals in combination.

Visual Lead Time

The concept of *visual lead time* refers to the driver/operator scanning far enough ahead of the apparatus, for the speed it is being driven, to ensure that appropriate action can be taken if it becomes necessary. For example, if the driver/operator is concentrating on vehicles that are 100 feet (30 m) in front of the apparatus and based on the speed the apparatus is being driven it would take 200 feet (60 m) for it to stop or perform an evasive maneuver, a collision is likely to occur. The driver/operator needs to learn to match the speed he is traveling with the distance ahead of the vehicle he is surveying. Visual lead time interacts directly with reaction time and stopping distances. By "aiming high in steering" and "getting the big picture," it is possible to become more keenly aware of conditions that may require slowing or stopping.

Braking and Reaction Time

A driver/operator should know the total stopping distance for a particular fire apparatus. The *total stopping distance* is the sum of the driver/operator reaction distance and the vehicle braking distance (Figure 4.34). The *driver/operator reaction distance* is the distance a vehicle travels while a driver/operator is transferring his foot from the accelerator to the brake pedal after perceiving the need for stopping. The *braking distance* is the distance the vehicle travels from the time the brakes are applied until the apparatus comes to a complete stop. Tables 4.6 a and b show driver/operator reaction distances, vehicle braking distances, and total stopping distances for different sizes of vehicles. These tables indicate approximates for vehicles, and the statistics may vary for different fire apparatus. Each department should conduct braking distance tests with its own appa-

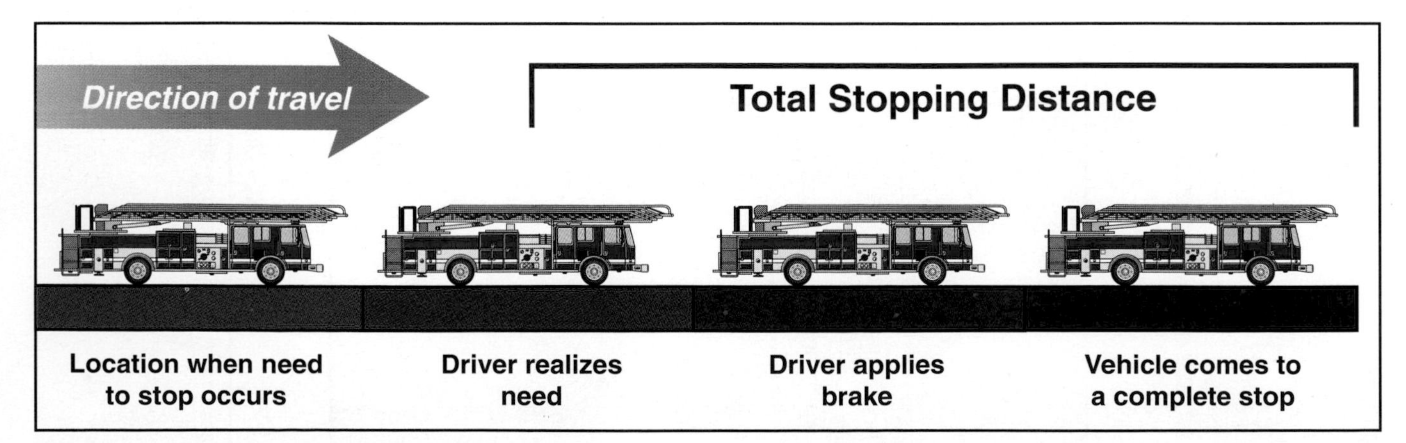

Figure 4.34 The total stopping distance is the distance a vehicle travels from the time the driver/operator realizes the need to stop and when the apparatus is actually stopped.

Table 4.6a (US)
Braking and Stopping Distances (dry, level pavement)

Speed (mph)	Average Driver Reaction Distance (feet)	Braking Distance (feet) Vehicle A	Vehicle B	Vehicle C	Vehicle D	Total Stopping Distance (feet) Vehicle A	Vehicle B	Vehicle C	Vehicle D
10	11		7	10	13		18	21	24
15	17		17	22	29		34	39	46
20	22	22	30	40	50	44	52	62	72
25	28	31	46	64	80	59	74	92	108
30	33	45	67	92	115	78	100	125	148
35	39	58	92	125	160	97	131	164	199
40	44	80	125	165	205	124	169	209	249
45	50	103	165	210	260	153	215	260	310
50	55	131	225	255	320	186	280	310	375
55	61	165	275	310	390	226	336	371	451
60	66	202	350	370	465	268	426	436	531

Typical Brake Performance
A–Average automobile
B–Light two-axle trucks
C–Heavy two-axle trucks
D–Three-axle trucks and trailers

Table 4.6b (metric)
Braking and Stopping Distances (dry, level pavement)

Speed (km/h)	Average Driver Reaction Distance (meters)	Braking Distance (meters) Vehicle A	Vehicle B	Vehicle C	Vehicle D	Total Stopping Distance (meters) Vehicle A	Vehicle B	Vehicle C	Vehicle D
16	3.4		2.1	3	4		5.5	6.4	7.3
24	5.2		5.2	6.7	8.8		10.4	11.9	14
32	6.7	6.7	9.1	12.2	15.2	13.4	15.8	18.9	21.9
40	8.5	9.4	14	19.5	24.4	18	22.6	28	32.9
48	10.1	13.7	20.4	28	35.1	23.8	30.5	38.1	45.1
56	11.9	17.7	28	38.1	48.8	29.6	39.9	50	60.7
64	13.4	24.4	38.1	50.3	62.5	37.8	51.5	63.7	75.9
72	15.2	31.4	50.3	64	79.2	46.6	65.5	79.2	94.5
80	16.8	39.9	68.6	77.7	97.5	56.7	85.3	94.5	114.3
88	18.6	50.3	83.8	94.5	118.9	68.9	102.4	113.1	137.5
96	20.1	61.6	106.7	112.8	141.7	81.7	129.8	133	161.8

Typical Brake Performance
A–Average automobile
B–Light two-axle trucks
C–Heavy two-axle trucks
D–Three-axle trucks and trailers

ratus. Apparatus manufacturers may also be able to provide this information for specific apparatus.

There are a number of factors that influence the driver/operator's ability to stop the apparatus:

- Condition of the driving surface
- Speed being traveled
- Weight of the vehicle
- Type and condition of the vehicle's tires and braking system
- Whether the vehicle is being driven uphill or downhill

A dry, paved road provides the optimal stopping ability from a driving surface standpoint. The ability of the apparatus to stop is negatively affected by wet, snowy, icy, or unpaved roads. Driver/operators must compensate for these conditions by reducing their speeds by an appropriate amount to match the conditions.

The correlation between vehicle weight and speed and stopping distance should be obvious to anyone. At an equal speed, it will take a greater distance to stop a three-axle quint than it will a brush pumper (Figures 4.35 a and b). It will also take a greater distance to stop a vehicle that is going 50 mph (80 km/h) than the same vehicle when it is traveling 30 mph (48 km/h).

The type and condition of the tires and braking system have a tremendous impact on the ability to stop the fire apparatus. Several serious fire apparatus accidents have been traced to poor maintenance of the braking system or worn tires. Obviously, a vehicle that has a properly maintained braking system and good tires will stop faster than one that does not.

Weight Transfer

The effects of weight transfer must be considered in the safe operation of fire apparatus. Weight transfer occurs as the result of physical laws that state that objects in motion tend to stay in motion; objects at rest tend to remain at rest. Whenever a vehicle undergoes a change in velocity or direction, weight transfer takes place relative to the severity of change. Apparatus driver/operators must be aware that the large weight carried on most aerial apparatus can contribute to skidding or possible rollover due to excessive weight transfer. These hazardous conditions can result from too much speed in turns, harsh or abrupt steering action, or driving on slopes too steep for a particular apparatus.

Use only as much steering as needed to keep weight transfer to a minimum. Steering should be smooth and continuous. Also, maintain a speed that is slow enough to prevent severe weight transfer from occurring. This is particularly important on curves.

Combating Skids

Avoiding conditions that lead to skidding is as important as knowing how to correct skids once they occur. The following are some of the most common causes of skids involving driver error:

- Driving too fast for road conditions.
- Failing to properly appreciate weight shifts of heavy apparatus.
- Failing to anticipate obstacles (these range from other vehicles to animals).
- Improper use of auxiliary braking devices.

Figure 4.35a Large three-axle quints may require a sizeable stopping distance. *Courtesy of Joel Woods, Maryland Fire & Rescue Institute.*

Figure 4.35b The stopping distance for this brush pumper would be much less than a three-axle quint traveling at the same speed.

- Improper maintenance of tire air pressure and adequate tread depth. Tires that are overinflated or lacking in reasonable tread depth make the apparatus more susceptible to skids.

Most newer, large fire apparatus are equipped with an all-wheel, antilock braking system (ABS). The power for the braking ability comes from air pressure. These systems are effective in that they minimize the chance of the vehicle being put into a skid when the brakes are applied forcefully. ABS works using digital technology in an onboard computer that monitors each wheel and controls air pressure to the brakes, maintaining optimal braking ability. A sensing device located in the axle monitors the speed of each wheel. The wheel speed is converted into a digital signal that is sent to the onboard computer. When the driver/operator begins to brake and the wheel begins to lock up, the sensing device sends a signal to the computer that the wheel is not turning. The computer analyzes this signal against the signals from the other wheels to determine if this particular wheel should still be turning. If it is determined that it should be turning, a signal is sent to the air modulation valve at that wheel, reducing the air brake pressure and allowing the wheel to turn. Once the wheel turns, it is braked again. The computer makes these decisions many times a second until the vehicle is brought to a halt. Thus, when driving a vehicle equipped with an ABS, maintain a steady pressure on the brake pedal (rather than pumping the pedal) until the apparatus is brought to a complete halt.

Keep in mind that in the case of air brakes, there is a slight delay (sometimes referred to as brake lag) in the time from which the driver/operator pushes down on the brake pedal until sufficient air pressure is sent to the brake to operate. This must be considered when determining total stopping distance.

When an apparatus that is not equipped with an antilock braking system goes into a skid, release the brakes, allowing the wheels to rotate freely. Turn the apparatus steering wheel so that the front wheels face in the direction of the skid. If using a standard transmission, do not release the clutch (do not push in the clutch pedal) until the vehicle is under control and just before stopping the vehicle. When the skid is controllable, gradually apply power to the wheels to further control the vehicle by giving traction.

Proficiency in skid control may be gained through practice at facilities having skid pads. These are smooth surface driving areas that have water directed onto them to make skids likely (Figure 4.36). All training should be done at slow speeds to avoid damaging the vehicle or injuring participants. Some jurisdictions choose to use reserve apparatus or other older vehicles for this part of the training process (Figure 4.37). If a skid pad is not available, it may be possible to use an open parking lot.

Auxiliary Braking Systems

In addition to the engine retarders mentioned earlier in this chapter, the apparatus may be equipped with one or more other types of auxiliary braking systems. The driver/operator needs to understand these systems so that they can be used properly; in some cases, their use should be avoided completely.

Figure 4.36 The skid pad may be used for learning to control vehicles on slippery surfaces.

Figure 4.37 Some jurisdictions use old police vehicles for skid pad training. This saves wear and tear on in-service fire apparatus.

The first type of auxiliary braking system is the front-brake limiting-valve type. These were commonly installed on apparatus built before the mid 1970s but are also found on some newer apparatus. These were more commonly known as the "dry road/slippery road" switches (Figure 4.38). These devices were intended to help the driver/operator maintain control of the apparatus on slippery surfaces. This was accomplished by reducing the air pressure on the front steering axle by 50 percent when the switch was in the slippery-road position. This would prevent the front wheels from locking up, allowing the driver/operator to steer the vehicle even when the rear wheels were locked into a skid.

In reality, these systems were not overly effective or safe. With the switch in the slippery-road position, the braking capabilities were actually reduced by 25 percent. If the braking system was not in optimum condition to begin with, say it was only working at 80 percent of its designed capability, then using this switch would drop it 25 percent more or to 55 percent of the designed braking ability. After the adoption of the Federal Motor Vehicle Safety Standard 121 of 1975 by the United States government, few trucks were built with this switch installed. *IFSTA recommends that apparatus equipped with this switch have it placed in the dry-road position and disconnected.*

Another type of auxiliary braking system is the interaxle differential lock, also known as a power divider or third differential (Figure 4.39). This is another type of switch that may be activated from the cab of an apparatus that has tandem rear axles. It allows for a difference in speed between the two rear axles, while providing pulling power from each axle. This is intended to provide greater traction for each axle.

Under normal operating conditions, the interaxle differential switch should be in the unlocked position. Move the switch to the locked position when approaching or anticipating slippery-road conditions to provide improved traction. Always unlock the switch again when road conditions improve. You must lift your foot from the accelerator when activating the interaxle differential lock. Do not activate this switch while one or more of the wheels are actually slipping or spinning because damage to the axle could result. Also, do not spin the wheels with the interaxle differential locked because damage to the axle could result.

Some vehicles equipped with an ABS are also equipped with automatic traction control (ATC). ATC turns itself on and off; there is no switch for the operator to select. A green indicator light on the dash illuminates when the ATC is engaged. The engine speed decreases, as needed, until traction is acquired to move the chassis.

ATC helps improve traction on slippery roads by reducing drive-wheel overspin. ATC works automatically in two ways: First, when a drive wheel starts to spin, the ATC applies air pressure to brake the wheel. This transfers engine torque to the wheels with better traction. Second, when all drive wheels begin to spin, the ATC reduces the engine torque to provide improved traction.

Some vehicles equipped with ATC have a snow-and-mud switch. This function increases available

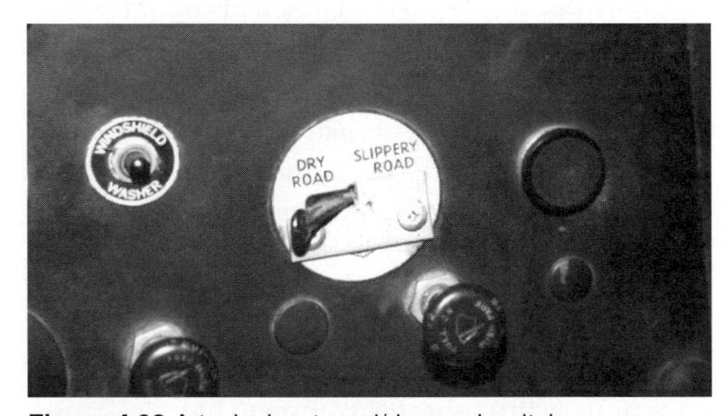

Figure 4.38 A typical wet road/dry road switch.

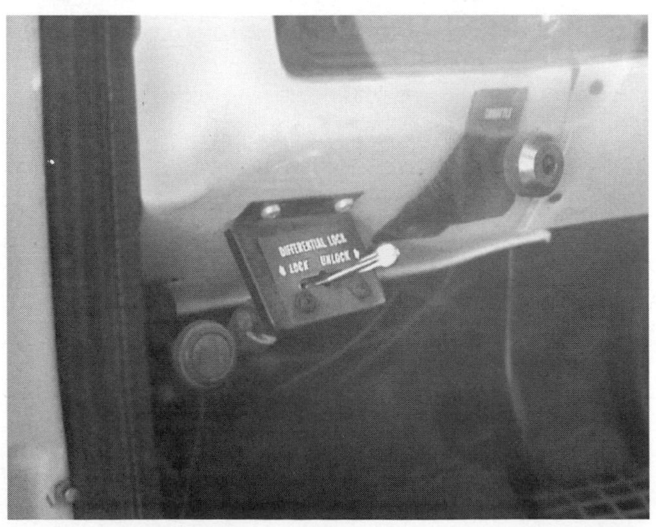

Figure 4.39 The interaxle differential may be controlled by a switch on the dashboard.

traction on extra soft surfaces. When this switch is activated, the ATC indicator light flashes continuously. When normal traction is regained, deactivate this feature. To deactivate the switch, press the switch a second time and turn off the vehicle ignition. If you desire to "rock" an apparatus out of a particular spot and the ATC has cut the throttle, activate the mud-and-snow switch. Use caution when activating this switch because if the apparatus regains traction suddenly, axle damage may occur.

Consult the apparatus manufacturer's operations manual for detailed information on operating the auxiliary braking system(s) on your apparatus.

Passing Other Vehicles

In general, it is best to avoid passing vehicles that are not pulling over to yield the right-of-way to the fire apparatus. However, in some instances, the need to pass will occur, and the driver/operator must be prepared to do it in the safest manner possible. The following guidelines should be used to ensure safe passing:

- Always travel on the innermost lane on multilane roads (Figure 4.40). Wait for vehicles in front of you to move to the right before proceeding.
- Avoid passing vehicles on their right (Figure 4.41). Most civilian drivers have a natural tendency to move to the right when an emergency vehicle is approaching. Thus, they could turn into your path if you are passing on the right. Some departments have strict SOPs prohibiting this practice.
- Make sure that the opposing lanes of traffic are clear of oncoming traffic if you must move in that direction.
- Avoid passing other emergency vehicles if at all possible. However, in some cases, it may be desirable for a smaller, faster vehicle (such as a chief's vehicle) to pass a larger, slower vehicle (such as an aerial apparatus). In these cases, the lead vehicle should slow down, turn off its warning lights, and move to the right to allow the other vehicle to pass. This maneuver should be coordinated by radio if possible.
- Flash your high beam lights to get the driver's attention when passing.

Adverse Weather

Weather is another factor to consider in terms of safe driving. Rain, snow, ice, and mud make roads slippery. A driver/operator must recognize these dangers and adjust apparatus speed according to the crown of the road, the sharpness of curves, and the condition of road surfaces. The driver/operator should decrease speed gradually, slow down while approaching curves, keep off low or soft shoulders, and avoid sudden turns. The driver/operator should recognize areas that first become slippery such as bridge surfaces, northern slopes of hills, shaded spots, and areas where snow is blowing across the roadway.

Because the stopping distance is greatly increased on slippery road surfaces, it is sometimes a good policy to try the brakes while in an area free of traffic to find out how slippery the road is. Speed must be adjusted to road and weather conditions so that the apparatus can be stopped or maneuvered safely.

Snow tires or tire chains will reduce the stopping distance and will considerably increase starting

Figure 4.40 Travel on the inside lane of multilane streets.

Figure 4.41 Avoid passing other vehicles on the right side of the road.

and hill-climbing traction on snow or ice. Apparatus may be equipped with the traditional, manually applied tire chains or the newer automatic variety. Automatic tire chains consist of short lengths of chain that are on a rotating hub in front of each rear wheel (Figure 4.42). The hubs swing down into place when a switch on the dashboard is activated (Figure 4.43). The rotation of the hub throws the chains underneath the rolling tires. These chains tend to lose their effectiveness in snow that is deeper than 8 inches (200 mm).

During slippery road conditions, the safe following distance between vehicles increases dramatically. Remember that it takes 3 to 15 times more distance for a vehicle to come to a complete stop on snow and ice than it does on dry concrete.

Figure 4.42 A typical automatic tire chain arrangement.

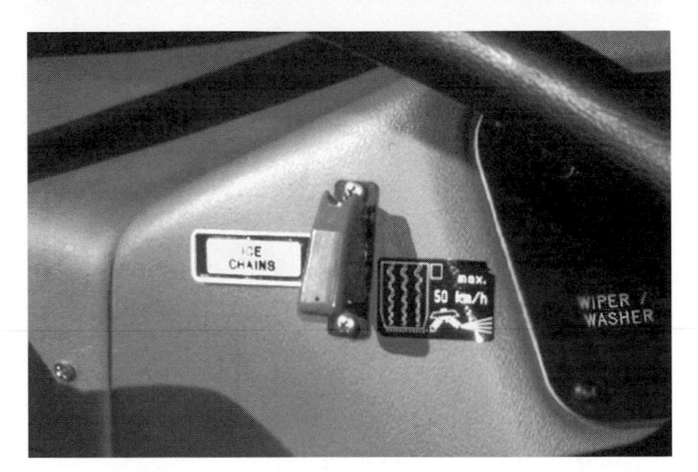

Figure 4.43 Controls for automatic tire chains are found on the dashboard.

Warning Devices and Clearing Traffic

All fire apparatus are equipped with some combination of audible and visible warning devices. Audible warning devices may include air horns, bells, mechanical sirens, or electronic sirens (Figures 4.44 a through d). Studies have shown that civilian drivers respond better to sounds that change pitch often. Short bursts with the air horns and the constant up-and-down oscillation of a mechanical or electronic siren are the surest ways to catch a driver's attention. At speeds above 50 mph (80 km/h), an emergency vehicle may "outrun" the effective range of its audible warning device. A study conducted by the staff of Driver's Reaction Course concluded that a siren operating on an emergency vehicle moving at 40 mph (64 km/h) can project 300 feet (90 m) in front of the vehicle. At a speed of 60 mph (97 km/h), however, the siren is only audible 12 feet (3.7 m) or less in front of the vehicle. Driver/operators must operate within the effective range of their audible warning devices.

Warning devices are of no value to the driver/operator or the public if they are not used. Warning devices should be operated from the time the apparatus begins its response until it arrives on the scene. The driver/operator needs to use some discretion in the use of sirens. When responding to sensitive situations, such as psychiatric emergencies, it is often better to turn off the siren as the apparatus nears its destination. The sudden use of an audible warning device immediately behind another vehicle may unnerve the driver. If this happens, the driver may suddenly stop or swerve, causing a collision. The use of warning devices is essential, but it does not give the driver/operator the right or permission to disregard other drivers.

The use of warning devices should be limited to emergency response situations. Each department should have standard operating procedures on which types of calls are considered emergencies and which are not. In general, certain types of calls for service, such as water shutoffs or covering an empty fire station, are not really emergencies. Warning devices and emergency driving tactics should not be used on these types of responses. Some fire departments have further extended their list of nonemergency responses to include such calls as automatic fire alarms, carbon monoxide detector alarms, smoke odor investigations, and trash fires. Be aware that emergency organizations have lost lawsuits when they were deemed responsible for causing collisions while driving in an emergency manner to nonemergency calls.

Figure 4.44a An air horn.

Figure 4.44b A bell.

Figure 4.44c A mechanical or coaster siren.

Figure 4.44d An electronic siren speaker.

Some fire departments have standard operating procedures that require driver/operators to turn off all warning devices and proceed with the normal flow of traffic while driving on limited-access highways and turnpikes. In reality, the apparatus will probably not be capable of driving as fast as many of the vehicles on that roadway, and the use of warning devices is unnecessary. Warning lights may be turned back on when the apparatus reaches the scene and parks.

There have been numerous collisions involving fire apparatus and other emergency vehicles. It is not always possible to hear the warning devices of other emergency vehicles when the audible warning device on your apparatus is being sounded. When more than one emergency vehicle is responding along the same route, units should travel at least 300 to 500 feet (90 m to 150 m) apart (Figure 4.45). Some fire departments rely upon designated response routes. This practice can be hazardous if a company is delayed or detoured for some reason. Standard operating procedures should call for radio reports of location and status, particularly when you are certain that you are approaching the same intersection as another emergency vehicle. Regardless of the system or pattern used, always take precautions to ensure a safe, collision-free response. This includes coming to a complete stop at any intersection that has a stop sign or a red signal light.

White lights can be readily distinguished during daylight hours. For this reason, headlights should

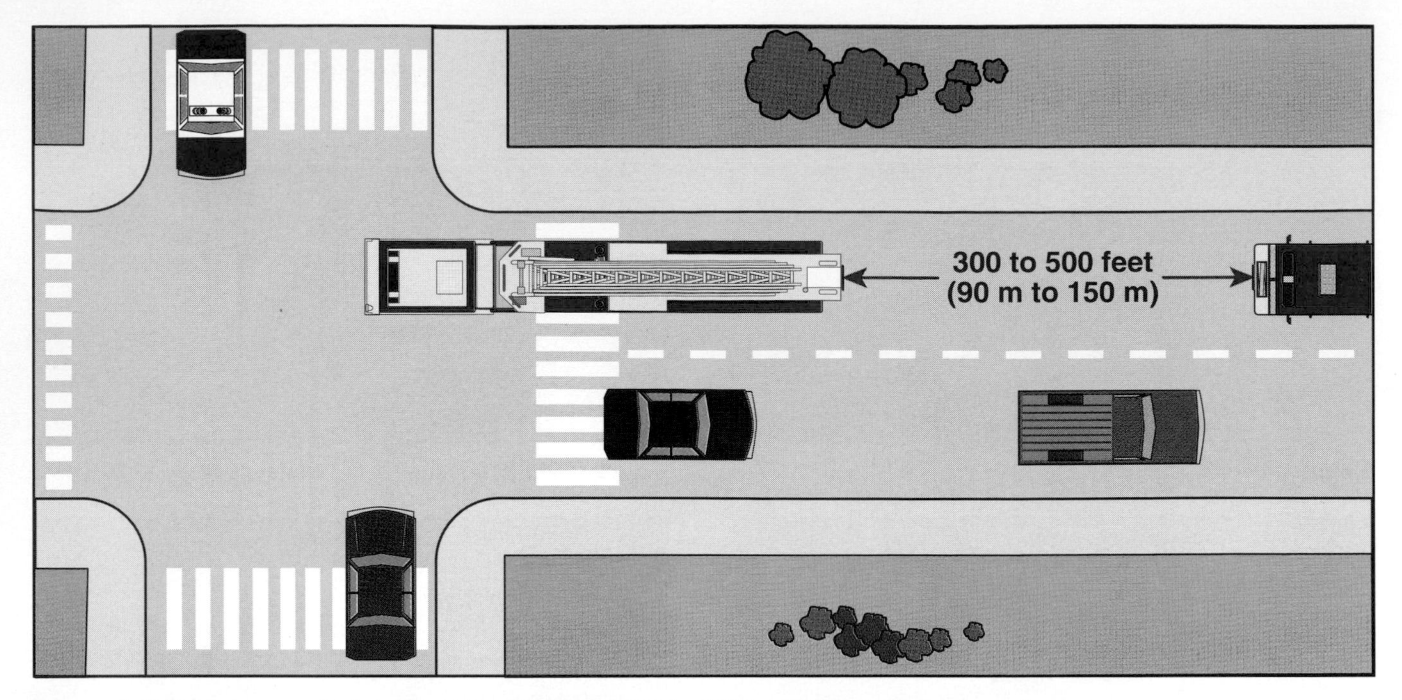

Figure 4.45 Emergency vehicles should travel at least 300 to 500 feet (90 m to 150 m) apart.

be turned on while responding (Figure 4.46). Some departments use white warning lights in conjunction with red or other colored lights, but some state and provincial laws prohibit the use of flashing white lights. A spotlight moving across the back window of a vehicle rapidly gains the driver's attention. The spotlight should not be left shining on the vehicle, however, because this blinds the driver. Headlights should be dimmed and spotlights turned off in situations where they may blind oncoming drivers, including the drivers of other apparatus that are approaching the scene. Headlight flashers are an inexpensive and effective warning device; however, some states (provinces) do not allow them or consider them warning lights. Do not drive with high beam headlights on constantly because they tend to drown out the other warning lights and blind oncoming drivers.

Figure 4.46 Always have the headlights turned on during daylight responses.

The trend in recent years has been to equip apparatus with large amounts of visual warning devices. While this quantity has proven to be effective during the response, studies have shown that they may actually pose hazards for the firefighters once the apparatus is parked on the scene during nighttime operations. The large quantity of warning lights, combined with on-scene floodlights overpower the effectiveness of the reflective trim on the firefighters' protective clothing or vests. This makes the drivers of approaching vehicles unable to see firefighters who are standing in the street. They also tend to attract the attention of drivers, often causing them to drive directly toward the lights, resulting in collision. In these situations, it is desirable to turn off some of the warning lights on the apparatus once it is parked. Some fire departments have equipped their apparatus with one or two small yellow warning lights or flashing yellow directional arrows that are turned on when the apparatus is parked on the scene. This allows the approaching vehicles' headlights to more effectively illuminate the reflective trim worn by firefighters.

Traffic Control Devices

Some jurisdictions use traffic control devices to assist emergency vehicles during their response. The driver/operator must be aware of the traffic control devices used in his jurisdiction and how they operate. One of the simplest involves placing a traffic signal in front of the fire station to stop the flow of traffic so that the apparatus can exit safely (Figure 4.47). This signal may be controlled by a button in the station or by the dispatcher. It may also be activated when the station is dispatched. Some jurisdictions have systems that control one or more traffic lights in the normal route of travel for fire apparatus. Again, these may be controlled from the fire station, remote controls on the fire apparatus, or from the dispatch center.

Another common system for controlling traffic signals for fire apparatus is the Opticom™ system, which involves the use of special strobe lights, called emitters, on the fire apparatus and sensors mounted on the traffic lights. The emitter generates an optical signal that is received by the sensor on the traffic light as the apparatus approaches (Figure 4.48). The sensor converts this signal to an electronic impulse that is routed to the phase selector in the traffic light control cabinet (Figure 4.49). The phase selector then provides a green light for the direction that the apparatus is traveling and red signals in all other directions (Figure 4.50). In some jurisdictions, the traffic light standard may be equipped with a white light that indicates to the driver/operator that the signal has been received and a green light is forthcoming. On some apparatus, the emitter is wired into the parking brake system. When the parking brake is set, the emitter will be turned off. On apparatus that do not have this feature, the driver/operator should remember to turn off the emitter when the apparatus is parked on the scene of an emergency.

Figure 4.47 Some jurisdictions have traffic control lights in front of their stations.

Figure 4.48 An Opticom™ emitter.

Figure 4.49 The Opticom™ sensor is located somewhere on the traffic light standard.

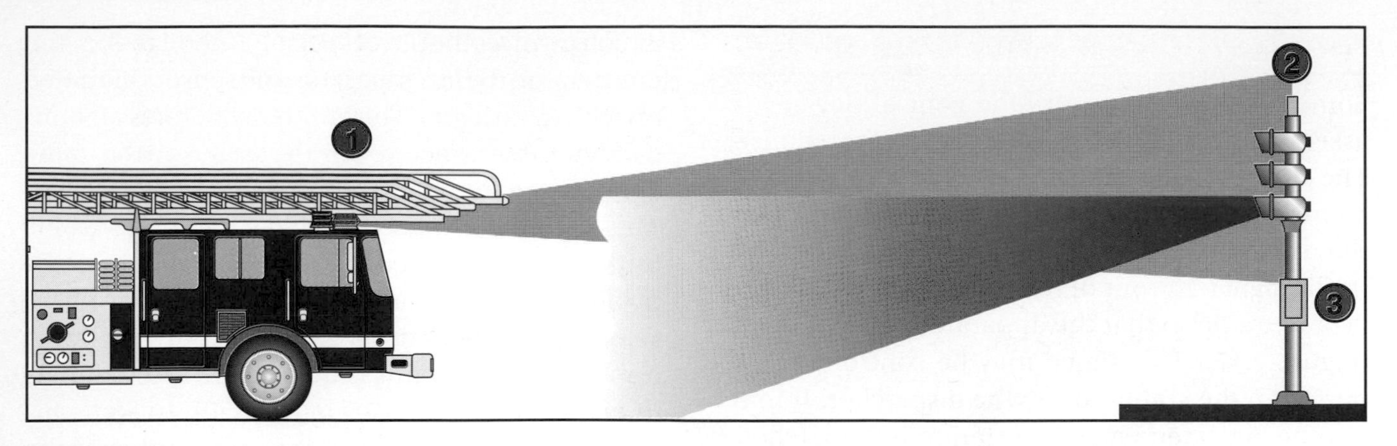

Figure 4.50 How the Opticom™ works: (1) the emitter generates an optical signal to the detector (2), which converts the optical signal to an electronic impulse. The phase selector (3) in the control cabinet processes and then manipulates the controller to provide a green signal for the emergency vehicle and red signals in all other directions. *Courtesy of 3M Safety and Security.*

Otherwise, the emitter could affect any traffic signals that are within reach and disrupt the normal flow of traffic.

A newer type of traffic control system is the SONEM 2000 system, which is activated by the emergency vehicle's siren as it approaches an intersection. A microphone on the traffic signal "hears" the siren and sends a signal to the traffic signal controller, ordering a preemption of the current traffic signal. The microphone may be adjusted to order the preemption from distances of anywhere from a few hundred feet (meters) to about one-half mile (1 km). Intersections equipped with this system will have 3-inch (76 mm) white and blue lights in each direction of travel, somewhere to the side of the regular traffic signals. As soon as the microphone sends the preemption signal to the signal controller, the direction of travel for the emergency vehicle gets a white light indicating that the signal was received and that a green traffic light is forthcoming. All other directions of travel get a blue light that indicates an emergency vehicle coming from one of the other directions has gained control of the signal first. This is extremely important when emergency vehicles are approaching the intersection from more than one direction. Driver/operators of vehicles getting the blue light know that they will have to come to a stop because a green signal is not immediately forthcoming in their direction of travel.

Regardless of which types of traffic control devices are used in any jurisdiction, they are not substitutes for using proper defensive driving techniques. When traversing an intersection, the driver/operator must maintain a speed that will allow the signal preemption to occur before the apparatus reaches the intersection and for evasive actions in the event another vehicle enters the intersection. If for any reason the fire apparatus does not get a green signal, the driver/operator should bring the vehicle to a complete stop at a red signal. Keep in mind that if two apparatus equipped with Opticom™ devices approach the same traffic signal from different directions, only the apparatus whose sensor affects the signal first will get a green light. The later-approaching apparatus gets a red signal. Do not assume that just because you did not get a green light that the system is not working. Approach the intersection with caution and come to a complete stop. Also keep in mind that at some intersections the traffic control system may give traffic approaching from the opposite direction a green light as well as you. Thus, if you need to make a left turn at that intersection, you should not automatically assume that approaching traffic has a red light.

Tillering

No activity in the fire service catches the interest of observers more quickly than an aerial ladder truck weaving in and out of traffic with the tiller operator negotiating turns that appear impossible. The tiller axle permits the tractor-trailer apparatus to be maneuvered in and out of traffic and around turns that would be difficult or impossible for other vehicles. There is, however, much more to the job than steering and maneuvering the apparatus. All previously discussed driving practices concerning starting the engine, shifting gears, and braking apply to tractor-trailer apparatus. Note, however, that the added weight increases the stopping distance of a tillered vehicle.

The Tiller Operator

The tiller operator must also be qualified to operate the aerial ladder and be familiar with the duties assigned to all truck company personnel. The tiller operator's driving assignments include straight-line driving, turning and backing, and proper placement of the trailer at fires to ensure the ready removal of ladders and the safe operation of the aerial ladder. Tiller operators must be particularly aware of such factors as the distance of the trailer from the base of the building involved, the angle of trailer placement in regard to the position of the tractor, proper overhead clearance, side and rear obstructions, and grades and slopes of the working area.

The tiller operator shares the driver's responsibility for the safety of the public and of other vehicles when responding to or returning from alarms. Efficient tiller operation demands a thorough knowledge of prescribed routes to first alarm assignments and an awareness of conditions to be anticipated.

Operator Training

Tillering, like other specialized operations, requires considerable practice and training. A beginner must be closely supervised by an experienced tiller operator. Instructors should be in a position to take over the tillering instantly, if necessary. Officers and instructors should stress the importance of the following:

- Using good signal practices
- ALWAYS trailing in-line on a straightaway
- Quickly bringing the trailer into line again as soon as a turn is completed
- Adequately observing the trailer overhang on turns
- Avoiding rough and jerky maneuvers
- NOT OVERTILLERING (overcorrecting)
- Keeping both hands on the wheel and giving undivided attention to the job

Tiller Control

Leaving the Fire Station

When preparing to drive a tractor-trailer ladder truck from a fire station, the tiller operator has several items to check before giving or returning the signal to proceed. On approaching the vehicle, the operator should check the position of the trailer wheels for alignment with the tractor. If they are not in line, be prepared to bring them into line as soon as the apparatus starts to move. The elevated seat position provides the tiller operator with a vantage point from which to check the readiness and safety of the crew and the proper security of ladders and equipment. The tiller operator must be secured in the bucket seat **WITH SEAT BELT FASTENED** before giving the signal to proceed.

As the truck starts to move, the tiller operator should center the trailer in the station doorway. When leaving some stations, the tractor must make a sharp turn because of a narrow street or obstruction. For example, if a sharp right turn is started by the driver, it may be necessary to turn the tiller wheel slightly to the left to keep centered in the doorway. Clearance must be maintained on the right side of the trailer ahead of the tiller operator. As the trailer wheels clear the doorway and the tractor continues to turn further to the right, the trailer pivots on its wheels and moves the trailer overhang toward the left side of the station door. This pivoting action must be corrected by steering to the right to keep the trailer overhang in the center of the doorway. If the reverse of this situation occurs and the ladder truck is turned sharply to the left out of the doorway, the trailer pivots on its wheels and moves the trailer overhang toward the right side of the station door. This pivoting action must be corrected by steering to the left to keep the trailer overhang in the center of the doorway.

Traveling Forward

Whenever possible, the tiller operator should keep the trailer wheels parallel with the frame of the trailer and in a direct line with the tractor. This is especially important when streets are wet or icy because there is a greater danger of skidding. When the tractor and trailer are not in line, the width of the apparatus is greatly increased. This also increases the chance of being involved in an accident.

When the tractor and trailer are in line, little movement of the tiller wheel is necessary while traveling forward. Under this condition, it is normal for the tiller operator to control only the incidental drift to one side or the other, similar to that which occurs when driving a passenger car. A good rule to remember: When moving forward, place a hand at the *top or front* of the tiller steering wheel, and the rear of the trailer moves in the direction the hand is moved.

When making a turn, or when weaving in and out of traffic on an emergency response, the tiller operator should keep the trailer in line with the tractor as much as possible. The operator should turn the tiller wheel only enough to maintain adequate clearances. In most tight turns, caused by traffic congestion or narrow streets, the tiller operator may have to turn the wheels to keep the trailer aligned with the tractor and to maintain adequate clearance for the trailer and the trailer overhang. This particular maneuver must also be compensated for afterward by a smooth recovery turn of the trailer wheels to maintain a proper tractor-trailer alignment.

When the apparatus is underway, it is essential that the tiller operator be alert to conditions on other streets ahead of the apparatus. The tiller operator should anticipate situations and plan compensating measures to prevent accidents. Particular attention should be given to trucks and buses because their height presents an added hazard to the trailer overhang and may require the tiller operator to take precautionary maneuvers. It is the duty of all members of the truck company, as well as that of the driver and tiller operator, to keep a sharp lookout for any overhanging obstructions that could injure the tiller operator or damage the apparatus.

Turning at Intersections

The tiller operator must exercise care at intersections. If the intersection is clear and wide, a right or left turn should not be difficult if the driver steers the tractor into a wide turn. If traffic is heavy or streets are narrow, extra precautions and good judgment must be used. The tiller operator must maintain a constant awareness of the vehicles on each side and to the rear of the apparatus. Even after starting a turn, developments may occur that require the tiller operator to maneuver the overhang of the trailer into the clear.

There are several methods of negotiating turns at an intersection, depending upon the conditions present and the immediate requirements of the situation. On most corners, the trailer tracks without an excessive amount of tillering. When approaching a turn at an intersection, the tiller operator should anticipate whether additional traffic lane space is required to make the turn. If necessary, the tiller operator can partially block a lane of traffic by steering the trailer slightly into whichever lane must be kept clear. (**CAUTION:** This is a low-speed maneuver! Below 20-25 mph [32 km/h to 40 km/h]).

When it is necessary to move out wide with the trailer while making a turn at an intersection, the tiller operator should start turning away from the corner about the time the front tractor wheels first enter the intersection. This maneuver prevents jackknifing with the tractor; however, it is necessary to compensate for the centrifugal force generated during the turn. *Jackknifing* occurs when the trailer and cab are at an angle less than 90° to each other. If it is necessary for the driver to turn sharply at a point near an intersection curb line, the tiller operator turns the tiller wheel quickly in the opposite direction to avoid striking or overrunning the curb.

It is essential that the tiller operator properly judge and maintain side clearances on all turns to sufficiently permit the safe passage of the trailer overhang. If at any time there is any doubt regarding side clearance, the operator should immediately signal the driver to stop.

DO NOT OVERTILLER ON TURNS! Avoid making a sudden swing-out that requires a sharp counter swing-in, unless it is necessary to avoid a collision. After a turn, the tiller operator should bring the trailer in line with the tractor *quickly* and *smoothly*.

Backing Maneuvers

When backing, steering the tiller wheels requires the opposite maneuvers of those used to steer the front wheels. For example, if the tiller wheel is turned to the right, the trailer travels sideways to the left; if the tiller wheel is turned to the left, the trailer moves to the right. This deviation from normal apparatus movement is one of the reasons relief drivers and tiller operators need to practice as often as possible. A good rule to remember: When backing, place a hand at the *bottom* or *rear* of the tiller steering wheel, and the trailer moves in the direction the hand is moved.

In any backing maneuver, the driver and the tiller operator must continue to signal each other. Tiller operators should be especially aware of the side of the trailer that is blind to the driver because of the jackknifing process. This area is forward of the trailer wheels up to the turntable, on the outside of the turn. It is very easy to inadvertently overjackknife the apparatus when backing if close cooperation is not maintained. It is also essential to station firefighters as guides to warn both the driver and the tiller operator of any inadequate clearances and to control traffic during the backing operation.

In all backing operations, the driver/operator must control apparatus movement smoothly and at low speed. The driver/operator must be prepared to stop the apparatus immediately and to steer the tractor so that it tracks the trailer in proper alignment. The tiller operator must pay particular attention to the position of the tractor, as well as to the clearance on both sides and to the rear of the trailer. The tiller operator must smoothly guide the trailer wheels to the parking spot and align the wheels with the trailer frame before the apparatus is stopped.

Driving Exercises and Evaluation Methods

After driver/operators have been selected and trained, their performance should be evaluated by some standard method. These evaluations should occur before the driver/operator is allowed to operate the apparatus under emergency conditions. NFPA 1002 provides some specific directions on how driver/operator candidates should be tested. These directions need to be followed by agencies that certify their personnel to the standard. Other agencies should at least follow the standard to avoid possible civil law liabilities should the driver/operator be involved in a collision. Most agencies use a combination of written and practical testing for driver/operator candidates.

All fire apparatus training and testing should follow the requirements contained in NFPA 1451, *Standard for a Fire Service Vehicle Operations Training Program (1997).*

Written Test

There are some facets of the driver/operator's job that are most easily tested through the use of a written exam. The written exam for driver/operators may include questions pertaining to the following areas:

- State and local driving regulations for emergency and nonemergency situations
- Departmental regulations
- Hydraulic calculations
- Specific operational questions regarding pumping
- Department standard operating procedures

Depending on local preference, the test may be open or closed book. The style of questions also vary according to local preference. They may be any combination of multiple-choice, matching, fill-in-the-blank, or essay-type questions.

Practical Driving Exercises

NFPA 1002 specifies a number of practical driving exercises that the driver/operator candidate should be able to successfully complete before being certified to drive the apparatus. The standard requires that driver/operators be able to perform these exercises with each type of apparatus they are expected to drive. For example, if your department has both straight chassis and tillered aerial devices, the driver/operator should be tested on both of them. Some jurisdictions prefer to have driver/operators complete these evolutions before allowing them to complete the road test. This ensures that driver/operators are competent in controlling the vehicle before they are allowed to drive it in public. The exercises that follow are those that are specifically required in the standard. Individual jurisdictions may choose to add other exercises that simulate local conditions. However, as a minimum, all these exercises should be completed.

NOTE: The descriptions for the exercises listed contain minimum dimensions for setting up these exercises. NFPA 1002 notes that these dimensions may not be reasonable for extremely large fire apparatus, such as aerial apparatus. This will require the local agency to modify the dimensions. The authority having jurisdiction should be able to justify any modifications that are made as being reasonable for local conditions.

Alley Dock

The alley dock exercise tests the driver/operator's ability to move the vehicle backward within a restricted area and into an alley, dock, or fire station without striking the walls and to bring the vehicle to a smooth stop close to the rear wall. The boundary lines for the restricted area should be at least 40 feet (12.2 m) wide, similar to curb-to-curb distance (Figure 4.51). Along one side and perpendicular is another simulated area 12 feet (3.66 m) wide and at least 20 feet (6.1 m) deep. The test procedure has the driver/operator moving past the alley (which is on his left), backing the apparatus, making a left turn in reverse into the defined area, and stopping. This exercise should then be repeated from the opposite direction. The driver/operator is considered to have successfully completed this exercise when able to back into the restricted area without having to stop and pull forward and without striking any obstructions or markers.

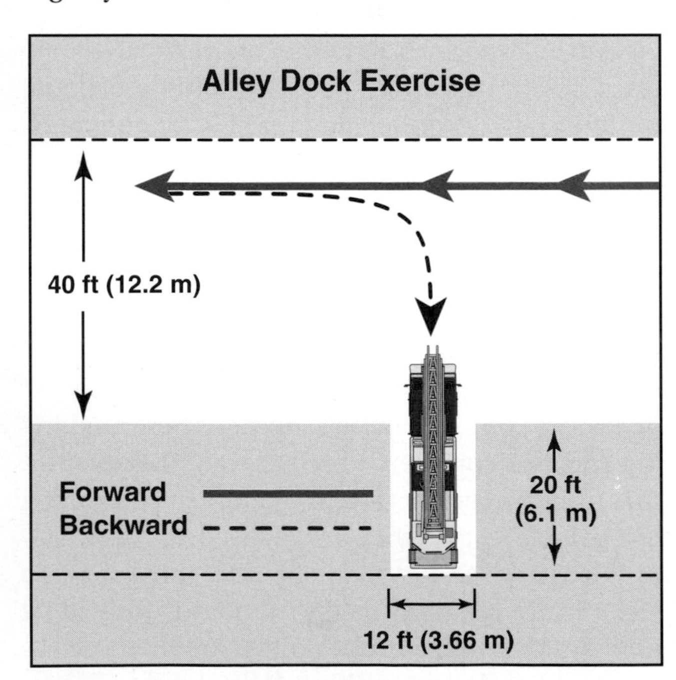

Figure 4.51 The alley dock exercise tests the driver/operator's ability to back into a tight spot.

As an alternative to the traditional alley dock exercise, the local jurisdiction may choose to substitute the apparatus station parking maneuver (Figure 4.52). In this exercise, an apparatus bay is simulated by allowing a 20-foot (6.1 m) minimum setback from a street that is 30 feet (9 m) wide. Sets of barricades are positioned 12 feet (3.66 m) apart at the end of the setback to simulate a garage door opening. A simulated apparatus bay should be constructed back from the "garage door" opening. This bay should be 10 feet (3 m) longer than the vehicle. A straight line may be provided inside the bay, and a traffic cone may be placed where the front left wheel is to stop. Again, the local jurisdiction may choose to alter these dimensions based on local conditions and apparatus size. The test procedure has the driver/operator moving past the setback area (which is on the left), backing the apparatus, making a left turn in reverse through the setback area, and into the apparatus bay area. This exercise should then be repeated from the opposite direction. The driver/operator is considered to have successfully completed this exercise when able to back into the apparatus bay without having to stop and pull forward and without striking any obstructions or markers.

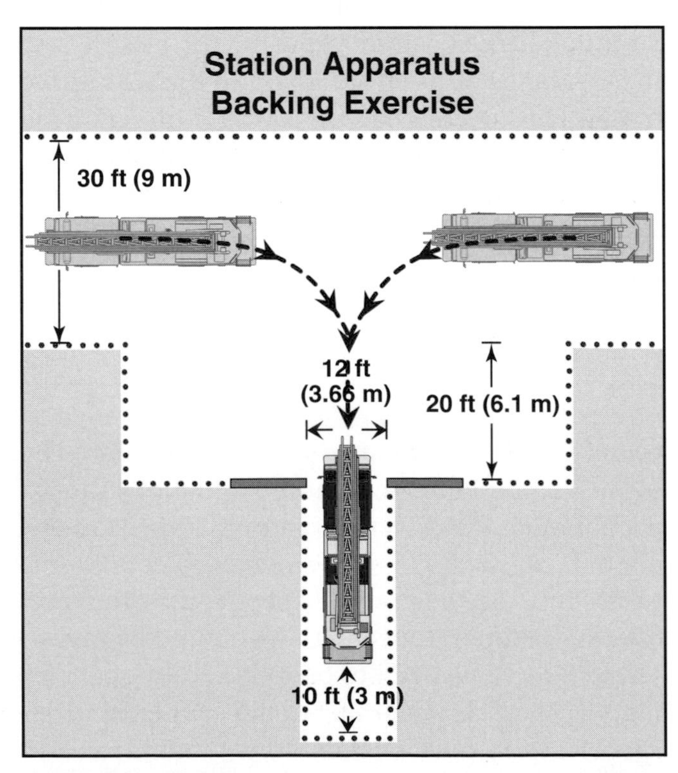

Figure 4.52 This exercise simulates backing the apparatus into the fire station.

Serpentine Course

This exercise simulates maneuvering around parked and stopped vehicles and tight corners. In the serpentine exercise, at least three markers are placed an equal distance apart in a line (Figure 4.53). The markers should be between 30 and 38 feet (9 m and 12 m) apart, depending on the size of the apparatus being used. Adequate space must be provided on each side of the markers for the apparatus to move freely. The driver/operator is required to maneuver the vehicle first backward through the cones and then forward. The course must be traveled in each direction in one continuous motion without touching any of the course markers.

First, the driver/operator is required to drive the apparatus along the left side of the markers in a straight line and stop just beyond the last marker. Then, the driver/operator must back the apparatus between the markers by passing to the left of No. 1, to the right of No. 2, and to the left of No. 3. At this point, the driver/operator must stop the vehicle and then drive it forward between the markers by passing to the right of No. 3, to the left of No. 2, and to the right of No. 1.

Confined Space Turnaround

The confined space turnaround exercise tests the driver/operator's ability to turn the vehicle 180° within a confined space. Although turning fire apparatus around may not be difficult in adequate space, it becomes more complicated in narrow streets or intersections.

This exercise may be performed in an area that is at least 50 feet (15.25 m) wide and 100 feet (30.5 m) long (Figure 4.54). The apparatus begins in the center of one end of the test area. The driver then pulls forward, moves toward one side or the other, and begins the turning process. There is no limit to the number of direction changes that are required before the apparatus is turned 180° and driven through the same opening it entered. A spotter may be used during the process of turning the vehicle

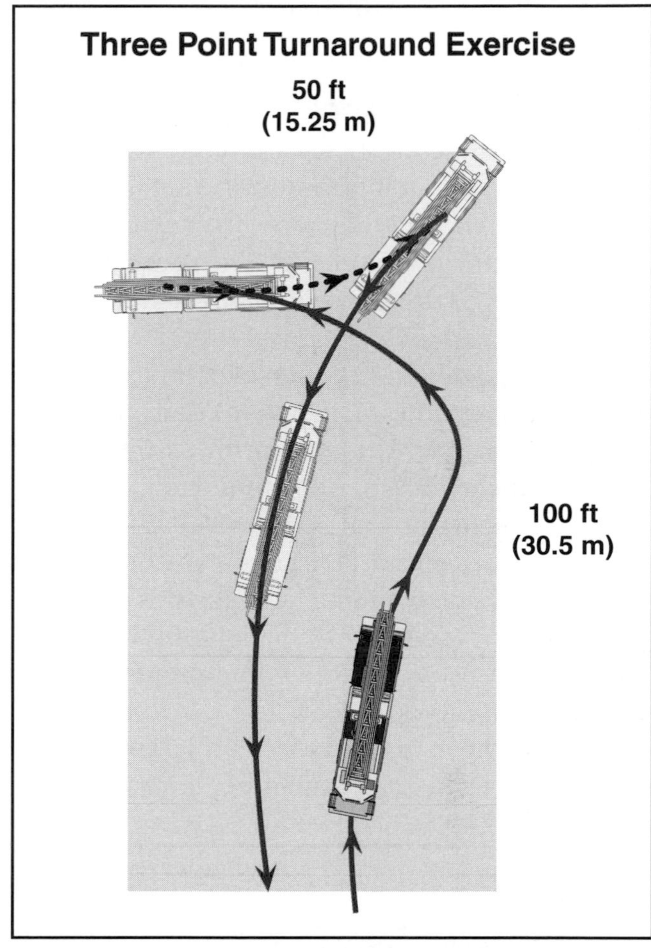

Figure 4.54 The turning around exercise simulates the need to change the direction of or to remove the apparatus from the scene.

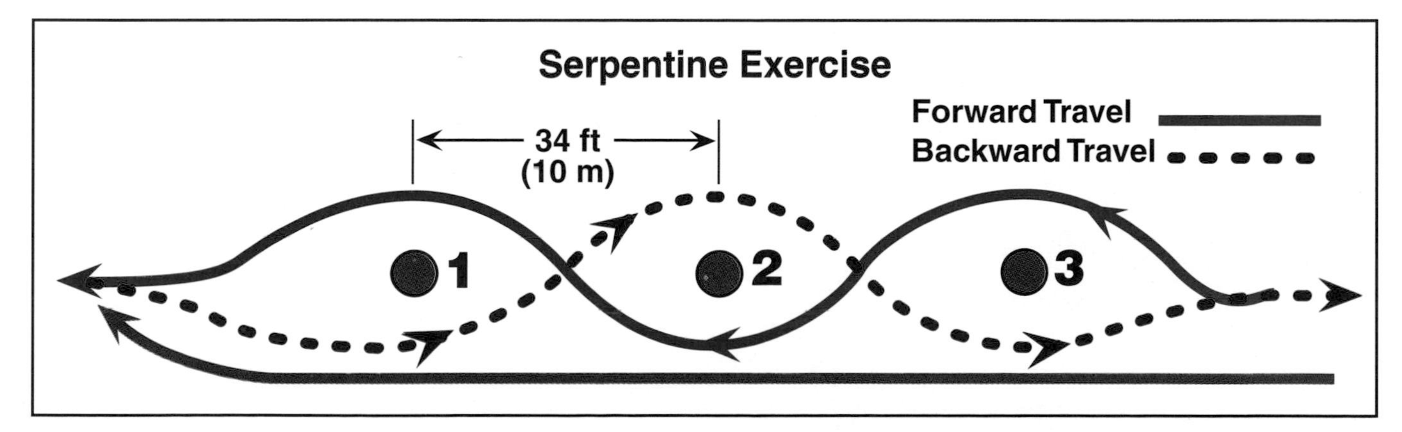

Figure 4.53 The serpentine exercise is intended to simulate maneuvering the apparatus in tight locations and around parked vehicles.

around. Successful completion of this exercise means that the apparatus has been turned 180° and driven through the original entrance point with no course markers being struck or without leaving the defined course.

If the apparatus is so small that it can complete a U-turn without stopping and backing in a course of the dimensions described here, make the course smaller so that backing is required.

Diminishing Clearance

The diminishing-clearance exercise measures a driver/operator's ability to steer the apparatus in a straight line, to judge distances from wheel to object, and to stop at a finish line. The speed at which the apparatus is driven is optional, but it should be fast enough to require the driver/operator to exercise quick judgment. The course for this exercise is arranged by two rows of stanchions that form a lane 75 feet (23 m) long. The lane narrows from a width of 9 feet 6 inches (2.9 m) to a diminishing clearance of 8 feet 2 inches (2.5 m) (Figure 4.55). The driver/operator must maneuver the apparatus through this lane without touching stanchions. At a point 50 feet (15 m) beyond the last stanchion, the driver/operator must stop with the front bumper within 6 inches (150 mm) of the finish line. Obviously, these dimensions need to be adjusted for larger vehicles such as full-sized quints and other large aerial apparatus.

Road Tests

Prior to being certified to drive emergency apparatus, driver/operators should demonstrate their ability to operate the apparatus on public thoroughfares. Driver/operators should only be allowed to operate on public thoroughfares after they have demonstrated the ability to control the apparatus they are driving. Each department will want to develop an established route that driver/operator candidates should follow. This route should cover all the usual driving conditions that can be expected within that jurisdiction. However, as a minimum, NFPA 1002 says that any road test that leads to certification should include at least the following elements:

- Four left and four right turns
- A straight section of urban business street or two-lane rural road at least one mile (1.6 km) in length
- One through intersection and two intersections where a stop must be made
- A railroad crossing
- One curve, either left or right
- A section of limited-access highway that is long enough to allow for at least two lane changes and includes a conventional on-ramp and off-ramp
- A downgrade that is steep enough and long enough to require gear changing to maintain speed
- An upgrade that is steep enough and long enough to require gear changing to maintain speed
- One underpass, low-clearance bridge

During testing, the evaluation of a driver/operator candidate on a road test is very subjective. In general, he should be evaluated on his adherence to posted traffic requirements and departmental policies, as well as his ability to safely control the vehicle.

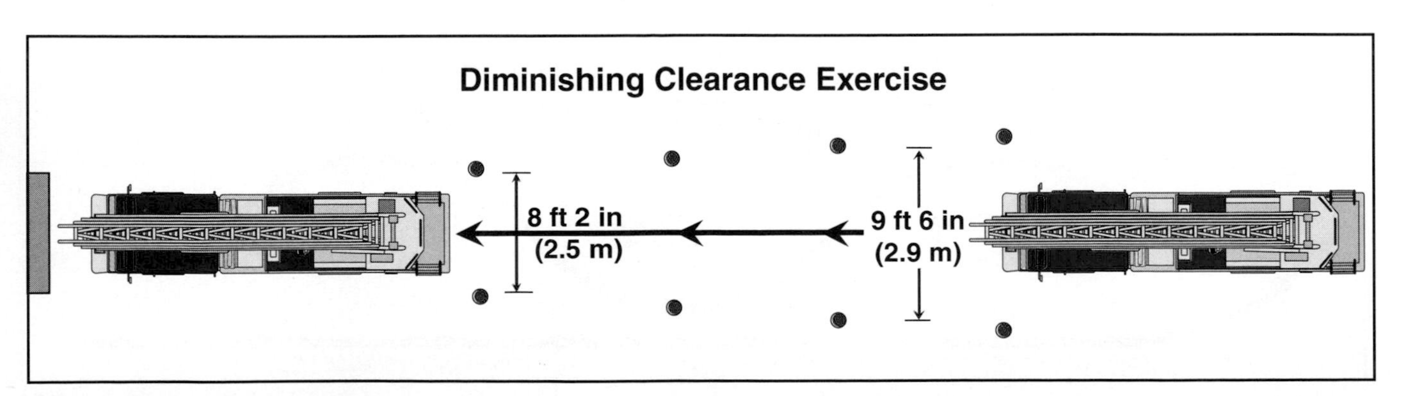

Figure 4.55 The diminishing clearance exercise measures the driver/operator's ability to steer the apparatus in a straight line and to judge the distance from the wheel to various objects.

Summary of Good Driving Practices

A collision or vehicular failure caused by irresponsible driving has many repercussions and is inexcusable. Lives that could have been saved may be lost. Property that should have been protected will be destroyed. Firefighters on the apparatus, innocent bystanders, or other drivers may be injured or killed, leaving the department and the driver/operator open to lawsuits and civil prosecution for such things as negligent homicide or manslaughter. The apparatus will be useless for an indefinite time, leaving citizens with less protection.

The following are some critical points to remember for safe operation and driving of fire apparatus:

- Remember that speed is less important than arriving safely at the destination.
- Slow down for intersections, and stop when faced with a red light or stop sign. Anticipate the worst possible situation.
- Drive defensively. Be aware of everything that is happening or likely to happen 360° around the apparatus.
- Expect that some motorists and pedestrians will neither hear nor see the apparatus warning devices.
- Be aware of the route's general road and traffic conditions. Adjust this expectation with the season, weather, day of the week, and time of day.
- Remember that icy, wet, or snow-packed roads increase braking distance.
- Do not grind the gears on manual transmission vehicles.
- Keep your foot off the clutch pedal until it is time to make a shift.
- Do not exceed 10 mph (15 km/h) when leaving the station.
- Do not race the engine when the apparatus is standing still. It is unnecessary and abuses the engine.
- Always use low gear when starting from a standstill. Using second or third gear and slipping the clutch damages the clutch and causes unnecessary, rapid wear.
- Keep the apparatus under control at all times.
- Take nothing for granted.

Chapter 5

Positioning Aerial Apparatus

Job Performance Requirements

This chapter provides information that will assist the reader in meeting the following job performance requirements from NFPA 1002, *Standard on Fire Apparatus Driver/Operator Professional Qualifications*, 1998 edition. Particular portions of the job performance requirements (JPRs) that are met in this chapter are noted in bold text.

4-2.1 Maneuver and position an aerial apparatus, given an aerial apparatus, an incident location, a situation description, and an assignment, so that the apparatus is properly positioned for safe aerial device deployment.

(a) *Requisite Knowledge*: Capabilities and limitations of aerial devices related to reach, tip load, angle of inclination, and angle from chassis axis; effects of topography, ground, and weather conditions on safe deployment; and use of the aerial device.

(b) *Requisite Skills*: The ability to determine the appropriate position for the apparatus, maneuver apparatus into proper position, and avoid obstacles to operations.

5-2.3 Maneuver and position an aerial apparatus equipped with a tiller from the tiller position, given the apparatus operating instructions, an incident location, a situation description, and an assignment, so that the apparatus is properly positioned and stabilized to safely accomplish the assignment.

(a) *Requisite Knowledge*: Principles of positioning and stabilizing the apparatus from the tiller position.

(b) *Requisite Skills*: The ability to determine the appropriate position for the tiller, maneuver the tiller into proper position, and avoid obstacles to operations.

For efficient and safe fire control, apparatus driver/operators must work together to ensure that aerial apparatus are positioned for maximum use and minimum stress to the aerial devices. In most cases, there are no hard and fast rules for the positioning of aerial apparatus. This is due to the wide variety of aerial apparatus, departmental standard operating procedures, weather conditions, road design and conditions, obstructions, building designs, and fire conditions. Though the basics of proper positioning can be formed into loose standard operating procedures, the process of positioning the apparatus at an actual fire incident will be at the discretion of the incident commander, truck company officer, or the driver/operator.

It can be argued that the fireground placement of aerial apparatus is the most critical of all the types of apparatus. The primary reason for this is that the aerial apparatus is equipped with an aerial device of fixed maximum length. On the other hand, most fire department pumpers carry in excess of 1,000 feet (300 m) of fire hose. It is almost always possible to add an extra section of hose to make an attack line long enough to do the job. You cannot add a section to the aerial ladder. Thus, it is more crucial to give the optimum scene position to the aerial apparatus than to the pumpers.

It is important that driver/operators be versed in the basics of proper positioning so that they can apply them when faced with a particular situation. This chapter discusses the various considerations in determining aerial apparatus placement. The first part of the chapter discusses some basic standard operating procedures that departments can establish for the placement of aerial apparatus. Placement is then discussed in terms of the specific function required for rescue, access to upper levels, ventilation, and fire suppression or exposure protection operations. Lastly, a variety of special incident positioning priorities are highlighted.

◆ Standard Operating Procedures for Positioning

Each incident is unique and may require the use of slightly different positioning techniques for aerial apparatus. Many fire departments have developed standard operating procedures (SOPs) that assist with the orderly placement of aerial apparatus on the fireground. Generally, these standard operating procedures apply to aerial apparatus that are assigned to the initial response to an emergency. Later-arriving aerial apparatus at large incidents will be positioned according to the incident action plans developed at the scene by the incident management personnel. In this sense, placement of the apparatus is tied to the apparatus staging procedures used by the department. Staging is described in more detail later in this chapter.

In many cases, the development of apparatus placement procedures should be a function of pre-incident planning. During pre-incident planning for target hazards (e.g., high rises or hospitals), special consideration should be given to aerial apparatus in terms of access to various portions of the occupancy, overhead obstructions, and any other factors that may influence the function of the apparatus (Figure 5.1). This information can be used to determine the best positions for the aerial apparatus depending on the conditions at the time of the incident.

Figure 5.1 Overhead obstructions that may impede aerial device deployment should be noted in pre-incident planning. *Courtesy of Ed Prendergast.*

Many different procedures are in use by fire departments for positioning aerial apparatus assigned to the initial response. The following are three examples:

- When two aerials respond to a given location, the first-arriving aerial takes the front of the building and the second goes to the rear or side, depending on building access.
- When two aerials respond to a given location, the first aerial's position is based on the present conditions, and the second aerial stages one block away or in accordance with departmental SOPs and awaits instructions.
- When a single aerial apparatus responds to a given location, the apparatus takes a strategically sound location in front of the fire building unless otherwise directed by the incident commander. For example, the apparatus may be centered in front of the involved portion of the building for offensive operations and at a corner of the building (out of the collapse zone) for defensive operations.

Another example of a standard procedure used by some departments eliminates the problem of having early-arriving engine companies positioning in a spot best suited for aerial apparatus. Of course, this procedure must be taught to engine company driver/operators as well as aerial driver/operators. The driver/operator of each type of apparatus should be trained in where to position the vehicle depending on the height of the building. If the fire building is less than five stories (about 60 feet [20 m]), engine companies should park on the side of the street closest to the building, and aerials park on the outside (Figure 5.2). The philosophy here is that the building is low enough to be reached by the aerial device even if it has to go over the closer engines. If the building is greater than five stories (about 60 feet [20 m]), the engines take the outside position, and the aerials park next to the building (Figure 5.3). This allows the aerial's maximum reach. Keep in mind that this procedure will only work in jurisdictions that have sufficiently wide streets and in which the buildings are not set back too far from the street. It also assumes that the department is operating aerial devices with a reach of 100 feet (30 m). It may have to be adjusted for jurisdictions that operate aerials with reaches of less or greater than 100 feet (30 m).

While the previously mentioned examples do provide some direction, each department must develop standard operating procedures and pre-incident plans that suit its local conditions. Once developed, all driver/operators must be trained to follow these SOPs on each and every response.

Tactical Considerations Affecting Aerial Apparatus Positioning

For any given situation, the proper distance between the objective and the aerial apparatus is the distance that affords maximum stability, the best climbing angle, and allows for adequate extension. This should be consistent with the planned use of the ladder and the conditions at the emergency. Long extensions at low angles place the maximum amount of stress on an aerial device and, in some cases, reduce the load-carrying capacity of that device (Figure 5.4). Whenever possible, long extensions at low angles should be avoided. This can be done by getting as close to the desired objective as safely possible. Because of sidewalks, parked cars, poorly-positioned early-arriving engine companies, and other roadside obstacles, it is seldom possible to get the apparatus directly adjacent to the objective.

Another factor that affects the distance from the building that the aerial device may be positioned is

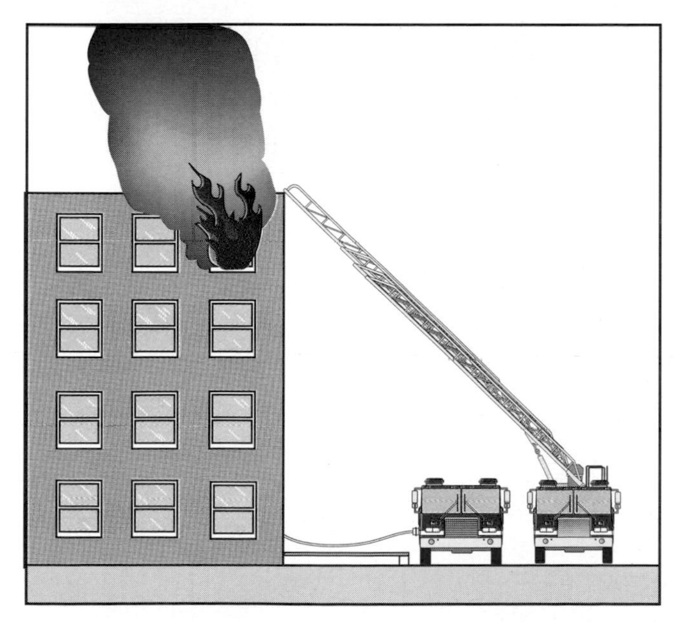

Figure 5.2 If the building is less than five stories, the aerial apparatus may be positioned outside the pumpers.

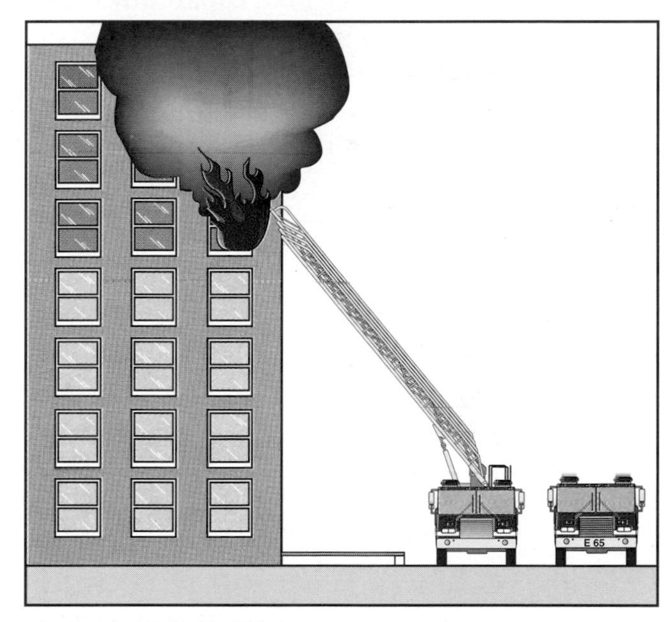

Figure 5.3 If the building is greater than five stories, the aerial apparatus should be positioned inside the pumpers.

Figure 5.4 Operating the aerial device at a long extension and low elevation places a tremendous amount of stress on the aerial device.

the condition of the fire building. If the building has been exposed to severe fire conditions or has otherwise been weakened, it is not desirable to park too close to the building. Positioning to avoid being caught by a collapsing building is discussed later in this chapter. Also, positioning to avoid being exposed to a sudden increase in radiant heat as a result of a collapse is discussed in further detail later in this chapter.

Possible stress to the aerial device can also have an impact on where the apparatus should be positioned. *Stresses* are those factors that work against the strength of the aerial device. Ladder and boom stress is a very important consideration in aerial operation. Stress may be imposed in both static (at rest) and dynamic (in motion) operation. The stress tends to be greater when the aerial device is in motion. Aerial device load capabilities vary from truck to truck and must be evaluated on an individual basis to determine the range of safe operation. Manufacturer's recommendations should be consulted and adhered to for maximum loading information. Aerial device stress can occur from one or a combination of the following conditions:

- Excessive degree of angle, both horizontal and vertical, measured from the truck's center line axis
- Operation in nonparallel positions (uphill, downhill, or lateral grades)
- Operation in supported vs. unsupported positions
- Length of aerial device extension
- Nozzle reaction from elevated master stream
- Weight and/or movement of hose, water, personnel, and/or equipment on the aerial device
- Wind reaction
- Improper operation of the aerial device (sudden starts and stops)
- Heat exposure (radiant and convected) (Figure 5.5)
- Ice on ladder or platform
- Impact with the building or another object
- Improper stabilization
- Wear caused by road travel

Figure 5.5 Occasionally, aerial devices are exposed to high levels of radiant heat. *Courtesy of Ed Prendergast.*

There are four main tactical uses for any aerial device (excluding water towers): rescue, access to upper levels, ventilation, and fire suppression. The following sections cover specific positioning considerations for each tactic.

Rescue

The best rescue approach is made from upwind, so consequently the apparatus should be parked so that the aerial device turntable is upwind of the target (Figure 5.6). If an approach is made from

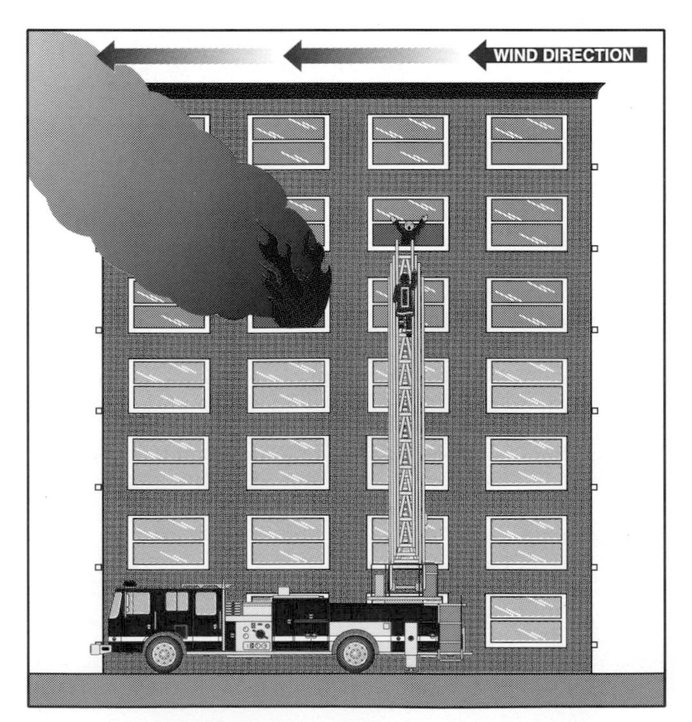

Figure 5.6 When possible, position the aerial device upwind of the rescue point.

downwind, the aerial device operator may have difficulty seeing the objective, and crews and victims will have to deal with the products of combustion or other toxic clouds. If possible, aerial apparatus used for rescue should be placed at the corner of the building (Figure 5.7). This allows rescuers to use the aerial device to reach victims on two sides of the building. This is also a less vulnerable position in the event that a structural collapse occurs. If a rescue is to be made from an area threatened by fire, hoselines can be used to protect the victims, rescuers, and the aerial device. However, use caution in fire stream selection if this is the case. Solid bore or straight stream master streams directed against an aerial device can place damaging load stresses on the device that could ultimately result in a collapse of the device. The preferred procedure is to use a wide-angle fog stream to push the heat or fire away from the aerial while the rescue operation is in progress. Caution must be used not to injure or knock people off the aerial device.

Placement of an apparatus equipped with an elevating platform device will be dependent on how you plan to get the victim into the platform. If the victim is to be lifted over the top rail, the apparatus should be placed so that the turntable is directly in line with the target. This allows the front of the platform to be placed squarely in front of the target. If the victim will be allowed to enter the platform through the hinged gated, the turntable must be spotted a little forward or behind the target. This is because the platform gates are typically on the side of the platform. Thus the device must be raised at an angle to the target to provide safe access to the gates.

Figure 5.7 Positioning on the corner of the building allows the aerial device to access two sides.

Access to Upper Levels

There are several operations, that require the use of aerial devices to give firefighters access to upper levels. Among them are performing interior truck work and using interior handlines off the aerial device. Aerial devices are used as a means of escape if interior or roof conditions become unsafe. Aerial devices are also used as a method to deploy portable equipment.

Whatever the situation, building coverage and aerial device reach should be maximized and upwind positioning used whenever possible. Apparatus position should afford the maximum degree of safety to the firefighters who are using the aerial device. This often involves positioning the apparatus on the side of the building opposite the fire (Figure 5.8). This position allows interior attack crews to advance hoselines toward the fire area from the unburned side. This is a standard fire fighting tactic used to avoid pushing the fire into uninvolved portions of the building. As well, firefighters who are retreating from unsafe positions will most likely move toward the unburned side. They can then escape down the aerial if necessary. When aerial devices are used for this purpose, they should not be repositioned unless approval is given by the firefighters who were deployed by it.

Figure 5.8 In some cases, it is advantageous to ladder the building on the side opposite the fire conditions.

Many departments choose not to use aerial devices for extending handlines or the other access reasons described in the previous paragraph. Their reasoning is that these uses tie up the aerial device and make it unavailable for deployment to other locations should the need arise. The choice of either policy is a local decision that must be clear in the department's SOPs.

Ventilation

Ventilation is a very important function of the truck company. Proper placement of the aerial device can make the ventilation process quicker and safer. If the aerial device is being used to provide access for ventilating a pitched roof, it may be possible to position the apparatus so that the firefighters may operate directly from the aerial device. This is beneficial in the event that a sudden roof collapse occurs.

When providing access for ventilating a flat roof, the aerial apparatus should be positioned on the unburned side of the structure, as close as possible to the area being ventilated (Figure 5.9). This minimizes the travel distance between the work area and the aerial device. This also could be important in the event of roof failure. Keep in mind that the aerial should be placed so that the vent holes will not be cut in the travel path back to the device. Position the apparatus so that the aerial device extends above roof level. This allows personnel to safely get on and off the aerial device and makes the aerial device easier to find once the firefighters are on the roof. Extend aerial ladders at least 6 feet (2 m) above the roof level. Extend the aerial platform so that the floor of the bucket is at roof level.

If the aerial device is being used to assist with horizontal ventilation, such as taking out windows, the turntable should be positioned so that the entire aerial device will be upwind of the ventilation point(s) and will have access to as many windows as possible. This rule is the same as that followed for ground ladder placement. Ultimately, the goal is to place the tip of the aerial device upwind of the window being opened and slightly higher than the opening (Figure 5.10). This allows the firefighter on the tip to open the window without having to deal with excessive heat, smoke, and falling glass.

All ventilation activities must be performed in a coordinated manner consistent with the incident plan established by the Incident Commander. Otherwise, excessive damage may be imposed on the structure, or the fire may be spread in an adverse manner. For more information on proper ventilation techniques, see the IFSTA **Fire Service Ventilation** or **Essentials of Fire Fighting** manuals.

Fire Suppression Activities (Elevated Master Streams)

The use of an aerial device for elevated master streams is common. Elevated master streams can be used in blitz attacks, defensive attacks, and exposure protection. When used in a blitz attack, position the apparatus to give the fire stream as much reach into the fire area as possible. Keep in mind that the ultimate goal will be to place the nozzle in the lower portion of the window opening so that the fire stream may be directed upward toward the ceiling. Make sure

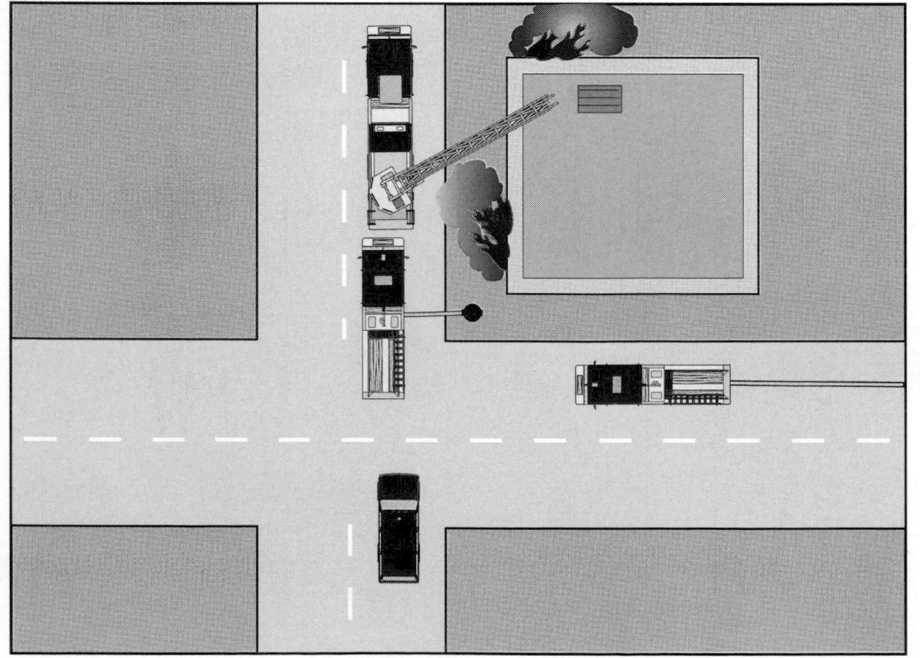

Figure 5.9 Position the aerial device as close to the area to be vented as possible. This reduces travel time to the aerial device in the event a hasty retreat is required.

Figure 5.10 The aerial apparatus must be positioned so that the tip of the aerial device can be positioned at the upper corner of the window on the upwind side of the opening.

Figure 5.11 Aerial devices can be used to make a blitz attack on some structure fires. *Courtesy of Joel Woods, Maryland Fire & Rescue Institute.*

Figure 5.12 Aerial devices can be used to make ground level attacks.

that the apparatus is close enough to the building so that the tip of the aerial device can be put into the desired position (Figure 5.11).

WARNING!

External fire attacks utilizing master streams must never be performed when crews are working inside the building. The disturbance of the interior thermal balance, the large volume of steam created, the possibility of being struck by the stream, and the additional weight added to the building by external master stream attacks pose a serious safety threat to firefighters working in the building.

In some cases, the aerial device may be used as a master stream close to ground level. This is particularly true of apparatus equipped with water towers and telescoping aerial platforms (Figure 5.12). If the aerial device is to be used in this way, position the apparatus so that the turntable is directly in line with the intended target (usually a window or door of some type). This will allow the fire stream to penetrate as far into the building as possible. As well, the driver/operator will need to judge the distance to the building accurately so that the nozzle will get close enough to be effective.

Elevated streams are most commonly used in defensive operations. In this type of operation, the elevated stream may be used to directly attack the fire, cool brands and gases within the thermal column, and protect exposures. When a defensive attack is employed, the chance for building collapse must be considered. Positioning the apparatus at the corners of, or an otherwise safe distance from, the building decreases the chances of damage to the truck in case of building collapse. It also decreases the chances of damage from radiant heat from the fire. However, this safe distance must be balanced with a close enough distance to allow the bulk of the fire stream to be able to hit the seat of the fire area or cover the exposures.

Spotting Considerations

In addition to the tactical and apparatus design positioning considerations, there are a number of other factors that the driver/operator must consider when determining the final operating position of the aerial apparatus. The following are some of these factors:

- Surface conditions (soft pavement or soil)
- Weather and wind conditions
- Electrical hazards and ground or overhead obstructions
- Angle and location of aerial device operation
- Fire building conditions

Surface Conditions

The surface condition of the spotting area must be considered when using an aerial device. Apparatus should be parked on soft surfaces only as a last alternative. If working on soft surfaces, operators must watch for settling. As we will learn in more detail in the next chapter of this manual, the stabilizer pads supplied with the apparatus are to be used whenever the stabilizer system is deployed (Figure 5.13). However, placement on soft surfaces may require additional cribbing or support materials may need to be placed under the stabilizer pads to further distribute the weight of the apparatus. When cribbing is used to supplement stabilizer stability, the material must be of adequate size and strength to withstand loading imposed by the aerial device. Ensure proper placement of cribbing to preclude slipping from underneath the stabilizer pads or stabilizer shoes.

The driver/operator must also be alert for surfaces that are stable when the incident begins but which may become unstable as the incident progresses. Frozen soil may be a stable surface as long as it stays frozen. However, apparatus exhaust and warm water from fire fighting operations may cause the ground to thaw and become unstable. Even during warm weather operations, dry, solid ground can become unstable from fireground runoff. Be alert for these conditions.

Thin-skinned paved surfaces may be as problematic as soft soil. These are especially common in parking lots. Most parking lots are not constructed with the same techniques and materials as public roadways. In some cases, a parking lot may consist of merely a thin layer of asphalt applied over dirt or a minimal gravel base. The thin surface may not provide a stable enough base for the apparatus stabilizers. If stabilizer pads or cribbing are not placed beneath the stabilizer shoes, the stabilizers may puncture the parking surface, and the truck may tip over when the aerial device is operated off the side of the vehicle (Figure 5.14). Parking lots that may be problematic for aerial apparatus operations should be identified in pre-incident plans.

Driver/operators must also be alert for areas that may contain vaulted surfaces. Vaults beneath the

Figure 5.13 Stabilizer pads must be used every time the stabilizers are lowered.

Figure 5.14 The use of stabilizer pads is especially crucial when operating on unstable parking lots.

ground may be created by underground parking structures, utility chases, drainage culverts, basements that extend under sidewalks, or underground transportation systems. These surfaces are unstable for the support of aerial apparatus. Through pre-incident planning, the driver/operator must be familiar with locations where the deployed stabilizer could possibly punch through the surface and cause the aerial apparatus to tip over.

When soft or otherwise unstable surfaces are present or even suspected, avoid the area if at all possible. If it is not possible to avoid the area, the apparatus stabilizers should be properly supported with stabilizer pads and supplemental materials (Figure 5.15). When possible, park the apparatus in a manner that will allow the aerial device to be operated directly over the front or rear of the apparatus (Figure 5.16). This lessens the possibility of a stabilizer settling into the surface. It also lessens the chance of the apparatus tipping over even if it does settle into the surface.

Figure 5.15 It some cases, extra cribbing may be required to achieve solid stabilization.

Figure 5.16 When possible, position the aerial apparatus so that the aerial device may be operated over the front or rear of the apparatus.

Weather Conditions

Weather conditions can affect the spotting of an aerial apparatus in several ways. These include cold, hot, or windy weather conditions. Driver/operators must have an understanding of how to deal with the weather extremes common to their jurisdiction.

The most common spotting consideration that faces driver/operators in cold weather conditions is the presence of ice or snow on the parking surface. When possible, the driver/operator should avoid parking the apparatus and deploying the aerial device on snowy or icy roads. However, since this is not always possible, the driver/operator must know how to make the operation as safe as possible. The driver/operator should monitor icy ground for melting conditions that could decrease stabilizer stability. In some cases, after the stabilizer has been lowered onto an icy or snow-packed surface, the ice/snow may melt or break away leaving the stabilizer off the ground (Figures 5.17 a and b). This

Figure 5.17a Sometimes setting the stabilizers on snow or ice is unavoidable.

Figure 5.17b Watch for melting that could result in a void beneath the stabilizer.

will require the stabilizer to be lowered further to ensure solid contact with the ground.

Icing of ladders, booms, and platforms requires additional caution during operation due to the weight of the ice. The amount of weight added to the aerial device by firefighters and equipment has to be decreased when icing conditions are present (Figure 5.18). Try to avoid parking the apparatus in a position where the raised aerial device will be subject to accidental contact with fire streams or overspray. If the aerial device is going to be used for elevated master stream operations, avoid parking it so that the stream will be operated against the wind. This will minimize overspray reaching the aerial device. If the aerial device does receive a buildup of ice, it may be necessary to spray it with a de-icing fluid so that normal operation may be resumed.

The primary spotting consideration associated with hot weather is that extreme heat may tend to weaken marginal or otherwise firm paved surfaces. For example, an asphalt parking lot that may provide a marginal level of support during more temperate weather conditions will become soft during periods of extreme heat (Figure 5.19). This may result in an unstable base.

Figure 5.18 Ice adds significant weight to the aerial device. *Courtesy of Ron Jeffers.*

Moderate to high winds impose a dynamic load on the aerial device and may reduce the overall stability. This is caused by the force of the wind blowing against the device and trying to move it in a direction for which it was not designed. The movement by the wind also magnifies the other loads placed on the ladder by personnel and equipment. When it is necessary to operate during high wind conditions, spot the apparatus in a manner that requires the aerial device to be raised only to the minimum extension needed. It is also helpful to position the apparatus so that the aerial device may be used over the front or rear of the vehicle, parallel to the wind. Driver/operators and company officers should also be aware of locations in the response district and sides of certain buildings that are more prone to gusting wind conditions and avoid them when possible. Always adhere to the manufacturer's recommendations for operations in windy conditions.

Electrical Hazards and Ground or Overhead Obstructions

Aerial driver/operators must continually be aware of overhead power lines (Figure 5.20). When parking the apparatus, it is just as important to look up as it is to look at the ground. If possible, the driver/operator should avoid spotting the apparatus in a position that will require a lot of aerial device maneuvering around the obstructions. Remember that

Figure 5.19 In warm weather conditions, apparatus may easily sink into the asphalt.

the goal is always to maintain a distance of at least 10 feet (3 m) between the aerial device and overhead electric lines.

Caution should be exercised around other types of overhead lines such as telephone and cable TV lines. Occasionally, these normally harmless lines will be in contact with electrical lines somewhere down the line and may also be energized. Articulating boom operators have two areas of the apparatus to monitor: the platform and the boom, particularly in the area of the boom joint or hinge (Figure 5.21). Personnel on the apparatus are generally considered to be susceptible to electric shocks regardless of whether they are in contact with the ground or not.

WARNING!

If it becomes necessary to exit an apparatus in contact with electrical lines, personnel need to jump clear of the energized apparatus to reduce the risk of being electrocuted (Figures 5.22 a and b). At no time should they be in contact with the truck and the ground at the same time.

Figure 5.20 Driver/operators must use care to avoid tangling the aerial device in overhead wires. *Courtesy of Ron Jeffers.*

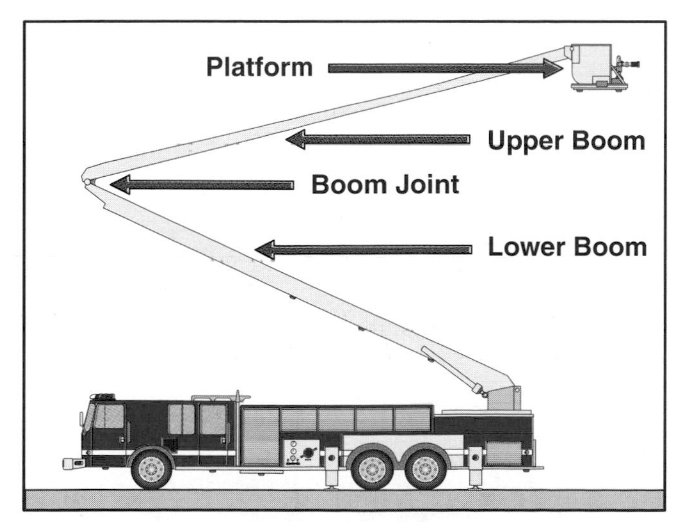

Figure 5.21 Driver/operators must be aware of the multiple contact points on an articulating aerial device.

Figure 5.22a If it becomes necessary to exit a potentially energized aerial device, the firefighters should jump clear of the apparatus.

Figure 5.22b Never contact the apparatus and the ground at the same time because there is a chance the aerial device is in contact with energized wires.

Trees, overhangs, parked vehicles, trash containers, and similar obstructions may also affect the operation of the stabilizers and/or aerial device and should be considered/avoided when positioning aerial apparatus.

Angle and Location of Aerial Device Operation

In many cases, principles of spotting the aerial apparatus are linked with stabilizing the apparatus. Stability of aerial apparatus can be improved by operating the aerial device in line with the longitudinal axis (apparatus body) (Figure 5.23). In other words, the aerial apparatus is most stable when the aerial device is operated directly over the front or rear of the vehicle. Increasing the angle of the aerial device away from the longitudinal axis of the truck decreases the amount of load that can be carried safely. An angle perpendicular to the apparatus is the least stable position available (Figure 5.24). By positioning the truck body in a line with the expected position of aerial use, the stability of the apparatus can be increased. Thus, if you have the entire parking lot at your disposal, nose or back the aerial apparatus into position rather than parking parallel to the objective. For rear-mounted aerial devices, backing the apparatus in is the preferred method, as this maximizes the reach of the aerial device. Nosing the apparatus in would shorten the possible reach by a distance equal to the length of the apparatus. The opposite would be true of midship-mounted aerials.

Tillered aerial apparatus may be positioned to increase stability by jackknifing the apparatus. *Jackknifing* involves turning the tractor at an angle from the trailer. Greatest stability occurs when this angle is approximately 60° from in-line, and the aerial device is extended away from this angle (Figures 5.25 a and b). Good stability occurs at angles up to 90°. Beyond 90°, stability decreases rapidly. The driver/operator must be familiar with the manufacturer's recommendations for that particu-

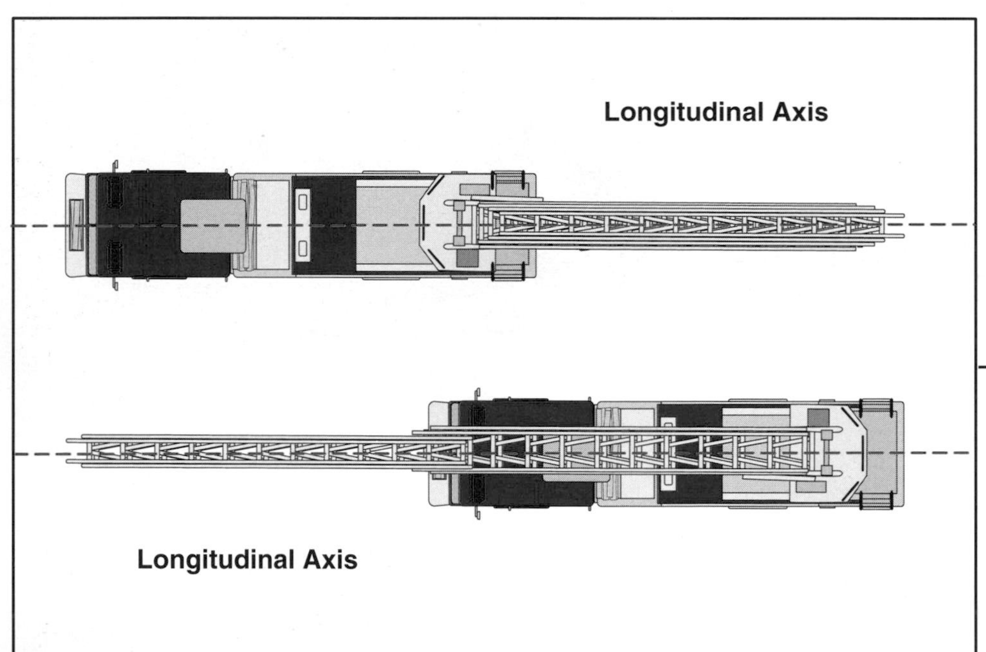

Figure 5.23 It is always most desirable to operate the aerial device in line with the longitudinal axis of the apparatus.

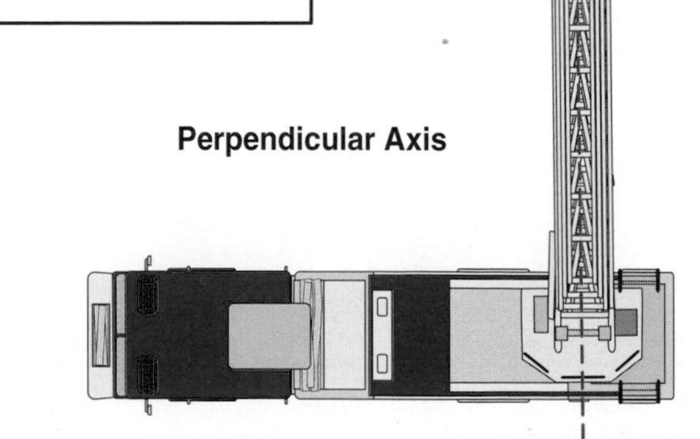

Figure 5.24 The aerial apparatus is least stable when the aerial device is operated perpendicular to the longitudinal axis of the apparatus.

lar apparatus. Modern tillered apparatus may have stabilizer systems that allow them to be adequately stabilized without jackknifing. However, even those apparatus become more stable when the apparatus is jackknifed.

Stress in aerial devices is also increased when the ladder rungs are operated nonparallel to the ground. This occurs when the apparatus is parked on an incline and the aerial device must operate off the side of the truck (Figure 5.26). These positions create a torsion or twisting action on the ladder or boom and the turntable. When an apparatus must operate off an incline, the operator can reduce these stresses by spotting the turntable downhill from the point of operation.

When approaching from the uphill side, the apparatus should be pulled past the building, and the aerial device should be operated off the back of the truck (Figure 5.27). When approaching from the downhill side, the apparatus should stop short of the building, and the aerial should be operated over the cab (Figure 5.28). Ideally, the truck should be operated in the uphill position with the aerial device directly in-line to reduce the stress.

In some cases, it is possible to level the truck somewhat by using the stabilizers to raise one side

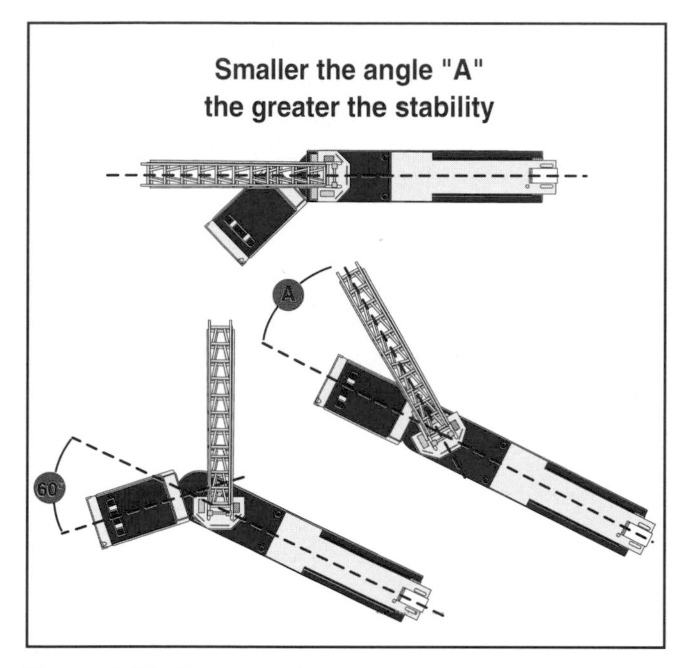

Figure 5.25a On some tiller apparatus, it is required that the apparatus be jackknifed in order to properly stabilize the apparatus for aerial device deployment.

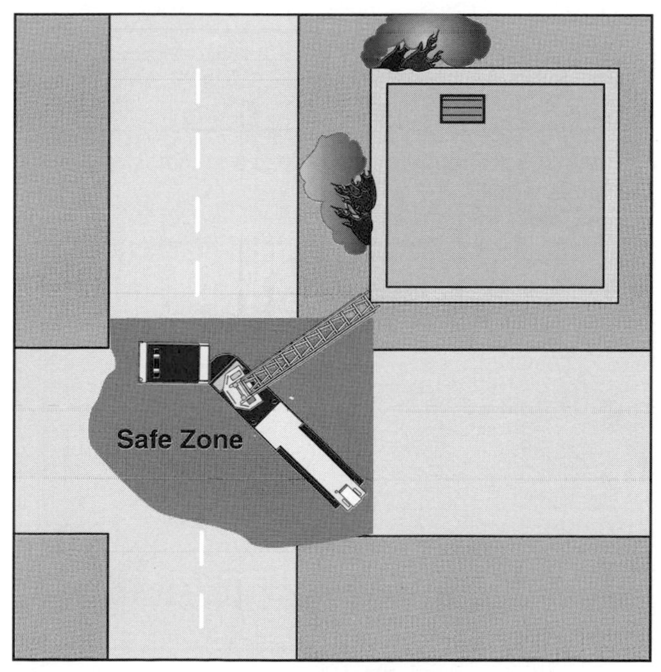

Figure 5.25b The proper position of a tiller apparatus on the corner of a building.

Figure 5.26 Note the twist in this aerial device. It is created by operating over the side of the apparatus while parked on a hill. *Courtesy of Ron Jeffers.*

Figure 5.27 When approaching from the uphill side, drive past the fire building and operate over the rear of the apparatus.

of the truck more than the other. Generally, this is only possible on grades that are perpendicular to the long centerline of the vehicle (Figure 5.29). The ability to do this depends on the type of stabilizer with which the truck is equipped. For the most part, it can only be done with single-chassis vehicles that are designed to be lifted completely off the ground.

Another spotting consideration is whether the aerial device is designed to be operated in either an unsupported (cantilever) or a supported (resting on a wall) position (Figures 5.30 a and b). *Follow the manufacturer's recommendations for either condition.* If the manufacturer recommends that its device be operated in a supported position, realize that the loading or the amount of extension must be reduced for low angles of elevation during unsupported operations. The maximum loading for any unsupported aerial device occurs when operated at angles between 70° and 80° from horizontal.

Figure 5.28 When approaching from the downhill side, stop short of the building and operate over the front of the apparatus.

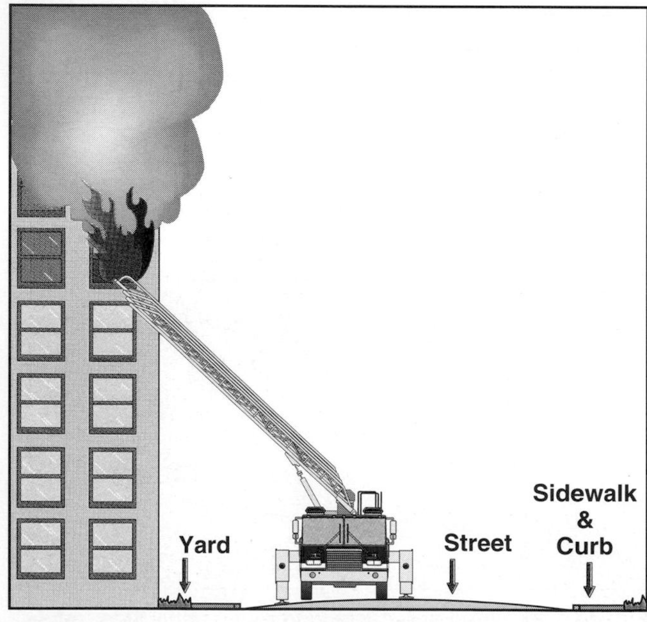

Figure 5.29 Crowned streets require the driver/operator to level the truck using the stabilizers.

The amount of extension affects aerial device stress. As extension increases, aerial loading must decrease. Aerials operating at a low angle of elevation and at long extensions are at their weakest operational position. This type of operating position should be avoided if at all possible. This can be done by spotting the apparatus as close to the intended target as safely possible.

The driver/operator must be familiar with the load limitations of the aerial device when flowing water and when not flowing water. Equally as important is knowing the range of motion that is acceptable for the aerial device under both conditions.

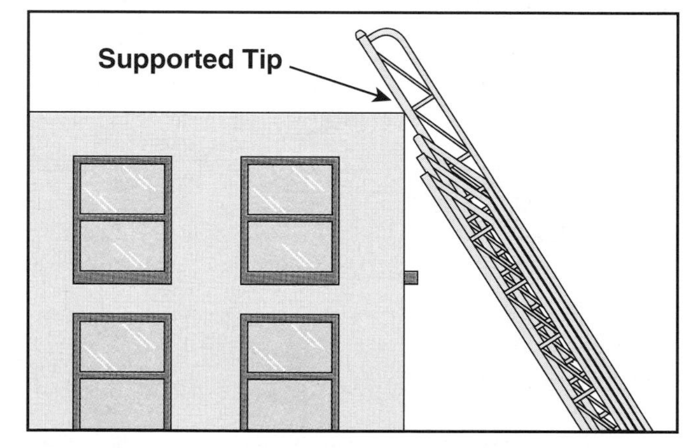

Figure 5.30a A supported aerial tip is one that is resting against its target.

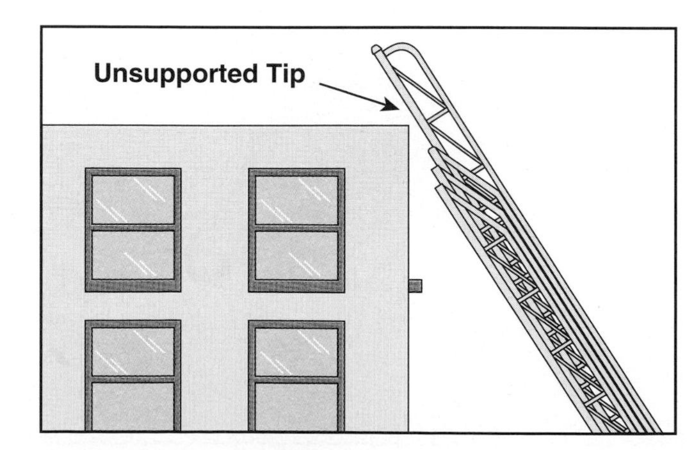

Figure 5.30b Most modern aerials are designed to be operated in the unsupported position.

Fire Building Conditions

The condition of the fire building, as well as other building-related concerns, must be considered when positioning the apparatus. Buildings that have been subjected to extensive fire damage or buildings in poor condition before the incidence of fire may be subject to sudden collapse. For this reason, apparatus should be parked far enough away so that they will not be in the collapse zone should a collapse occur. The collapse zone should be at least equal to one and one-half times the height of the building (Figure 5.31). Realize also that even if the apparatus is not struck by falling debris that it may be subjected to higher levels of radiant heat and smoke following a collapse. This must also be considered when spotting the apparatus. In many cases, the aerial apparatus is the most expensive exposure on the fireground. Keep this in mind when positioning the apparatus. It makes little sense to damage a $500,000 piece of apparatus while fighting a fire in a $15,000 garage.

There are many indicators that a building may become unstable. Signs that a serious exterior collapse may occur include: bulging walls; sagging roofs; large cracks in the exterior; falling bricks, blocks, or mortar; and interior collapses. Pre-incident planning aids in identifying buildings with a serious potential for collapse. Buildings that are old and poorly maintained should be targeted. The presence of ornamental stars or large bolts with washers at various intervals on exterior masonry walls indicates that reinforcement ties are in place to hold together otherwise unstable walls (Figure 5.32).

The intensity of the fire also dictates apparatus placement. Large hot fires require the apparatus to be spotted farther away from the fire building. Consideration must also be given to the fire's potential growth. If the fire has the potential to grow or spread to exposures, the apparatus must be placed so that it is not trapped by the advancing fire (Figure 5.33). ***Always leave a way out.*** Avoid making the apparatus an exposure hazard itself. If the apparatus is to be positioned in a dead-end access, back the apparatus into position if possible. This will make an escape faster if it becomes necessary.

Another consideration for spotting apparatus is the debris that can fall from the fire building. This is of particular concern at high-rise fire incidents. Large pieces of glass, roof-mounted signs, steel gates, and other debris may be falling from many stories above street level. This can pose a serious hazard to personnel operating off the apparatus and to the apparatus itself. In these situations, the apparatus should be spotted away from the area in which debris is falling, and all personnel should be kept safe of the falling debris zone.

Special Positioning Considerations

A variety of other scenarios and conditions affect apparatus placement. The driver/operator must be

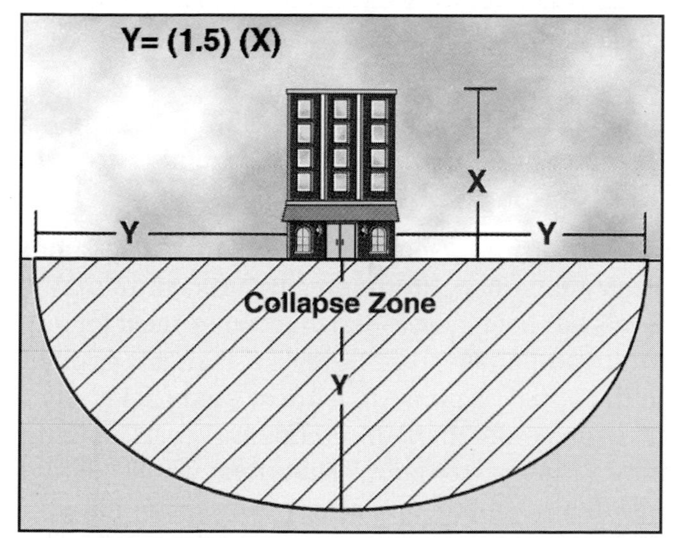

Figure 5.31 The collapse zone is equal to 1.5 times the height of the building.

Figure 5.32 The presence of stars or plates are an indicator that this building was unstable even before it caught fire.

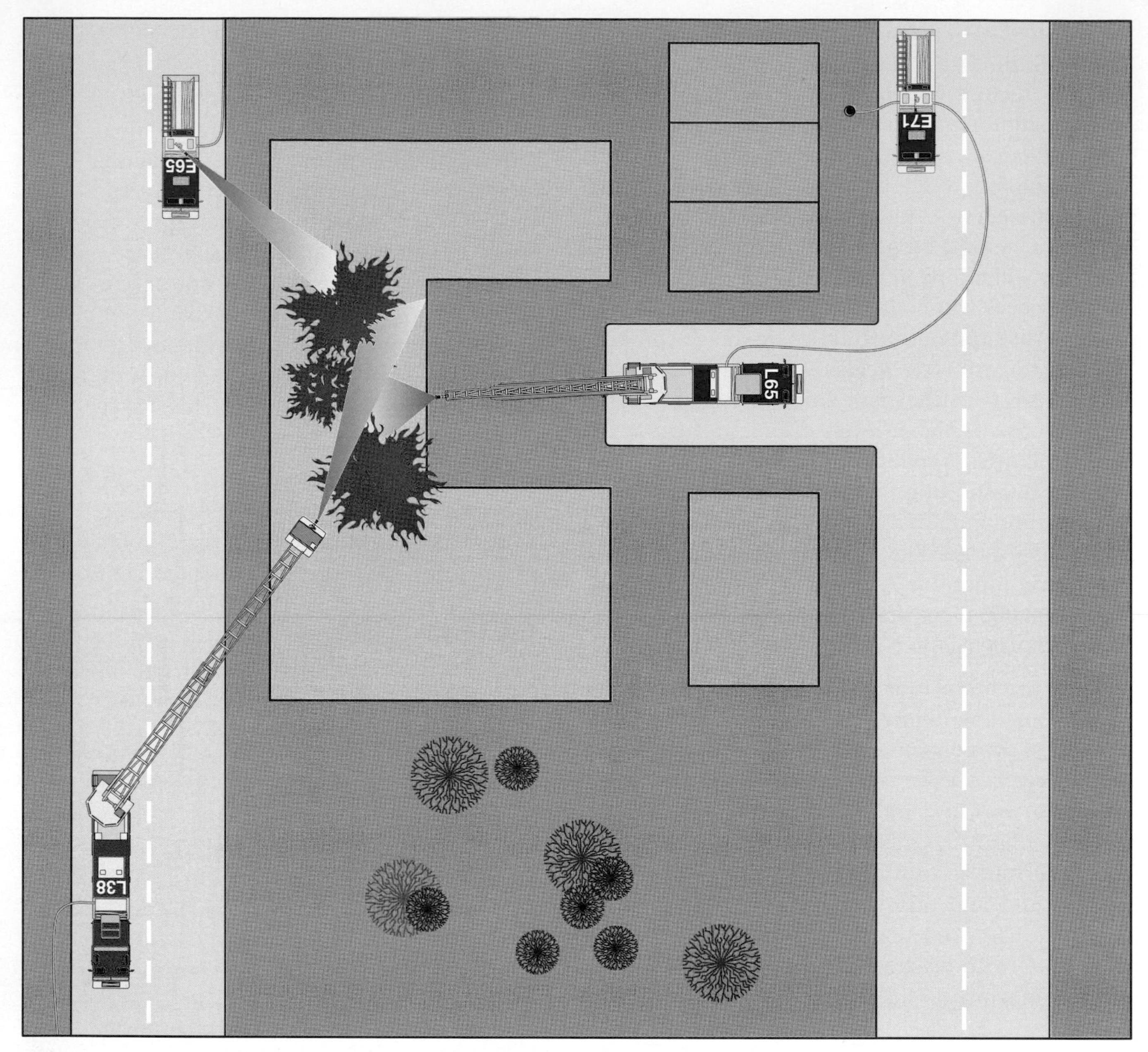

Figure 5.33 Always leave a suitably quick escape route for the apparatus when operating at large fires.

familiar with these circumstances so that the apparatus is placed in a safe, yet effective location for each situation.

Staging

Often, apparatus placement at the scene of a fire or medical incident is limited by the order in which responding apparatus arrive. A late-arriving aerial apparatus may be blocked from a better position by an earlier-arriving pumper or rescue vehicle. Standard operating procedures governing apparatus placement is one way to prevent this type of situation from occurring. An apparatus staging procedure facilitates the orderly positioning of apparatus and allows the Incident Commander to fully utilize the potential of each unit and crew.

Through improvements in incident management strategies, an apparatus staging procedure in two levels has been developed that can be used for any multicompany response. Level I staging is applied to the initial response to a fire or medical incident involving more than one company. Level II staging is used in greater alarm situations where a large number of emergency vehicles are responding to an incident. Level II staging procedures must be initiated by the Incident Commander when requesting additional alarms or by a dispatcher when a large initial response is called for.

Level I staging is used on every emergency response when two companies performing like functions are dispatched (for example, two truck companies). The first-due engine company, truck company, rescue or squad company, and command officer proceed directly to the scene. Later-arriving units park or stage at least one block prior to reaching the scene in their direction of travel or otherwise according to department SOPs (Figure 5.34). The Incident Commander may order the staged units to lay additional supply lines, send personnel to the scene, or proceed to the scene and set up. Engine or quint companies in jurisdictions that typically perform straight or forward hose lays should stage near hydrants or other sources of water. Staged apparatus should not allow their paths to the scene to become blocked.

Level II staging is used when numerous emergency vehicles will be responding to an incident. Incidents that require mutual aid or that result in multiple alarms require Level II staging. When additional units are requested by the Incident Commander, an apparatus staging area is designated. Companies are informed of the staging area location when they are dispatched and respond directly to that location. A parking lot or open field can serve as a staging area (Figure 5.35). Generally, the company officer of the first-arriving company at the staging area becomes the staging manager. On large-scale incidents, a chief officer may be assigned to the staging manager function. The stag-

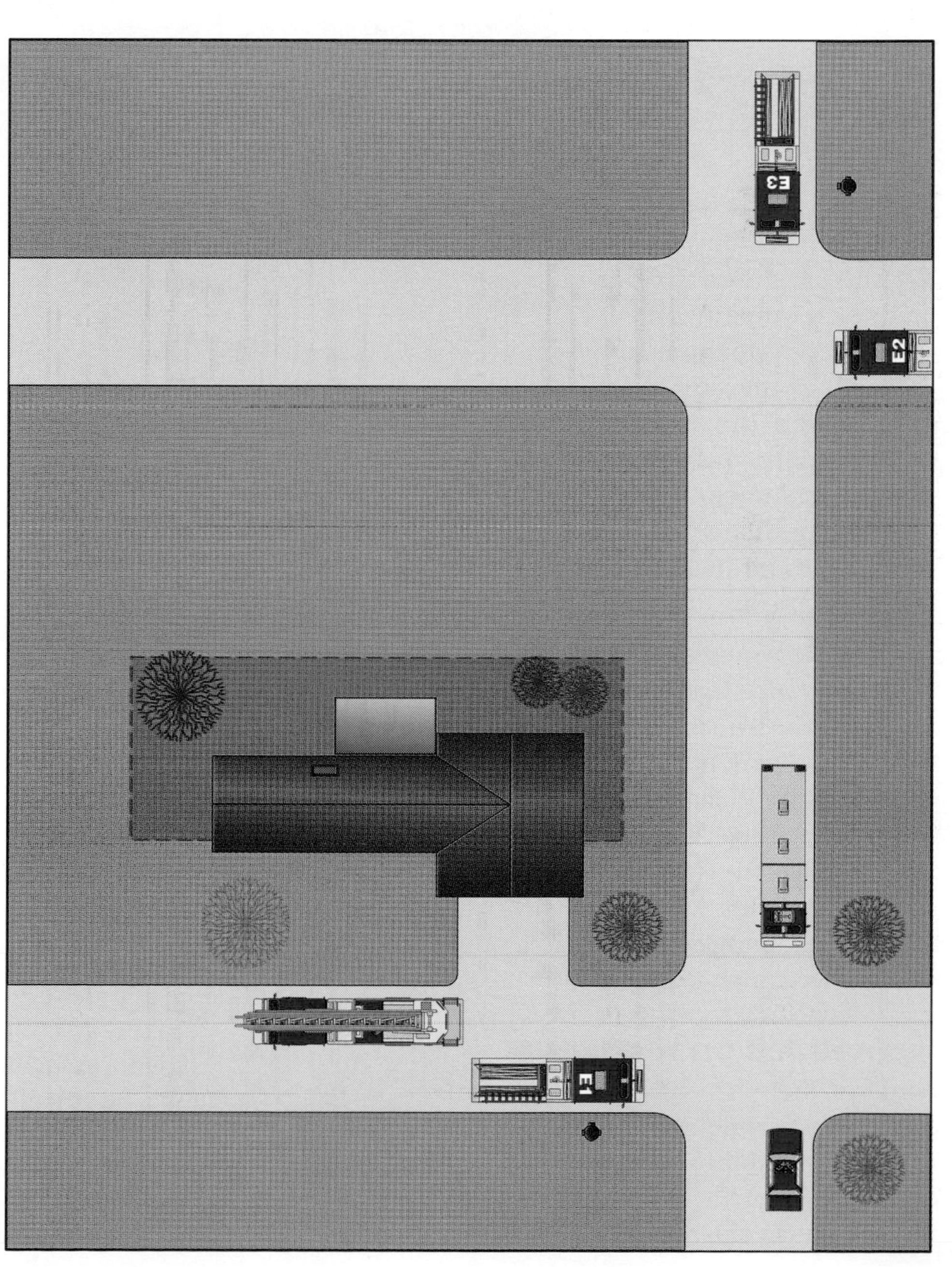

Figure 5.34 Level I staging.

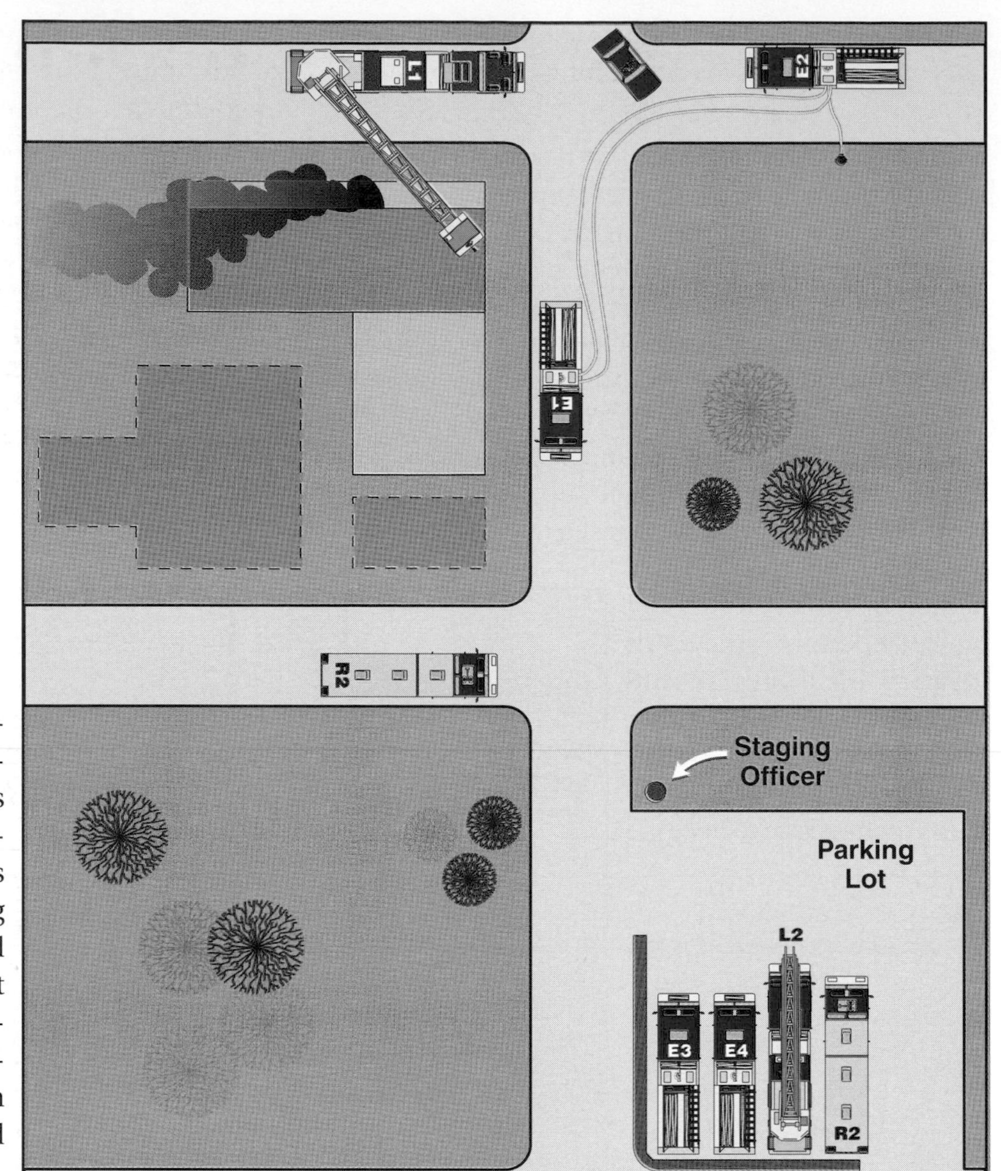

Figure 5.35 Level II staging.

ing manager should communicate available resources and resource needs to the Incident Commander. Company officers should report to the staging manager as they arrive and park. When the Incident Commander requires additional assistance, companies are summoned through the staging manager and sent to the scene.

The staging area should be a secure area that is free of nonemergency traffic. The apparatus belonging to the staging manager should park near the entrance to the staging area and should leave its emergency lights flashing/rotating (Figure 5.36). All subsequent apparatus arriving at the staging area should turn off their emergency lights when they park. This makes it easy for incoming companies to identify the location of the staging manager.

Operations on Highways

Some of the most dangerous scenarios faced by firefighters are operations on highways, interstates, turnpikes, and other busy roadways. The most common types of incidents on these thoroughfares are motor vehicle crashes and/or fires. The potential for multiple-injury crashes or crashes involving hazardous materials is also high. There are numerous challenges relative to apparatus placement, operational effectiveness, and responder safety when dealing with incidents on busy roadways. Because many jurisdictions use truck companies to perform extrication operations at motor vehicle crashes, the aerial apparatus driver/operator must understand the principles of positioning on highways.

Problems associated with simply accessing the scene can be a challenge to emergency responders. This is particularly true on limited-access highways and turnpikes. Apparatus may have to respond over long distances between exits to reach an incident. In some cases, apparatus will be required to travel

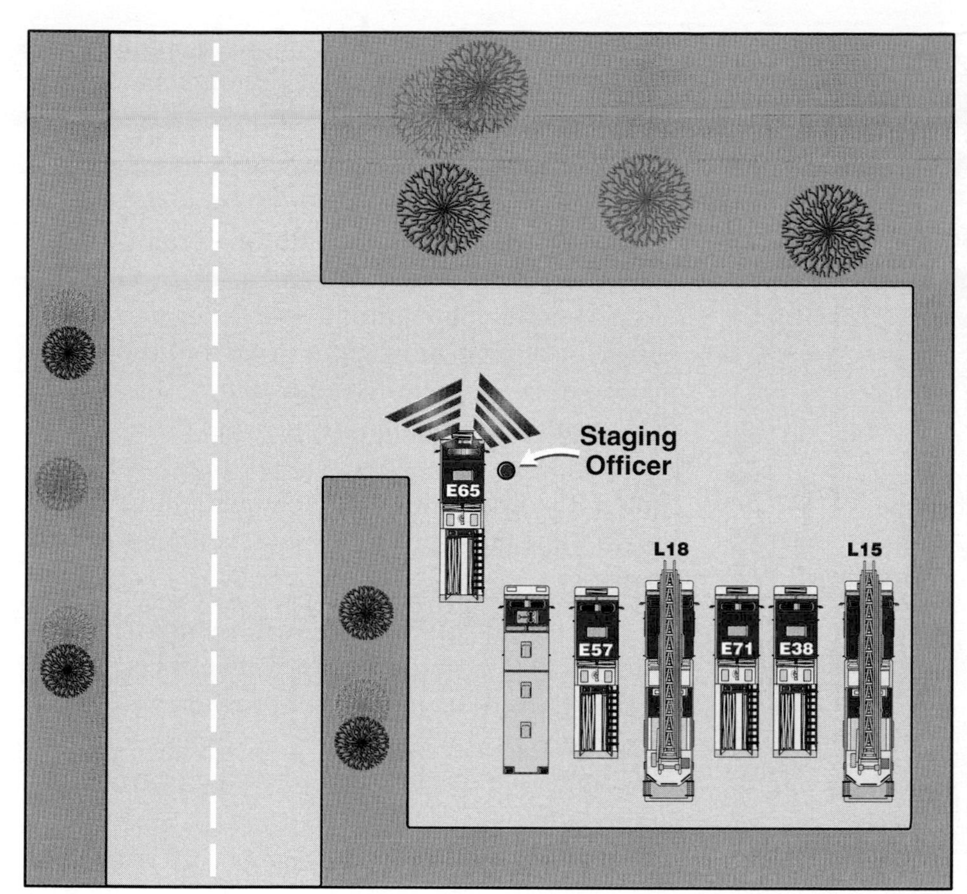

Figure 5.36 The staging officer's apparatus should be the only one in the staging area with its warning lights flashing.

a long distance before there is a turnaround that allows them the ability to get to the opposite side of the median if necessary. Apparatus should not be driven against the normal flow of traffic unless the road has been closed by police units. Incidents occurring on bridges or overpasses may require the use of aerial apparatus or ground ladders in order to reach the scene from below. Operating on bridges is discussed later in this chapter.

Water supply can be a problem on roadways that are in rural areas or even on limited-access highways in urban areas. Hydrant placement on highways may be infrequent or may not exist. It may be necessary to stretch hoselines or use an aerial device from an overpass or underpass to get water to the level of the highway.

The driver/operator should use prudence when responding to an incident on a highway or turnpike. A fire apparatus usually travels slower than the normal flow of traffic, and the use of warning lights and sirens may create traffic conditions that actually slow the fire unit's response. At nighttime incidents, a minimum number of warning lights should be used at the scene to prevent blinding other drivers or distracting them, possibly leading to another crash. Once on the scene, headlights should be on the low-beam setting and should not be flashing. However, all warning lights should *never* be turned off because that might compromise the safety of firefighters working at the scene.

Cooperation between police and fire department personnel at highway incidents is essential. At least one lane next to the incident lane should be closed. Additional or all traffic lanes may have to be closed if the extra lane does not provide a safe barrier. Fire apparatus should be placed at an angle between the flow of traffic and the firefighters working on the incident to act as a shield (Figure 5.37). Front wheels should be turned away from the firefighters working highway incidents so that the apparatus will not be driven into them if struck from behind. Also consider parking additional apparatus 150 to 200 feet (45 m to 60 m) behind the shielding apparatus to act as an additional barrier between firefighters and the flow of traffic.

If the aerial device is going to be deployed on the highway, there are a couple of safety considerations the driver/operator must keep in mind. Traffic moving by the area may cause vibrations that affect the stability of the apparatus, especially on

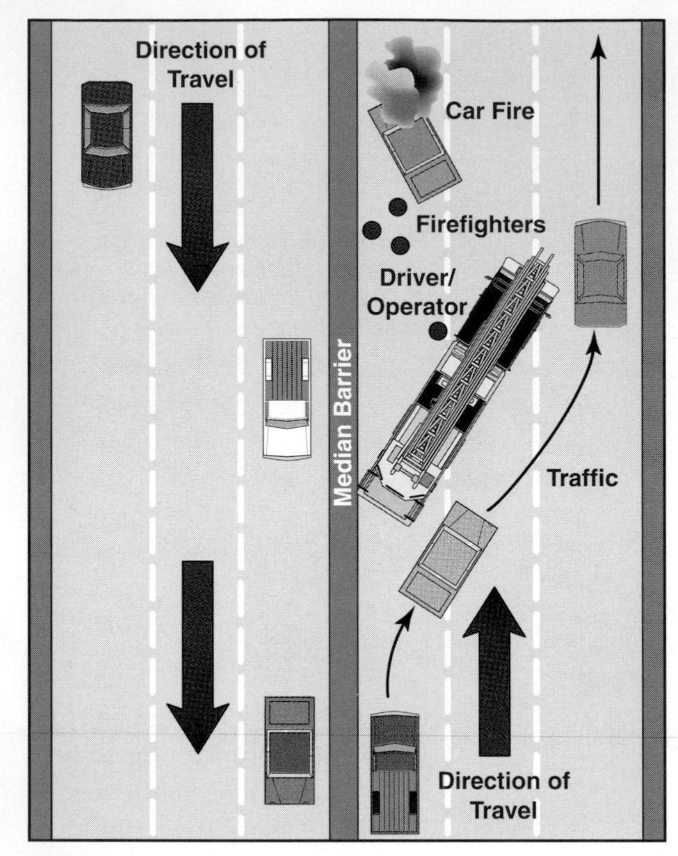

Figure 5.37 Use the apparatus to place a barrier between the firefighters and oncoming traffic.

elevated roadways. The driver/operator should frequently check the stabilizers to make sure that they are still in solid contact with the stabilizer pads and the ground. Also, the driver/operator should use caution when maneuvering the aerial device to make sure that it is not placed in a position where it may be struck by another apparatus or passing traffic. This is of particular concern when operating articulating apparatus that have the boom and knuckle behind the platform.

Hazardous Materials Incidents

Hazardous materials incidents are increasingly common in today's fire service. Emergencies involving hazardous materials may occur in a transportation setting or a fixed facility. The possibility that a hazardous material may be involved in an incident should be considered by the company officer and driver/operator on virtually every fire or transportation incident. It is not possible, in this manual, to cover all information needed to train the driver/operator in hazardous materials identification. However, there are a few general considerations that driver/operators should keep in mind when responding to a potential hazardous materials emergency.

If you are the driver/operator of the first-arriving apparatus, never drive directly into the scene without first attempting to identify the material involved. Failure to heed this advice could result in the apparatus becoming an ignition source for flammable gases (by driving into a vapor cloud) or in contamination of the apparatus and its crew. Always stop well short of the incident scene until the nature of the hazard is understood. Do not park over manholes or storm drains. Flammable materials flowing into the underground system could ignite and explode.

Try to obtain information on the wind speed and direction while en route to the scene. This may be obtained from the dispatcher or by observing the conditions as you respond. If at all possible, park and approach the incident from the upwind and uphill side.

Most jurisdictions use a series of control zones to organize the emergency scene. The zones prevent sightseers and other unauthorized persons from interfering with first responders, help regulate movement of first responders within the zones, and minimize contamination. Control zones are not necessarily static and can be adjusted as the incident changes. Zones divide the levels of hazard of an incident, and what a zone is called generally depicts this level. The three most common terms for the hazardous materials zones are the hot zone, warm zone, and cold zone (Figure 5.38).

- *Hot zone* (also called restricted zone, exclusion zone, or red zone) — an area surrounding the incident that has been contaminated by the released material. This area will be exposed to the gases, vapors, mists, dusts, or runoff of the material. The zone extends far enough to prevent people outside the zone from suffering ill effects from the released material.
- *Warm zone* (also called the contamination reduction zone, limited-access zone, or yellow zone) — an area abutting the hot zone and extending to the cold zone. The warm zone is used to support workers in the hot zone and to decontaminate personnel and equipment exiting the

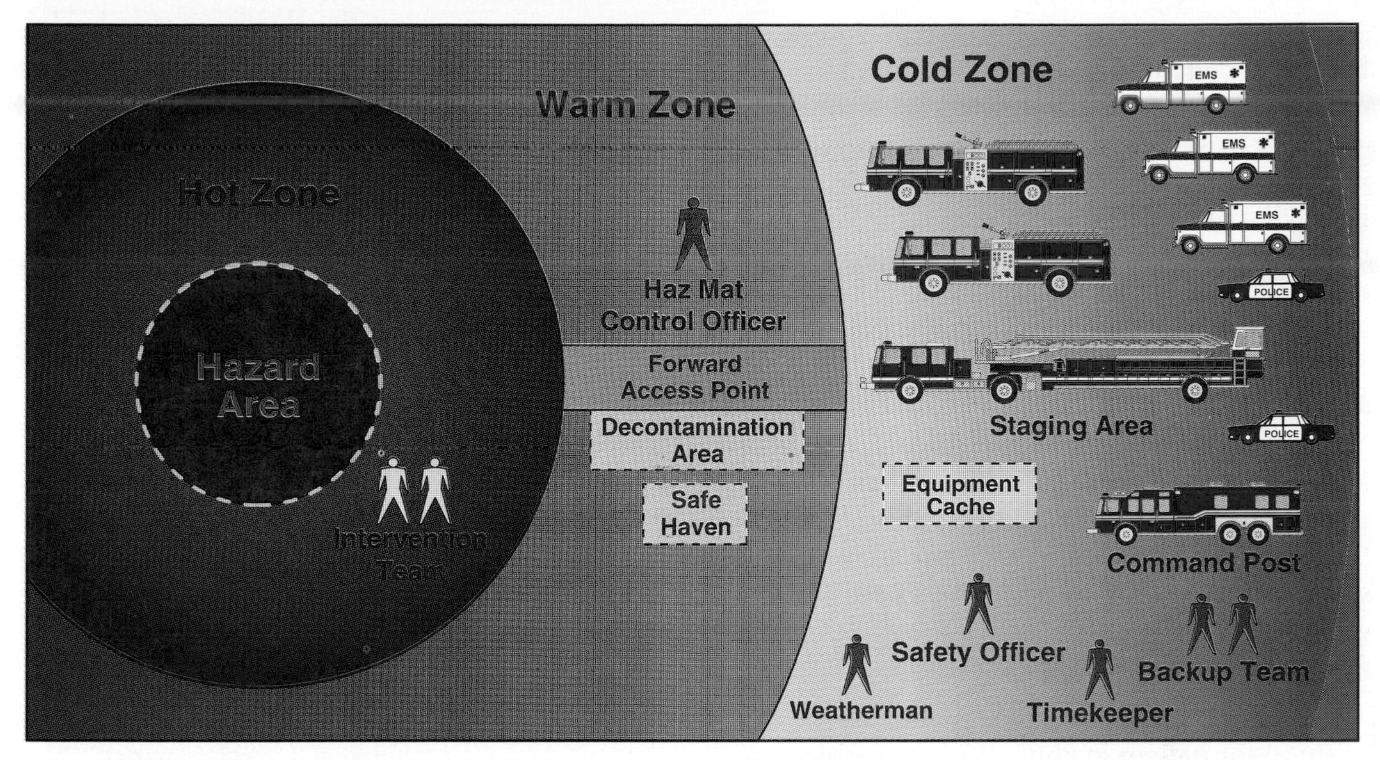

Figure 5.38 Control zones are used to coordinate hazardous materials scenes.

hot zone. Decontamination usually takes place within a corridor (decon corridor) located in the warm zone.

- *Cold zone* (also called the support zone or green zone) — encompasses the warm zone and is used to carry out all other support functions of the incident. Workers in the cold zone are not required to wear personal protective clothing because the zone is considered safe. The command post, the staging area, and the triage/treatment area are located within the cold zone. Most of the time the driver/operator and his apparatus are positioned in the cold zone.

If the hazardous materials incident involves ignited flammable or combustible liquids (or the potential for them to ignite), the driver/operator must be aware of the dangers of being exposed to high levels of radiant heat and position the apparatus accordingly. If the material involved is a corrosive, be mindful that any contact with the aerial device could result in severe damage.

Avoid staging all apparatus in the same location when responding to bomb threats and other potential terrorist incidents. It is possible that an explosive device could be placed in the staging location with the intent of harming emergency personnel. For more information on responding to hazardous materials emergencies, see IFSTA's **Hazardous Materials for First Responders** manual.

Operating Near Railroads

Occasionally, emergency incidents occur on or in close proximity to railroads. Driver/operators should understand the hazards associated with operating near railroads and take any measures possible to minimize those hazards. It is difficult to stop the movement of trains on the track during emergency operations. Always treat a railroad track as a potentially active line.

Never park the apparatus on the railroad tracks. As well, keep the apparatus far enough away from the tracks so that it (or a deployed aerial device) will not be struck by a passing train. When possible, park the apparatus on the same side of the tracks as the incident. This negates the need to raise the aerial device across the tracks or for personnel to be traversing back and forth between each side. Most railroad companies advise that vehicles be kept at least 25 feet (8 m) away from the tracks whenever possible.

If it becomes absolutely necessary to raise an aerial device across a railroad track, attempt to confirm from the rail company that train traffic has been halted on that set of tracks. Even when a halt confirmation has been received, keep the aerial device at least 25 feet (8 m) above the level of the rails as an added safety precaution. Use caution when operating aerial apparatus in the vicinity of rail lines that operate from high-voltage, overhead electrical lines. Always keep the aerial device at least 10 feet (3 m) away from all overhead lines.

Emergency Medical Incidents

Many of the calls that fire departments respond to are emergency medical incidents. Apparatus positioning on these incidents is not only important from a tactical standpoint, it is important from a safety standpoint. Many firefighters have been injured or killed when struck by traffic while operating at emergency medical incidents.

It is important to allow the ambulance the best position for patient loading. When possible, park the apparatus off the street, although this may be difficult with large aerial apparatus. This virtually eliminates any of the hazards associated with traffic. If locating off the street, make sure that the surface is stable enough to support the weight of the fire apparatus. Typically, residential driveways and yards will not safely support the weight of an aerial apparatus.

If it is not possible to locate off the street, use the apparatus as a shield between the work area and traffic. Park larger apparatus (such as an aerial apparatus) between smaller apparatus (such as an ambulance) and the oncoming flow of traffic (Figure 5.39). In particular, guard the patient loading area of the ambulance by shielding it with another vehicle. If possible, place traffic cones to direct traffic away from the apparatus.

Aircraft Incidents

Aerial apparatus may respond to incidents involving aircraft for a number of reasons. As mentioned earlier in this chapter, aerial apparatus frequently carry extrication equipment that may be useful for disentangling victims of an aircraft crash. Some incidents involving large-frame aircraft, such as the DC-10, 747, or 767, may require the use of an aerial device to access the interior of the plane (Figure 5.40). Driver/operators of aerial apparatus assigned to airport fire stations, or those who may respond to an airport facility, should be familiar with the procedures for approaching and positioning at an aircraft emergency.

There are two basic types of aircraft incidents to which aerial apparatus may respond. These include aircraft crashes and non-crash-related fires. In reality, there are few tactical uses for aerial de-

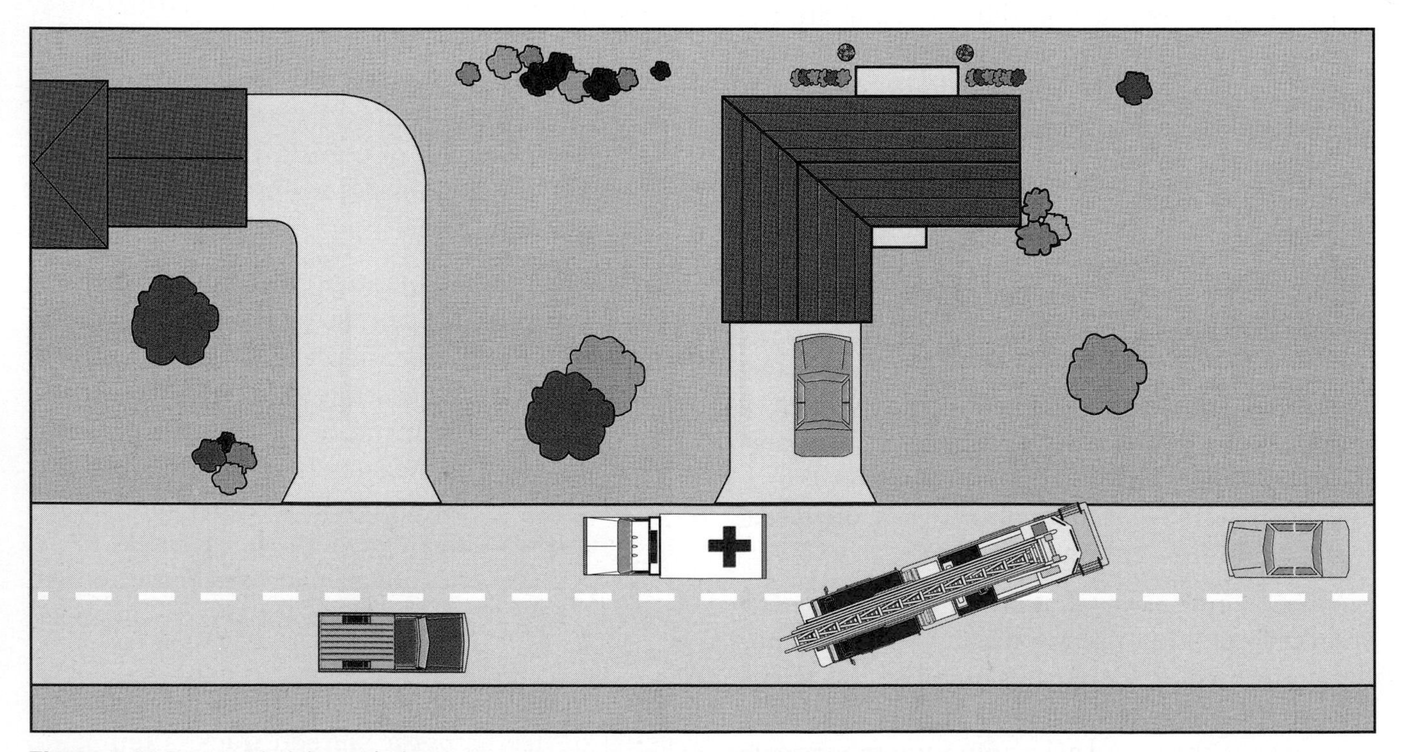

Figure 5.39 Park behind the ambulance to protect the patient loading zone.

vices at aircraft crash scenes. The apparatus will be primarily relegated to transporting firefighters and providing necessary portable equipment needed in handling these incidents. Many crashes occur away from runways or other paved surfaces. Aerial apparatus are typically not well suited for off-the-road operations. Thus, it may not be possible to get the apparatus very close to the actual scene or work area. In these cases, the apparatus should be parked as close to the scene as reasonably possible. It may be necessary to ferry equipment and personnel from the aerial apparatus to the work area using a more suitable vehicle.

Aerial apparatus are more likely to be useful at non-crash-related fires involving the types of large-frame aircraft previously mentioned. These incidents almost always occur on paved surfaces that allow the aerial apparatus direct access to the aircraft. At these incidents, the tactical uses for the aerial device are basically the same as those described earlier in this chapter for structural fire fighting. The aerial device may be used to provide access for ventilation, to rescue passengers, to deploy handlines to the interior of the aircraft, or for master stream operations. If the tip of the device is to be placed in an aircraft doorway, position the apparatus as described earlier for accessing the window of a building (Figure 5.41). If the objective is to provide access over a wing or remove victims from a wing, the positioning principles are basically the same as for providing access to the roof of a building (Figure 5.42).

There are a few special circumstances that driver/operators of aerial apparatus must remember when responding to a non-crash-related fire. This is particularly true if you have been standing by and waiting for the aircraft to land or come to a stop. Oftentimes the apparatus are right there when the plane comes to a halt. In these situations, the driver/operator should avoid spotting the apparatus in a position that hinders the deployment of the aircraft emergency slides. These slides inflate and drop from the door openings of the aircraft. On larger

Figure 5.40 This Boeing 747 is an example of a large frame aircraft.

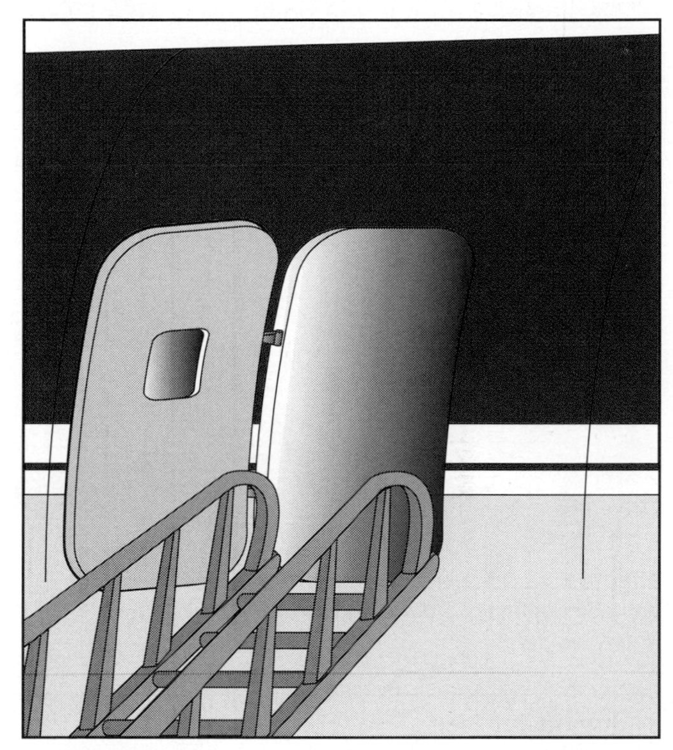

Figure 5.41 Position the tip of the aerial in the doorway of the aircraft.

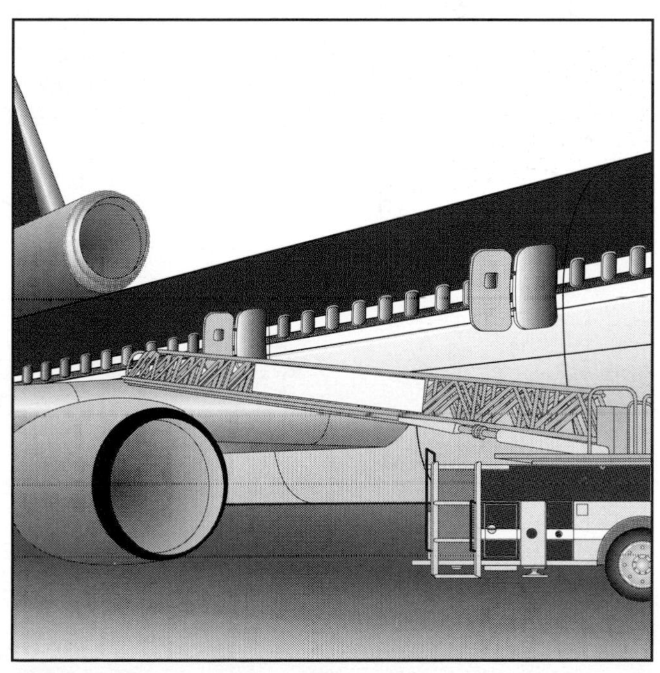

Figure 5.42 The aerial device may be used to ladder a tall wing structure.

aircraft, they inflate and drop from the rear of the wings near the emergency window exits (Figure 5.43). The driver/operator must be familiar with the type of aircraft and the location of the chutes. These chutes will allow the aircraft to be evacuated much more quickly than trying to remove victims down the aerial device.

Also, use caution when positioning near jet engines that are still running. The potential for injuries created by the intake or exhaust forces of large jet engines is substantial and should be avoided (Figure 5.44). Make sure that the apparatus is not positioned so that people operating on the aerial device will be forced to travel through these danger zones.

Figure 5.43 Don't block the area where escape chutes may be deployed. *Courtesy of Dallas-Fort Worth International Airport.*

When responding to any type of aircraft incident and positioning on the scene, keep a few general safety requirements in mind:

- Use caution when you approach the incident scene to avoid running over victims.
- Watch for pools of unignited jet fuel — do not drive through them or close enough to present an ignition source.
- Position upwind of any fire conditions or vapors from unignited pools of fuel.
- Watch for wreckage or other items that may flatten tires; they may be obscured by smoke or darkness.

For more information on responding to aircraft emergencies, see the IFSTA **Aircraft Rescue and Fire Fighting** manual.

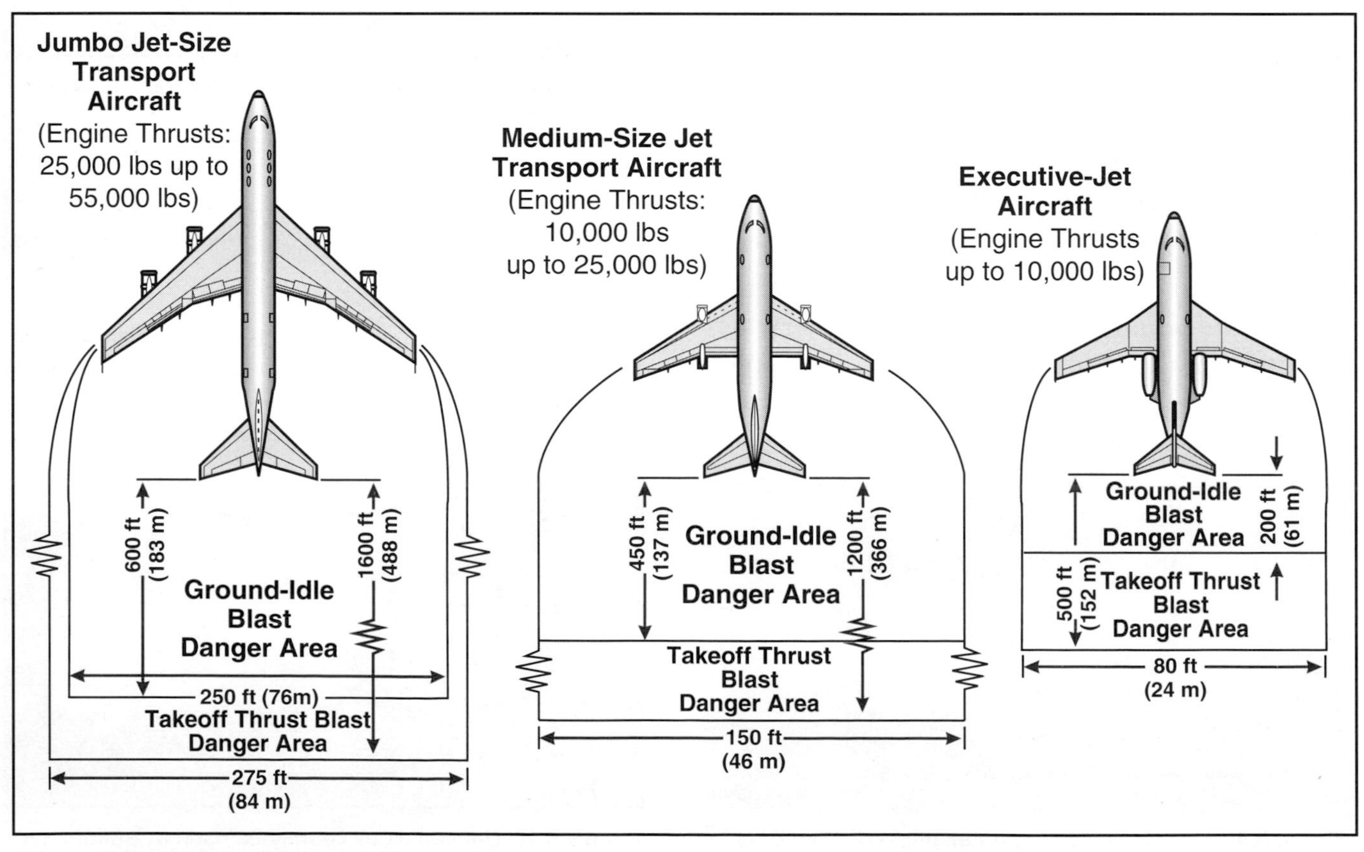

Figure 5.44 Avoid the danger areas around jet engines.

Operating on Bridges

In some cases, the aerial apparatus may be required to position on and operate from a bridge or overpass. This most commonly occurs when a large fire of some type occurs next to the bridge. In these cases, the Incident Commander may determine that it is tactically important to deploy elevated master streams from the bridge side of the fire. The following guidelines should be followed in these situations:

- Make sure that the bridge has a load capacity that makes it safe for the apparatus to drive on and operate from.
- Use caution when raising the aerial device if the bridge has a superstructure above the road surface. If electric lines are present, follow the standard 10-foot (3 m) safety distance from the lines.
- Make sure that the road surface on which the apparatus is setting up is in good repair. Avoid weak spots that might allow a stabilizer to punch through the bridge decking.
- Follow the principles discussed earlier in this chapter for setting up the aerial apparatus on an incline if the bridge has a significantly sloped roadway.
- Be alert for wind conditions on top of a large bridge that may be more severe than on either side of the access to the bridge. Keep in mind that when on the bridge, high-wind operating principles for aerial device operation may be necessary.
- Recognize that most large bridges are designed to move somewhat in response to forces placed upon them by wind, traffic, and water movement below. This movement or shaking will be somewhat amplified at the end of a raised aerial device. High winds will further affect the movement of the aerial device. When this movement becomes uncomfortable for firefighters or begins to place excessive lateral stress on the aerial device, minimize the extension of the aerial device as much as possible.
- Remember that bridges and elevated sections of roads will be the first road surfaces to freeze during cold weather. This can pose serious travel and positioning hazards later in the incident.

Operating at Petroleum Storage/ Processing Facilities

Aerial apparatus are frequently used to provide elevated master streams at fires involving large storage tanks or fuel/chemical processing facilities. The master streams may be used for exposure protection or fire attack. Departments that respond to these types of facilities must have solid standard operating procedures for handling these types of incidents. Apparatus positioning and placement must be part of these SOPs and should be a part of pre-incident planning for these target hazards. Pre-incident planning and training sessions between the industrial and municipal firefighters protecting a facility are crucial to the success of industrial emergency incidents.

When operating at storage tank fires, the apparatus should never be spotted inside the dike that surrounds the affected tank(s) (Figure 5.45). Always position the apparatus outside the dike's walls, unless the roadway is built on top of the dike. In that case, the apparatus may actually be set up on the dike itself, although a constant monitoring of the conditions of the dike must be maintained. An upwind location is most desirable if the aerial device is being used for direct fire attack. It is more efficient to discharge foam streams downwind. This also reduces the amount of heat and smoke to which the apparatus will be exposed. If the aerial device is being used to protect exposures, the specific needs of the incident will dictate the exact spot for the apparatus to park. However, this does not negate the need to keep the apparatus outside the dike.

Positioning the aerial apparatus to attack fires or protect exposures in processing facilities or refineries can be an extremely challenging task. It goes without saying that the nature of the fuels involved with these fires dictate that the apparatus be positioned upwind whenever possible. These facilities present a number of challenges to positioning large aerial apparatus. Some of these challenges include:

- Narrow driveways — It may be extremely difficult and time consuming to maneuver the apparatus into the necessary position.
- Dead-end accesses — Because incidents of this nature have the potential to become more severe

in a rapid fashion, the apparatus should never be placed in a position where a difficult reverse retreat would be required. Take the time to back the apparatus into position if this is the case.

- Overhead obstructions — Refineries and chemical processing facilities are typically a maze of overhead piping and conduits (Figure 5.46). Make sure that the final spot chosen for the aerial apparatus is one that will allow the aerial device to be deployed effectively without coming into contact with these overhead obstructions.

Technical Rescue Incidents

The proper positioning of aerial apparatus at the scene of a technical (trench, below grade, confined space) rescue operation will depend on whether the aerial device is being used to directly aid in the mitigation of the incident or the apparatus is simply serving in a support role on the scene. The types of technical rescue incidents to which an aerial apparatus may respond are too numerous to describe individually in this manual. However, there are some general principles that may be followed for these types of incidents.

If the aerial device is going to be part of the rescue effort, the apparatus should be placed in a position that will minimize the angle and extension to which the aerial device will be raised. These principles are much the same as those described earlier in the chapter for positioning the apparatus in order to operate the device to a window or roof. Avoid spotting the apparatus in a location that will require the apparatus to be stabilized on top of debris or otherwise unstable surfaces. Also be aware of dangling debris and unstable structures that could drop or collapse on top of the apparatus and the firefighters working around it (Figure 5.47).

If the aerial apparatus is going to be used in a support role, it should be parked in a manner that does not block other apparatus that need to be closer to the work area. However, it should not be parked so far away that it is difficult to carry ground ladders and other portable equipment to the scene if needed.

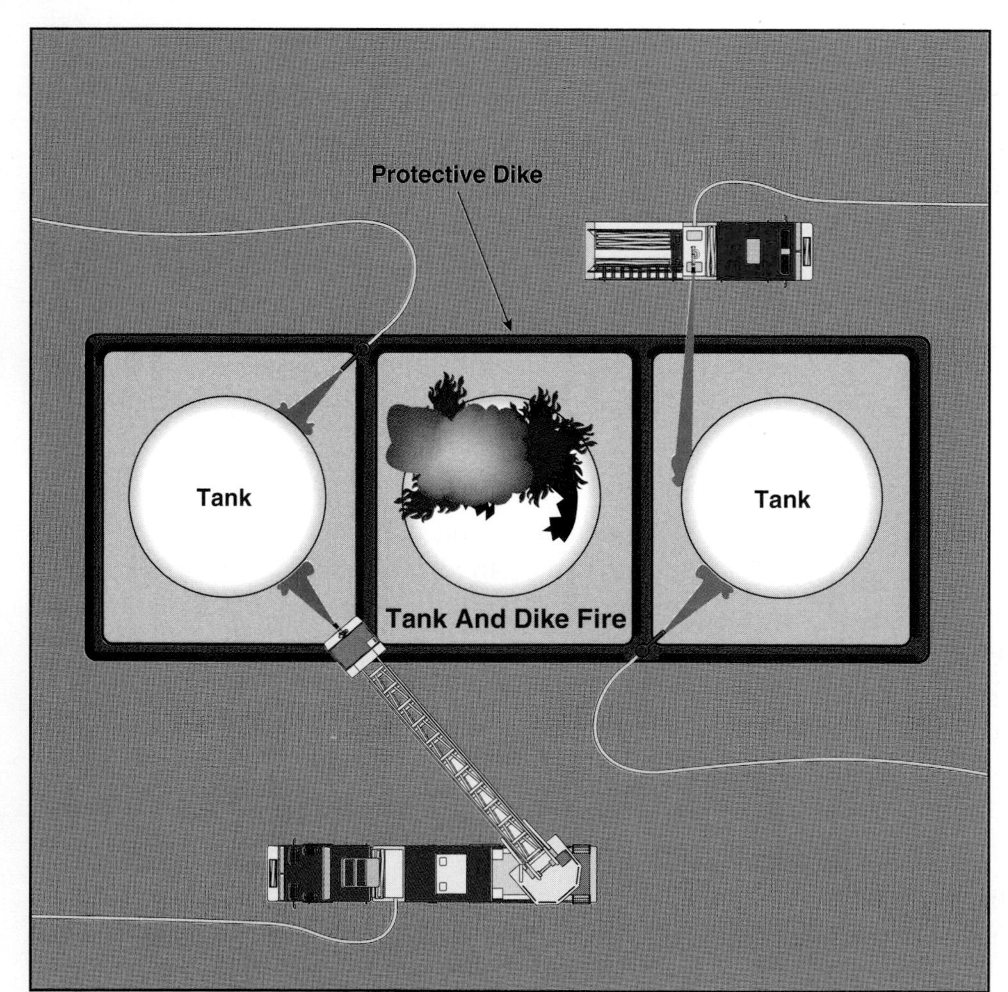

Figure 5.45 Keep all apparatus outside the dike area surrounding the storage tank.

Some other general guidelines for positioning the apparatus at technical rescue scenes are:

- Avoid parking in a position where apparatus exhaust fumes, noise, or vibrations will affect victims and rescuers.
- Stay an adequate distance away from trench walls and other unstable terrain.
- Do not block the scene access to later-arriving vehicles.

Additional information on using the aerial device to effect rescues is covered in Chapter 9 of this manual.

Figure 5.46 The aerial device must be used cautiously around overhead obstructions at the refinery.

Figure 5.47 Be alert for dangling debris that could fall onto the aerial device or apparatus. *Courtesy of Steve George.*

Chapter 6

Stabilizing The Apparatus

Job Performance Requirements

This chapter provides information that will assist the reader in meeting the following job performance requirements from NFPA 1002, *Standard on Fire Apparatus Driver/Operator Professional Qualifications*, 1998 edition. Particular portions of the job performance requirements (JPRs) that are met in this chapter are noted in bold text.

4-2.2 Stabilize an aerial apparatus, given a properly positioned vehicle and the manufacturer's recommendations, so that power can be transferred to the aerial device hydraulic system and the device can be safely deployed.

(a) ***Requisite Knowledge*: Aerial apparatus hydraulic systems, manufacturer's specifications for stabilization, stabilization requirements, and effects of topography and ground condition on safe stabilization.**

(b) ***Requisite Skills*: The ability to transfer power from the vehicle's engine to the hydraulic system and operate vehicle stabilization devices.**

Perhaps the most critical aspect of preparing an aerial device for use is the apparatus stabilization process. When an aerial device is raised and extended to either side of the chassis, the wheel span alone is not adequate to handle the torquing momentum created by the extended aerial device. It is, therefore, necessary to broaden or widen the base of the supporting chassis beyond the span of the wheels so that it resists the tendency to tip over when the aerial device is extended over the side of the apparatus. ***IFSTA recommends that stabilizers (also called outriggers) and stabilizer pads (also called jackplates) be fully deployed every time the aerial device is raised from its bed.*** NFPA 1914, *Standard for Testing Fire Department Aerial Devices,* says:

"The truck is considered to be in a state of stability when no sign of overturning is evident with the aerial ladder or elevating platform in operation. The lifting of a tire or outrigger on the opposite side of the vehicle from the load does not necessarily indicate a condition of instability. Instability occurs when an aerial device can no longer support a given load and overturning is imminent."

The principles related to stabilizing an aerial apparatus are actually fairly simple. To begin this understanding, you must first accept the fact that an aerial apparatus is a system of connected components. When parked, with the aerial device in its stowed position, the apparatus has a center of gravity located somewhere along the longitudinal midline of the chassis, between the front and rear axles. Should the aerial device be deployed without first setting the apparatus stabilizers, the base of stability would be limited to the rectangle formed at each corner by the tires of the apparatus (Figure 6.1).

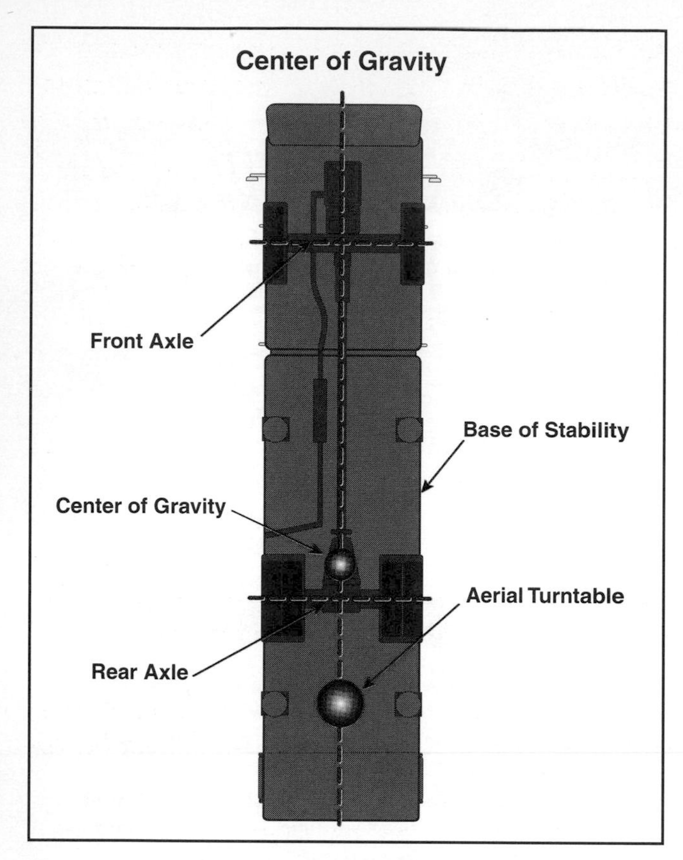

Figure 6.1 The center of gravity of the apparatus before the aerial device is deployed.

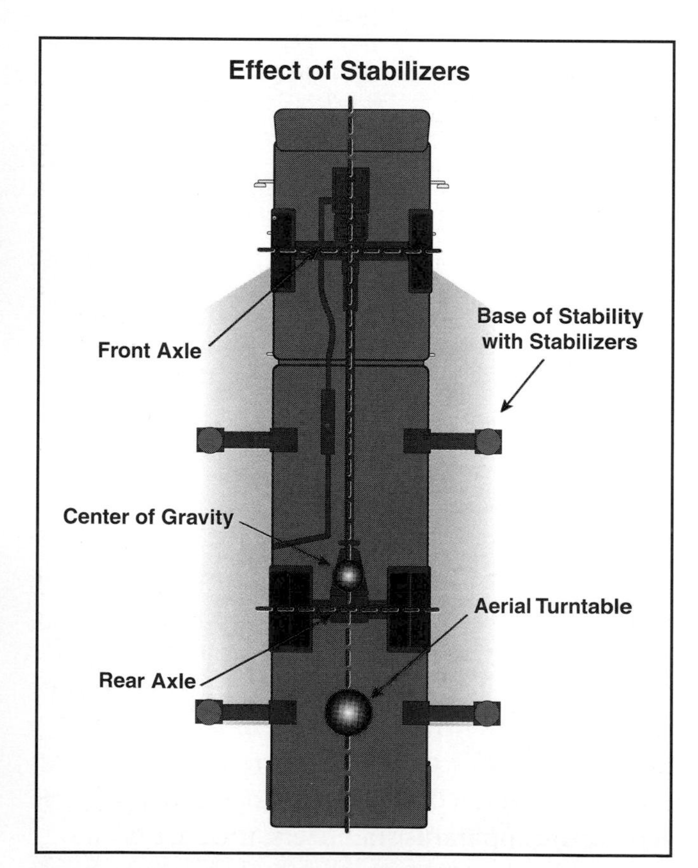

Figure 6.2 The base of stability is increased when the stabilizers are deployed.

By deploying the apparatus stabilizers, the base of stability for the aerial apparatus is effectively increased (Figure 6.2). Should the aerial device be extended straight over the front of the apparatus at this point, a corresponding forward shift in the center of gravity would also occur (Figure 6.3). If the driver/operator begins to rotate the extended aerial device, the center of gravity would move once again — this time away from the longitudinal midline of the apparatus (Figure 6.4). If the device were to be rotated 360°, a "gravity circle" would be traced out. As long as this gravity circle does not extend outside the base of stability, the apparatus should remain stable.

The effects of "short jacking" an aerial apparatus can be seen in Figure 6.5. The fact that the gravity circle extends beyond the base of stability on the short-jacked side indicates that overturning is likely should the aerial device be extended in that direction. Newer apparatus may be equipped with limit switches that prevent the movement of the aerial

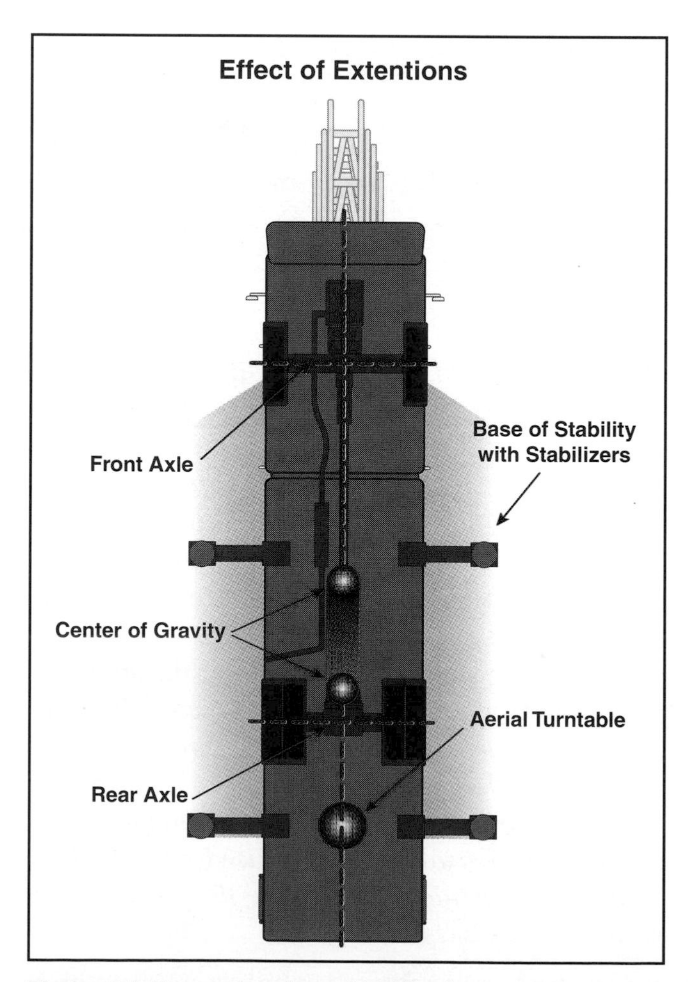

Figure 6.3 The center of gravity shifts forward if the aerial device is extended over the front of the apparatus.

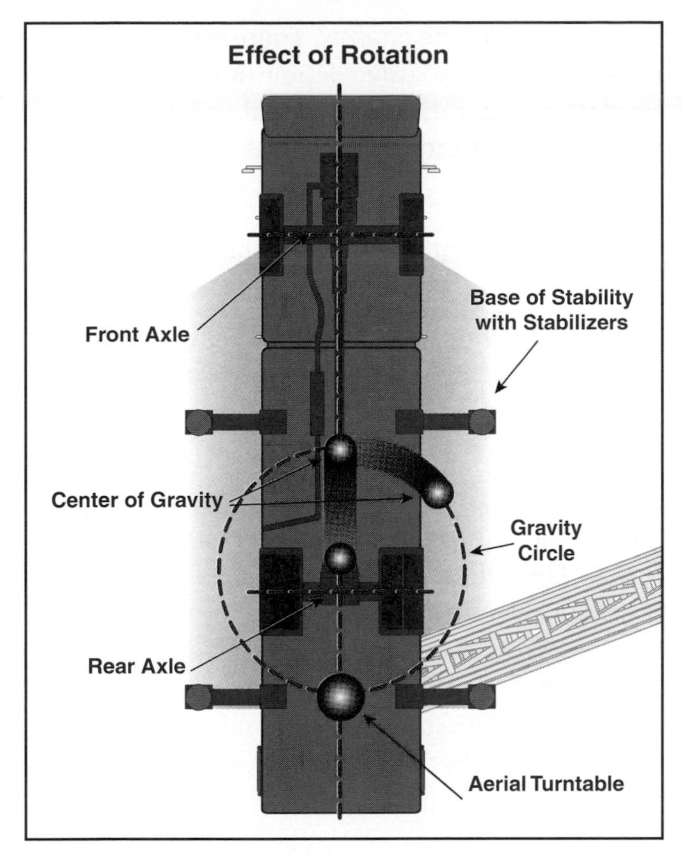

Figure 6.4 The center of gravity shifts as the aerial device is rotated.

device to the short-jacked side. However, the driver/operator should not rely on these automatic features to prevent a serious accident from occurring.

Lateral grades may also have an adverse effect on the stability of the apparatus. If the apparatus is parked on a lateral grade, the center of gravity (and the theoretical gravity circle) will be shifted to the lower side. If the aerial device is raised over the lower side, the gravity circle may extend beyond the base of stability (Figure 6.6).

Overloading the aerial device can also have an adverse effect on the stability of the apparatus, even on a level surface with the stabilizers fully deployed. Excessive loading on the aerial device will result in an expansion of the gravity circle (Figure 6.7). In some cases, the gravity circle may expand beyond the base of stability, particularly to either side of the apparatus. This would also be the case with older aerial devices that are extended too far at low angles of elevation.

Stabilizers (also called outriggers, ground jacks, etc.) are required to prevent the apparatus from tipping as the aerial device is extended away from

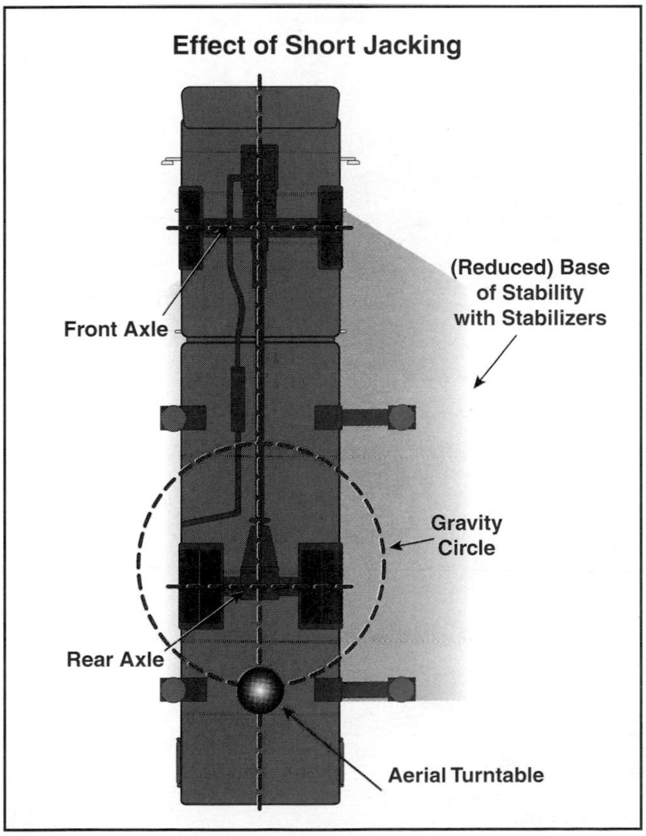

Effect of Short Jacking

(Reduced) Base
of Stability
with Stabilizers

Front Axle

Gravity
Circle

Rear Axle

Aerial Turntable

Figure 6.5 It is conceivable that the gravity circle will extend past the stabilizers if they are short jacked.

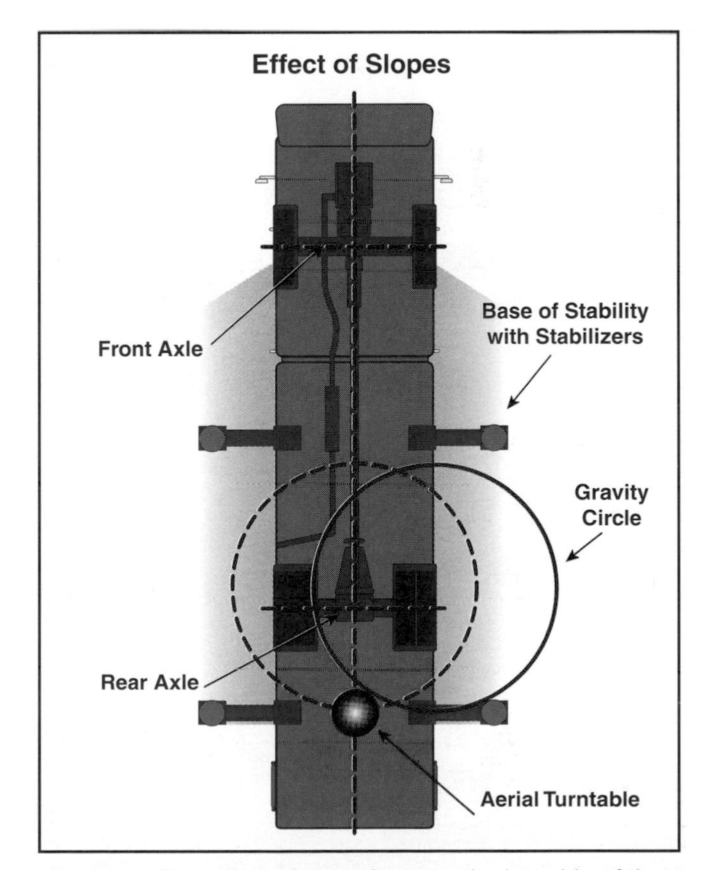

Figure 6.6 The effect of operating over the low side of the apparatus.

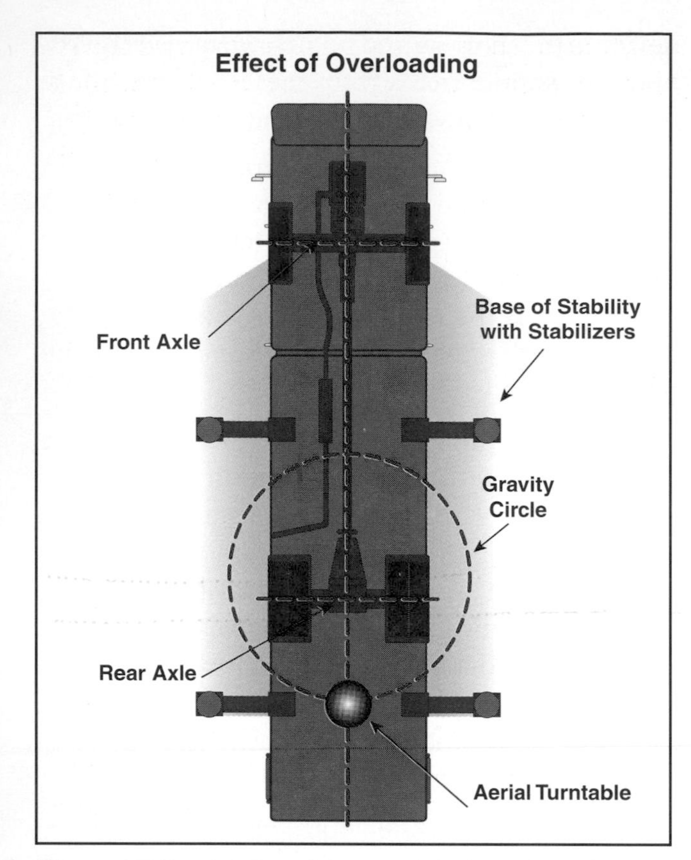

Figure 6.7 Overloading of the aerial device effects the stability circle.

Figure 6.8a A post-type stabilizer.

the centerline of the chassis. There are many stabilizing devices available, depending on the type, size, and configuration of the aerial apparatus. Some of the more common are (Figures 6.8 a through c):

- Straight jacks that extend down from the chassis.
- A-frame, scissor, or X-style stabilizers that extend down and away from the chassis at an angle.
- Box stabilizers that extend straight away from the truck and then have jacks that extend straight to the ground. These are sometimes referred to as H-style stabilizers.

This chapter begins by outlining the hydraulic system that controls the stabilizers and the aerial device. The remainder of the chapter provides a step-by-step outline of processes used to stabilize the apparatus. Included are specific directions for stabilizing the apparatus on even and uneven terrains. Information on using manual stabilizers and stabilizing tillered aerial apparatus is contained at the end of this chapter.

Figure 6.8b An A-frame stabilizer.

Figure 6.8c An H- or box-type stabilizer.

Basics of Aerial Device Hydraulic Systems

In the early days of aerial apparatus, aerial devices and stabilizers were operated manually by any one of several designs including spring-operated and gear-driven. Today, all aerial devices are hydraulically powered, as are virtually all stabilizers. Hydraulic fluid is the medium used within the system to transmit force. Hydraulic fluid is used because it is practically incompressible, and it allows force to be transmitted over a relatively large distance with little loss of power. Depending on the manufacturer of the aerial apparatus, the fluid may be under pressures of 3,500 psi (24 500 kPa) or more. It is for this reason that personnel searching for leaks in a hydraulic system under pressure should use extreme care and should never attempt to block a leak with any part of the body. Pinhole leaks at high pressure are capable of cutting through human tissue and may also cause severe burns.

WARNING!

Use extreme caution when searching for hydraulic fluid leaks. The extreme pressure behind these leaks may cause serious damage to human tissue.

Force is created on the fluid by a hydraulic pump (Figure 6.9). The pump is powered by a power take-off (PTO) arrangement off the vehicle's main engine and transmission. The pump used for the hydraulic system is a positive displacement pump. These types of pumps are explained in detail in Chapter 10 of the IFSTA **Pumping Apparatus Driver/Operator Handbook**. Each type pumps a fixed amount of fluid (measured in gpm or L/min) at a given pump speed (measured in rpm). This allows for predictable operation of the hydraulic components. The amount of fluid being pumped can be varied based on the speed at which the pump is operated.

The hydraulic fluid pumped into the hydraulic system is supplied from the hydraulic reservoir. Fluid displaced from the system flows back into the reservoir for storage before being recirculated through the system. The reservoir is designed to supply an adequate amount of fluid to operate the hydraulic system and to condition the fluid while it is stored in the tank. Baffles located in the tank slow the movement of the fluid through the tank. This allows air, heat, and foreign matter to be released from the fluid before it is reintroduced into the hydraulic system. Filters and exchange-type fluid coolers are also used to supplement the reservoir's ability to condition the fluid. Coolers, usually located in front of the vehicle's radiator, remove heat from the fluid as it returns to the reservoir from the hydraulic system.

The hydraulic fluid is supplied through the system by a series of tubing and hoses (Figures 6.10 a and b). Most manufacturers use steel tubing and aircraft-type, steel-braided hose (Figure 6.11). Hose must be rated to burst at a pressure that is at least four times stronger than normal operating pressure. Fluid is supplied to components above the rotation point of the turntable through a high-pressure hydraulic-swivel assembly to permit continuous rotational capability of the device.

In order to control the flow of hydraulic fluid through the system, a number of different valves are used. Valves start, stop, regulate, and direct the flow of fluid to control pressure in the system by allowing the pressure to be built up or released. Valves may be controlled manually, electrically, hydraulically, mechanically, or by a combination of these methods. There are numerous types of valves used in an aerial apparatus hydraulic system. *Check valves* prevent fluid from flowing backwards through a component and act as a safety feature in the event that a leak develops in the

Figure 6.9 Pressure is imparted on the hydraulic fluid by a hydraulic pump.

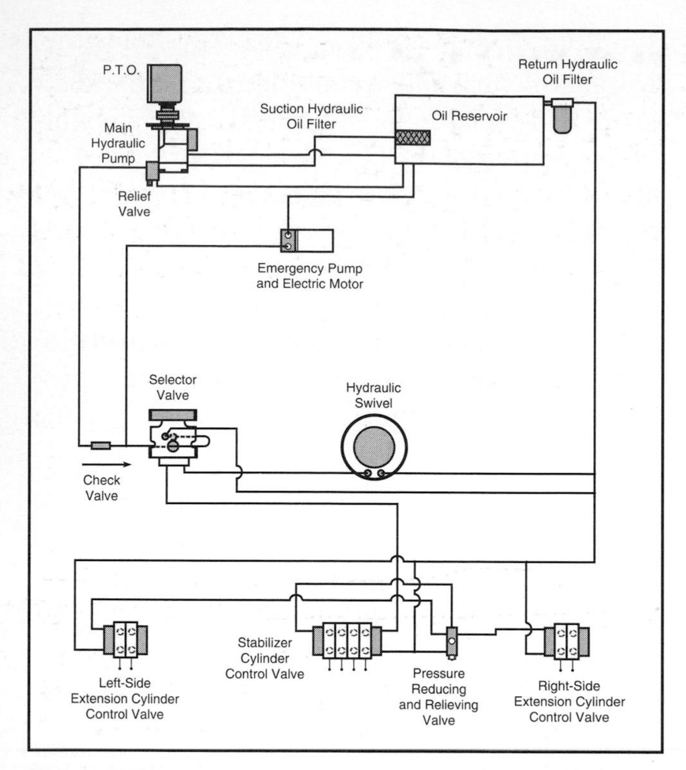

Figure 6.10a A schematic for a typical aerial apparatus hydraulic system.

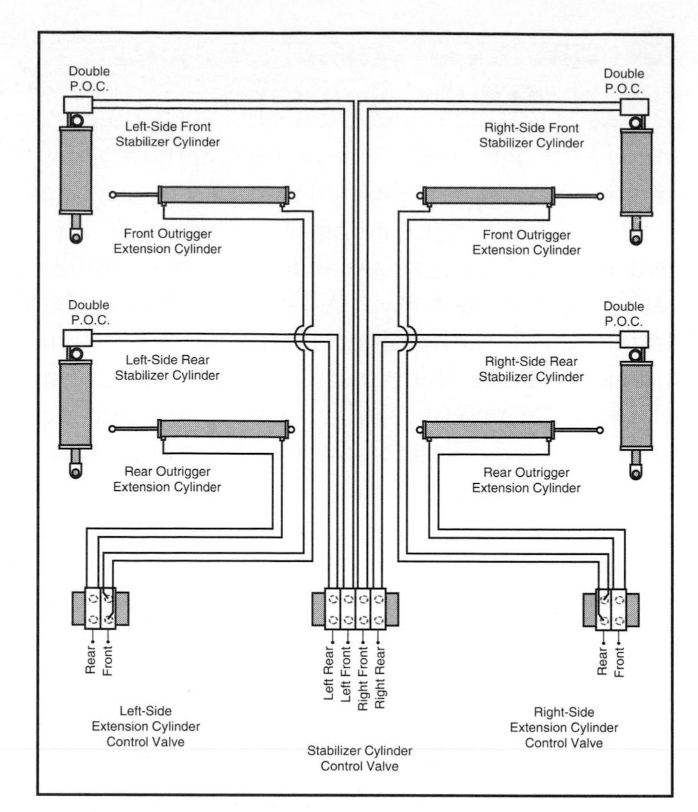

Figure 6.10b The parts of an aerial apparatus stabilizing system are shown in this diagram.

system. *Relief valves* limit the pressure built up in the system, thus preventing damage due to overpressurization. *Counterbalance valves* prevent unintentional or undesirable motion of the device from position.

One of the more important valves in the system is the selector valve (also known as the transfer or diverter valve). The *selector valve* is a three-way valve that directs fluid to either the stabilizer control valves or the aerial device control valves (Figure 6.12). A sliding spool in a housing directs fluid flowing through the valve to one work port or the other, at the same time blocking fluid flow through the opposite work port. By blocking fluid flow to the system not in use, the selector valve acts as an interlock to prevent both the stabilization and aerial device systems from operating at the same time. The selector valve also makes inadvertent operation of the stabilizers unlikely while the aerial device is deployed.

Once the fluid is directed into one system or the other, actuator valves, monitor valves, stack valves, and proportional directional control valves are used to direct and control the power in that system. These valves tend to be four- or five-way valves in order to accomplish two-directional control. This

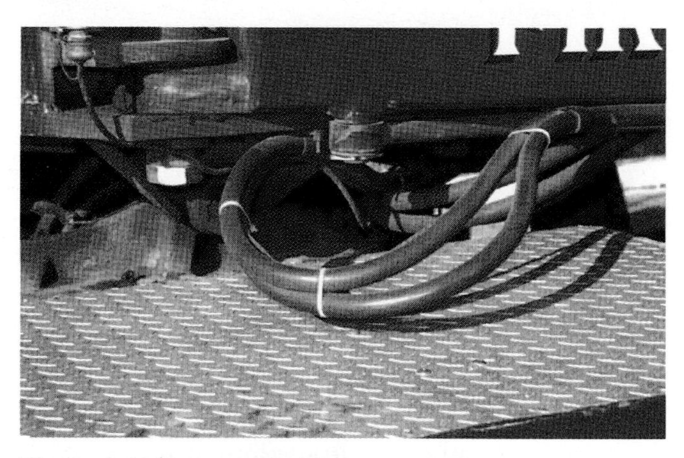

Figure 6.11 Hoses and tubing connect the various components of the hydraulic system.

Figure 6.12 A typical selector valve control.

final group of valves supplies fluid to the actuators, which are the devices that convert the fluid power developed in the system back into mechanical force. The actuators in these systems are hydraulic cylinders or motors (Figure 6.13). Hydraulic cylinders convert the energy in the system into linear mechanical force or motion. This is accomplished when pressurized hydraulic fluid is directed into a chamber created by fitting a piston into a cylindrical barrel (Figure 6.14). Hydraulic cylinders used for elevation and telescopic control of the aerial device and for operating the stabilizers are double acting. Double-acting cylinders are capable of receiving fluid under pressure from both sides of the piston so that force can be created in either direction. These cylinders have a piston sleeve that passes through the entire cylinder body and extends/retracts from both ends of the cylinder.

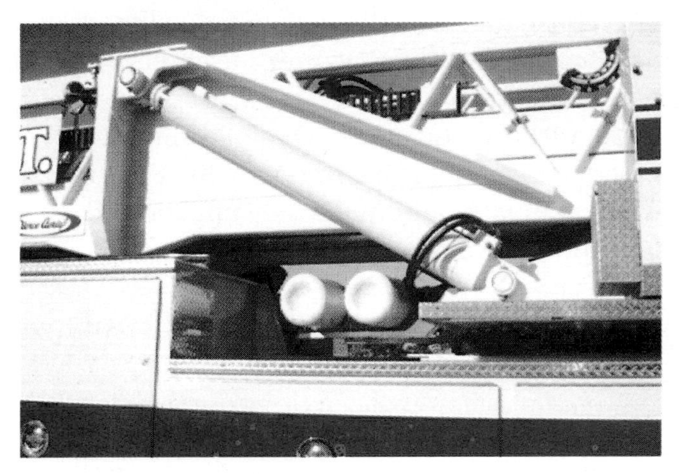

Figure 6.13 A hydraulic cylinder.

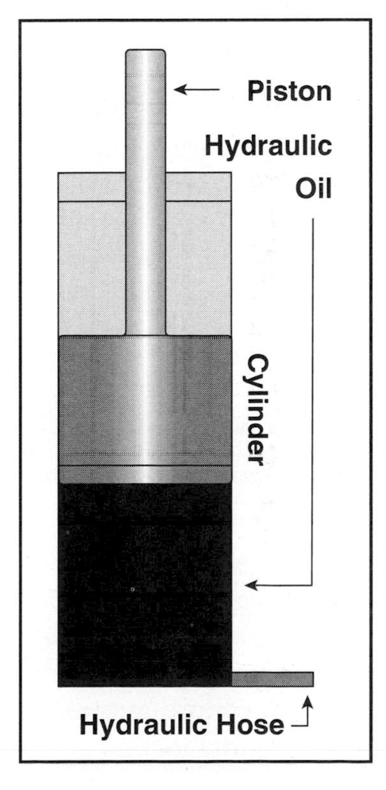

Figure 6.14 The parts of a hydraulic cylinder.

All manufacturers of aerial devices provide an auxiliary hydraulic pump for use in the event of a failure of the main hydraulic pump (Figure 6.15). The auxiliary pump is either a 12-volt DC electrically operated pump connected directly to the vehicle's battery or a 110-volt AC pump designed to be supplied from an invertor, generator, or ground shore. Either of these arrangements allows the pump to be operated even if the main vehicle engine goes down. These auxiliary pumps provide all the same motions as the main pump but at a reduced speed. However, the use of these pumps should be limited to bedding the aerial device after a main system failure occurs. They are not designed nor intended to be used for long periods of time. In fact, they should be operated only for intervals of about one minute, with an equal amount of rest

Figure 6.15 The auxiliary hydraulic pump is used if the main hydraulic system fails.

Figure 6.16 The aerial device may be manually rotated using a hand crank.

Figure 6.17 The PTO control is used to energize the aerial device and stabilization hydraulic system.

between operations. Auxiliary motors are subject to overheating if they are operated for longer periods of time. Some aerial devices — depending on the manufacturer of the device — are also equipped with hand cranks that rotate the turntable during hydraulic system failures (Figure 6.16). They must then rely on gravity to retract the aerial device.

Transferring Power to the Hydraulic System

In order to deploy the stabilizers and the aerial device, it will be necessary to transfer the power generated by the vehicle's engine from the road transmission to the hydraulic system. In order to accomplish this, the driver/operator will have to activate the power take-off (PTO) system. The PTO should not be engaged until the aerial apparatus has been properly positioned in accordance with the conditions on the scene. Chapter 5 of this manual discusses positioning in depth.

The PTO system provides the power to the hydraulic pump, which in turn creates a pressure on the fluid in the hydraulic system. The process of engaging the PTO is fairly similar for all types of vehicles. The engagement switch for the PTO system is usually located on the dashboard of the vehicle or between the driver and passenger/officer seats (Figure 6.17). The type of PTO activation switch varies from manufacturer to manufacturer and may be pneumatic, electric, hydraulic, or a combination of these three. Pneumatic systems require anywhere from 50 to 140 psi (350 kPa to 980 kPa) air pressure for the PTO to be operated, depending on the manufacturer. Regardless of the type of system used, the overall process for engaging the PTO is the same. The following sections describe the process based on a vehicle equipped with a manual or automatic transmission and with or without a fire pump.

Automatic Transmission Without Fire Pump

Use the following procedures for engaging and disengaging the PTO when the apparatus is equipped with an automatic transmission but no fire pump.

Engaging the PTO

Step 1: Set the parking brake and engage the tiller axle brake (if present) (Figure 6.18). If vehicle is equipped with a Jacobs engine brake, and if recommended by the manufacturer of the aerial device, turn the activating switch to the **OFF** position.

Step 2: Place transmission selector in the appropriate drive gear recommended by the manufacturer (Figure 6.19).

Step 3: Activate the PTO selector switch (Figure 6.20).

Step 4: Place the transmission selector in neutral. *If the PTO transfer is complete, the PTO indicator light should come on at this time* (Figure 6.21). On most makes of apparatus, completion of this procedure automatically releases the bed ladder locks.

Figure 6.18 Set the parking brake.

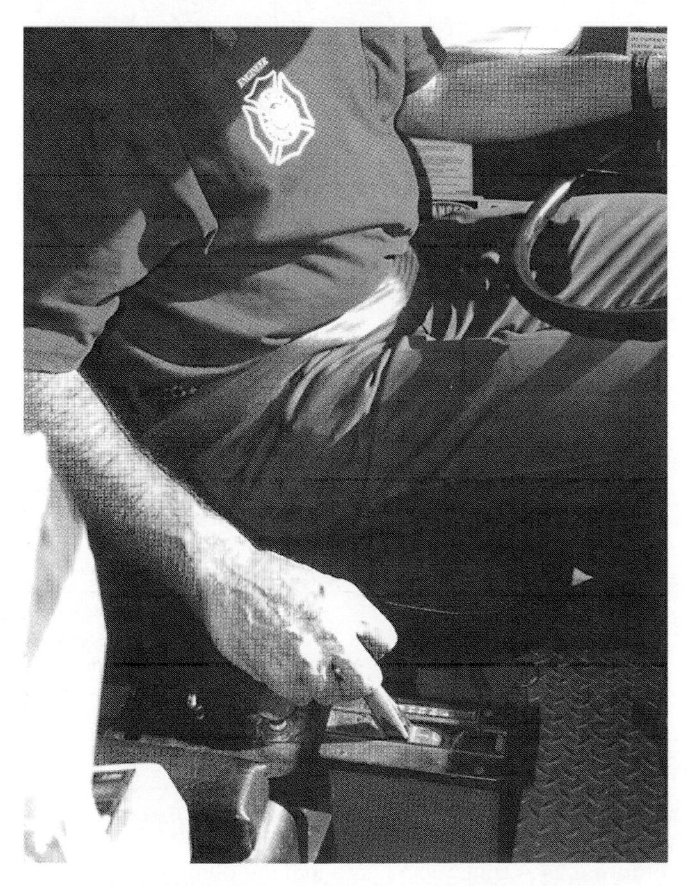

Figure 6.19 Place the transmission in the proper gear.

Figure 6.20 Activate the PTO control.

Figure 6.21 The PTO indicator light should illuminate if the transfer is complete.

Disengaging the PTO

Step 1: Return the engine to idle speed.

Step 2: Deactivate the PTO control switch while the transmission is in neutral.

Manual Transmission Without Fire Pump

Use the following procedures for engaging and disengaging the PTO when the apparatus is equipped with a manual transmission but no fire pump.

Engaging the PTO

Step 1: Set the parking brake and engage the tiller axle brake (if present). If vehicle is equipped with a Jacobs engine brake, and if recommended by the manufacturer of the aerial device, the activating switch should be turned to the **OFF** position.

Step 2: Depress the clutch pedal (Figure 6.22).

Step 3: Place the transmission in neutral.

Step 4: Activate the PTO selector switch.

Step 5: Release the clutch pedal. If transfer was complete, the PTO indicator light is lit. On most makes of apparatus, completion of this procedure automatically releases the bed ladder locks.

Disengaging the PTO

Step 1: Return the engine to idle speed.

Step 2: Depress the clutch pedal.

Step 3: Deactivate the PTO control switch.

Step 4: Release the clutch pedal slowly.

Automatic Transmission with Fire Pump in Use

Use the following procedures for engaging and disengaging the PTO when the apparatus is equipped with an automatic transmission and a fire pump.

Engaging the PTO and Fire Pump

Step 1: Engage the PTO. Use Steps 1 - 4 from procedure for Automatic Transmissions Without a Fire Pump. ***The PTO must always be fully engaged before engaging the fire pump.***

Step 2: Once the transmission is in neutral and required air or hydraulic pressure is present, operate the pump shift control as directed by the pump manufacturer (Figure 6.23).

Step 3: Place the transmission selector into the appropriate pumping gear. If the shift is complete, the pump control light should be lit and there should be a minimal reading on the speedometer (Figure 6.24).

Disengaging the PTO and Fire Pump

Step 1: Allow the engine to return to idle speed.

Step 2: Return the transmission selector to neutral.

Step 3: Deactivate the pump control switch.

Step 4: Deactivate the PTO control switch. The vehicle is now ready to drive.

Figure 6.22 Depress the clutch pedal with your left foot.

Figure 6.23 Operate the pump shift control if the fire pump will be used.

Figure 6.24 A light will indicate that the pump shift has been successfully made.

Manual Transmission with Fire Pump in Use

Use the following procedures for engaging and disengaging the PTO when the apparatus is equipped with a manual transmission and a fire pump.

Engaging the PTO and Fire Pump

Step 1: Engage the PTO. Use Steps 1 - 5 from procedure for Manual Transmissions Without a Fire Pump. The PTO must always be fully engaged before engaging the fire pump.

Step 2: Once the transmission is in neutral and required air or hydraulic pressure is present, operate the pump shift control as directed by the pump manufacturer.

Step 3: Depress the clutch pedal.

Step 4: Place the main transmission into the pumping gear (the direct drive gear).

Step 5: Release the clutch pedal. If the shift is complete, the pump control light should be lit and there should be a minimal reading on the speedometer.

Disengaging the PTO and Fire Pump

Step 1: Allow the engine to return to idle speed.

Step 2: Depress the clutch pedal.

Step 3: Return the transmission to neutral.

Step 4: Deactivate the pump control switch.

Step 5: Deactivate the PTO control switch. The vehicle is now ready to drive.

Some newer models of aerial apparatus are equipped with what is called a "hot-shift" PTO system. These systems can be engaged when the main transmission is in neutral. Hot shifting should only be done when the operator is sure that the vehicle is equipped with such a system. Attempting to engage a PTO from neutral in a vehicle not equipped with a hot-shift system could result in serious damage to the PTO and the main transmission.

Setting the Stabilizers

Once the PTO system has been engaged, the driver/operator may leave the cab of the apparatus to begin the actual deployment of the stabilizers and the aerial device. Some manufacturers recommend that the wheels of the apparatus be chocked at this time. This is particularly true of those types of apparatus that are not designed to be totally lifted off the ground once the stabilizers are set. Manufacturers who specify that their apparatus should be chocked also specify whether the front or rear wheels are the ones to be chocked. Whichever the case, the wheels should have appropriate chocks placed both in front of and behind the tire, on both sides of the vehicle (Figure 6.25). If any lifting of the apparatus is performed, the chocks should be removed before lowering the apparatus. Otherwise, the chocks may become jammed under the tires and will be impossible to remove without relifting the apparatus. Consult the specific manufacturer of your aerial apparatus to determine whether chocking is required. If the collapsible type of chocks are being used, make sure they are locked and properly deployed.

Preliminary Checks

Before the actual deployment of the stabilizers, the driver/operator should make several quick observations to ensure that some potential problems are

Figure 6.25 On level surfaces, the tires should be chocked front and back.

avoided. The first two checks that are necessary should have already been made: making sure that the PTO engagement light is lit before leaving the cab, and making sure that the wheels are properly chocked, if required. Next, the driver/operator should anticipate the expected travel path of the stabilizers to ensure that they do not strike anything as they are being deployed. The driver/operator should be observant of other vehicles, utility poles, hoselines, fences, signs, and *other personnel* operating on the scene (Figures 6.26 a and b). If there is any doubt as to the room for the stabilizers to deploy, a short pike pole or closet hook can be used to estimate the travel path and distance. If the apparatus was positioned properly, any fixed objects will be clear of the stabilizers. However, it is always possible for other fire personnel or civilians to be standing in the travel path of the stabilizers. If possible, a second firefighter should be positioned in the area of stabilizer deployment to keep people away from the stabilizers (Figure 6.27).

Once the stabilizers are in place, the driver/operator should check the surface they are resting on. Care should be taken to ensure that no loose objects, such as large rocks, fire hose, tools, underground utility access covers, or broken pavement, are in the area where stabilizer boots (sometimes also called the footplate or stabilizer pad) rest (Figure 6.28). If present, these objects should be moved, and the area receiving the stabilizer boots should be made as stable as possible. In cold weather conditions, the driver/operator should be alert for icy patches or frozen ground as a potential problem. Ice reduces the stabilizer boot's ability to make solid contact with the ground. Avoid ice, if possible. As ice melts, it leaves air between the stabilizer boot and the ground, compromising the stability of the apparatus. Therefore, the driver/operator must constantly monitor the stabilizers and continue to lower them for maximum contact with the ground. Frozen ground can also thaw from apparatus exhaust or water runoff and leave an unreliable surface for the

Figure 6.26a and b There are many potential obstacles to stabilizer travel that the driver/operator may encounter.

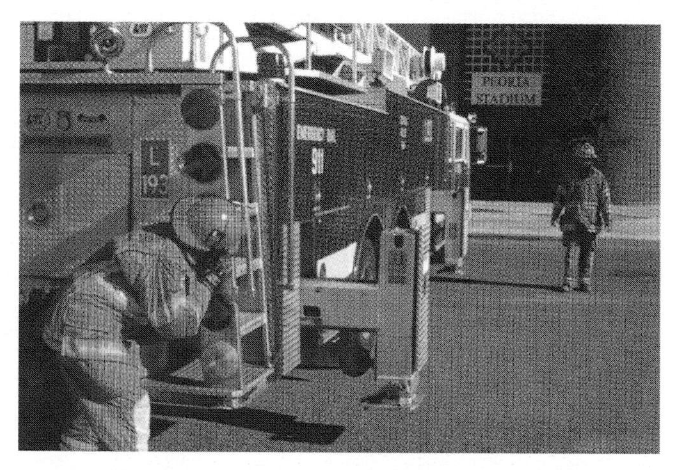

Figure 6.27 If possible, have a firefighter positioned to warn other people in the area of the deploying stabilizer.

Figure 6.28 Avoid placing the stabilizer on objects such as a large rock or hoseline.

Figure 6.29a Some apparatus have the selector valve located on the rear step area of the apparatus.

Figure 6.29b This apparatus has the selector valve located in a midship position.

stabilizer. In this case, it is necessary to add additional cribbing to the underside of the stabilizer boots to further distribute the weight load.

Operating the Selector Valve

Once the driver/operator is assured that the preliminary activities are successfully completed, the selector valve may be operated to provide hydraulic power to the stabilizing system. The selector valve is located either on the rear step area or midship on the driver's side of the vehicle, depending on the manufacturer of the apparatus (Figures 6.29 a and b). The physical operation of the valve varies, depending on the manufacturer (Figures 6.30 a through c). Whichever the type, the controls should be clearly marked as to the proper positions for each operation. At this time, the driver/operator should move the control to the stabilization position. This might also be called the neutral, outrigger, jack, or leveling position. On some makes of apparatus, it is the operation of the selector valve that automatically releases the aerial device bed locks.

Once the selector valve has been moved to the proper position, the apparatus is ready for stabilization to begin. The following sections highlight the principles of stabilization on both even and uneven terrain.

Stabilizing on Even Terrain

Stabilization is ideally performed on even terrain. Even terrain allows for the maximum stability of the apparatus and for the greatest range of safe movement of the aerial device. The stabilization

Figure 6.30a A typical selector valve.

Figure 6.30b A typical selector valve.

Figure 6.30c A typical selector valve.

controls are most commonly found on the rear step area of the apparatus; however, some manufacturers locate these controls midship on one or both sides of the vehicle (Figures 6.31 a and b). Generally, there is one control for each of the stabilizers, and more than one control can be operated at a time.

A few general principles of stabilizer controls should be pointed out before operation. Many of these controls are lever-type valves that move up to about 90° (Figure 6.32). The engine idle speed is regulated automatically by operating these controls or by a fast-idle toggle switch. This provides the power necessary to perform the desired action. Thus, when the driver/operator operates the control, an increase in engine speed (noticeable due to the increase in the sound level) occurs. Because operation of these valves also affects the vehicle's engine speed, rough, jerky operation of the valves should be avoided.

There are three basic ways that stabilizers move into position. There are those that extend straight down from the apparatus, fulcrum-types that swing down into position, and box-types that first extend outward and then down. Each operates a little differently.

The box-type is most common on newer apparatus (Figure 6.33). Box stabilizers are basically two-part devices. The first part is an arm that extends

Figure 6.31a These stabilizer controls are located on the rear step area of the apparatus.

Figure 6.31b Some stabilizer controls are located on the side of the apparatus.

Figure 6.32 A typical stabilizer control.

Figure 6.33 Box stabilizers typically raise the entire apparatus off the ground.

straight out from the side of the truck, parallel to the ground. Attached to the end of this extension arm is a stabilization jack that extends down to the ground to raise the vehicle. When operating this type of stabilizer, the first operation is to extend the parallel extension arms to their maximum travel distance, if possible (Figure 6.34). On some apparatus, the aerial device cannot be raised unless the extension arms are fully extended. If the design of the apparatus allows for partial extension when operating in restricted areas, as a minimum, the extension arms on the side that the aerial device is going to be operated over should be fully extended. Extension arms on the opposite side should be extended as far as possible. Remember that the load and range of safe movement capabilities are significantly decreased if the extension arms are not extended to their maximum positions. Consult your apparatus manufacturer's directions for recommended procedures in these situations.

Once the extension arms have been extended to their final positions, place the portable stabilizer pads directly beneath the stabilizer boots so that they will be in position when the jacks are lowered (Figure 6.35). Though some manufacturers or authorities may state that stabilizer pads are only necessary on unstable ground, ***IFSTA recommends stabilizer pads be used beneath the stabilizer boots every time the stabilizers are deployed.*** Once the pads are in position, the stabilizing jacks may be lowered onto them. When operating on level ground it makes no difference which side is lowered first. Lower both the stabilizers on one side of the vehicle first. The stabilizers should be lowered onto the pads until the slight amount of pressure applied lifts that side of the truck slightly (Figure 6.36). One good way to judge appropriate deployment is to note the tires on that side of the apparatus. When the bulge is removed from the tires, the stabilizers are resting on the pads properly. Once this is accomplished, the stabilizers on the other side of the vehicle can be lowered to the same point.

When the stabilizers have made contact with the ground, the driver/operator should alternate between the sides to raise the vehicle to its working position, as close to level as possible. Again, the amount it is to be raised depends on the manufacturer. Some stabilizers are designed to have all apparatus tires lifted from the ground, some to lift only the rear tires, and yet others are designed so that all tires remain on the ground.

Figure 6.35 Place the stabilizer pad beneath the stabilizer boot.

Figure 6.34 Extend the box stabilizer as far as it will go.

Figure 6.36 Slight pressure is applied to the stabilizers on one side of the apparatus.

Table 6.1 lists numerous manufacturers and their lifting requirements. Make sure that ***all stabilizers*** are in firm contact with the ground and bearing weight. Some types of apparatus may have safety interlocks that automatically engage once proper stabilization has been achieved.

Stabilization using fulcrum-type stabilizers or those that extend straight down from the chassis are similar to those previously described with the box type stabilizers. The primary difference is that these types of stabilizers can be deployed only in the fully extended position and cannot be deployed at all if obstructions are present. The method of deployment is the same for these types of stabilizers as it was for the box-type, with one exception. These stabilizers should be lowered to within a few inches of the ground prior to the placement of the stabilizer pads (Figure 6.37). This allows for correct positioning of the pads. The remainder of the procedure for stabilizing the apparatus is the same.

Table 6.1
Manufacturer's Lifting Requirements for Stabilization

Manufacturer	Recommendation
Baker	All wheels off ground
Bronto	All wheels off ground
Calavar	All wheels 10 inches (250 mm) off ground
Emergency One	All wheels 1-2 inches (25 mm to 50 mm) off ground
Hahn	Fully extended stabilizers
LTI 75-foot (23 m)	Rear wheels slightly off ground
LTI full-sized units	All wheels slightly off ground
Pierce/Smeal	All weight off axles
Snorkel	Raise rear step 2-3 inches (50 mm to 75 mm)
Sutphen	Front tires slightly raised, rear tires on ground
Thibault	All weight off axles

Stabilizing on Uneven Terrain

In some situations, setting up on uneven terrain is unavoidable. When this is necessary, the driver/operator is required to make the apparatus as stable as possible, realizing that the operation of the aerial device may be somewhat limited as compared to operation on level ground. There are two ways that the truck can be uneven: laterally (side to side) or longitudinally (end to end) (Figures 6.38 a and b). Of the two, the easiest to correct is lateral unevenness.

Figure 6.37 Place the stabilizer pad beneath the boot when it becomes apparent where the boot will come to rest.

Correction of lateral unevenness can be achieved on grades of up to 5 or 6 percent. A 5 percent grade means that there is a rise of 5 feet (5 m) for every run of 100 feet (100 m) (Figure 6.39). Driver/operators must be familiar with the requirements for their specific make of apparatus. Most manufacturers of aerial devices recommend that the stabilizers on the high side of the vehicle be lowered first (although some may specify the opposite). These stabilizers should be lowered only until solid contact is made with the ground (Figure 6.40). Then, the low side stabilizers should be lowered until they raise the low side of the truck level with the high side (Figure 6.41). The driver/operator must make sure that the truck has been raised enough to activate the interlocks, if the truck is equipped with them. If the interlocks have not activated, it is necessary to raise both sides of the truck until they do. After the apparatus is stabilized, it is preferable that the aerial device be operated over the high side of

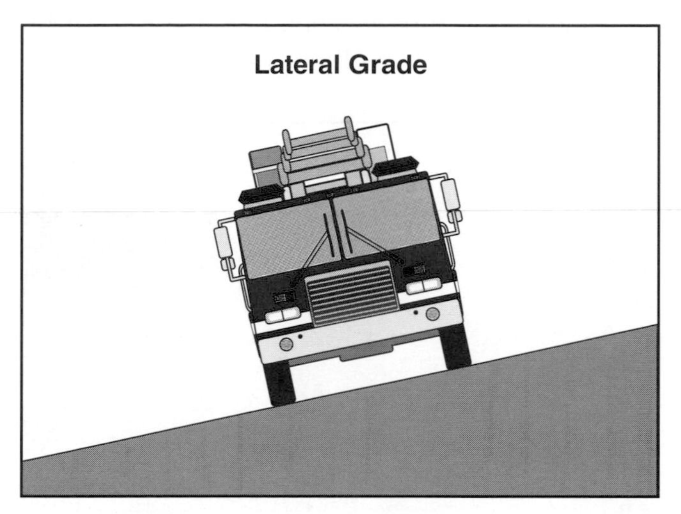

Figure 6.38a A lateral grade.

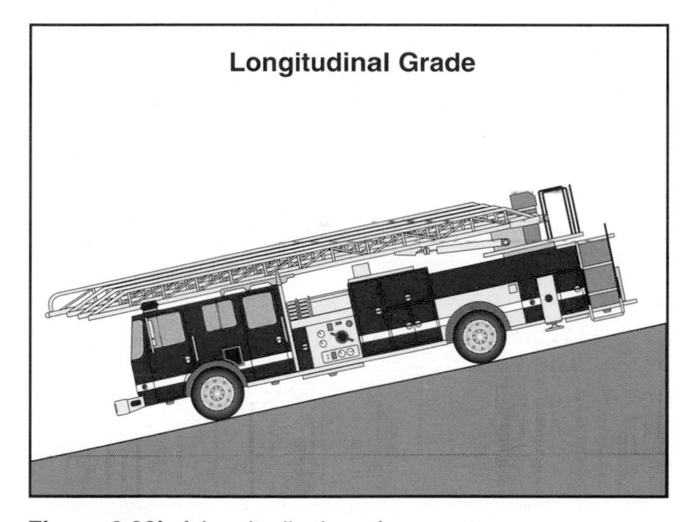

Figure 6.38b A longitudinal grade.

Figure 6.40 Lower the stabilizers on the high side of the apparatus until they are just in touch with the ground.

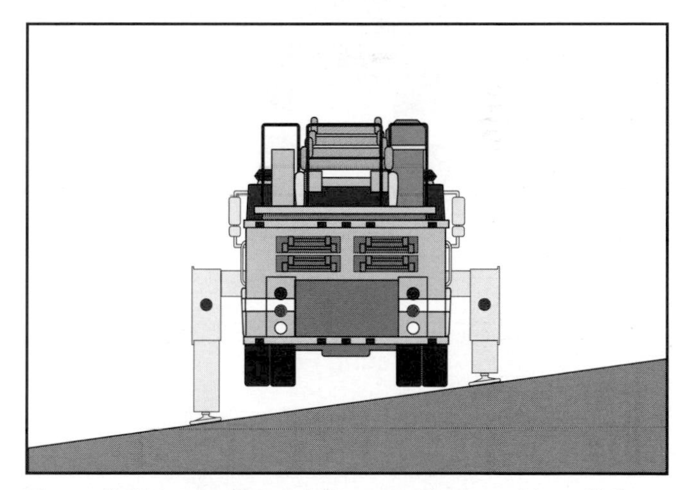

Figure 6.41 Lower the stabilizers on the low side until the apparatus is brought to level.

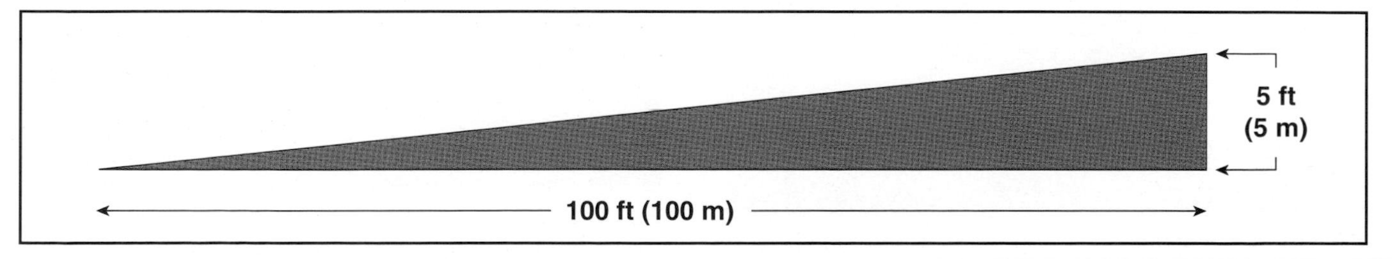

Figure 6.39 This example shows a 5 percent grade.

the apparatus (Figure 6.42). If this is not possible, remember that lesser amounts of aerial device extension and loading have to be used, based on manufacturer's recommendations.

Operating on a longitudinal (end-to-end) grade is a little trickier than operating on a lateral grade. If the aerial turntable is not leveled, the rungs of the aerial device may not be exactly parallel to the ground if the aerial device is operated over the side of the apparatus. This will generate lateral and torsional stresses in the aerial device when it is raised and loaded (Figure 6.43). This can be destructive to the ladder and very dangerous to personnel working on or around it. Each manufacturer has different requirements and maximum longitudinal grades for its apparatus. These vary from 6 to 14 percent. The driver/operator must consult the instructions for the particular device being operated to determine its maximum grade.

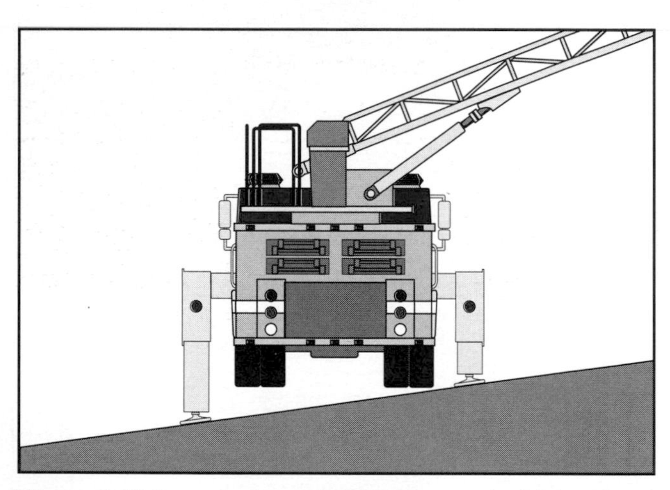

Figure 6.42 Operate the aerial device over the high side of the apparatus.

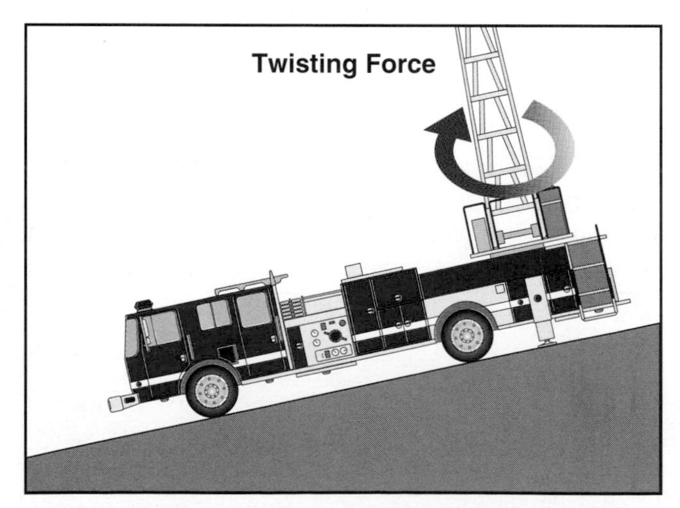

Figure 6.43 Operating on a longitudinal grade causes twisting forces on the aerial device.

When setting up on a longitudinal grade, the apparatus should be positioned to minimize the stresses on the aerial device due to the grade. The aerial device should be operated either directly over the front or rear of the apparatus, depending on whether the apparatus stopped short of the objective or pulled past it (Figure 6.44). The apparatus should not be placed crosswise on the grade. Articulating aerial devices should generally be operated off the rear of the vehicle whenever possible (Figure 6.45). The stabilizers should be used to level the apparatus as much as possible. It is more difficult to do this on longitudinal grades than on lateral

Figure 6.44 Telescoping aerial devices may be operated over the front or rear of the apparatus on a longitudinal grade.

Figure 6.45 Articulating aerial devices should be operated over the rear of the apparatus when parked on a longitudinal grade.

Figure 6.46 Do not set the stabilizer on a curb. In this case, it is much safer to simply move the truck away from the curb.

Figure 6.47 The stabilizer interlock light will illuminate when the truck is suitably stabilized.

grades. ***The aerial device should not be extended or retracted over the side of the apparatus when parked on a longitudinal grade.*** Check the manufacturer's requirements for aerial device loading on grades. Most have restricted loads, some as much as one-half the normal load, when operating on grades.

Stabilizing on Curbs or Other Obstructions

If possible, avoid placing stabilizers on curbs, sidewalks, parking blocks, water valve covers, or other obstructions. Unfortunately, sometimes such placement is unavoidable (Figure 6.46). These situations should be treated the same as setting up on a lateral incline. Placement of the stabilizers should be done in the same manner. Lateral incline load and extension restrictions also apply in this instance. Make sure that the object to be placed under the stabilizer is solid and capable of supporting the necessary weight. Be particularly cautious of crumbling curbs or parking blocks. Generally, crumbling curbs or parking blocks will not support the amount of weight necessary to stabilize the apparatus. The vehicle should be repositioned rather than deployed in an unsafe manner.

Locking the Stabilizers and Transferring Power to the Aerial Device

Once the stabilizers have been fully deployed, the driver/operator must ensure that they are locked and that they will stay in place. On most aerial

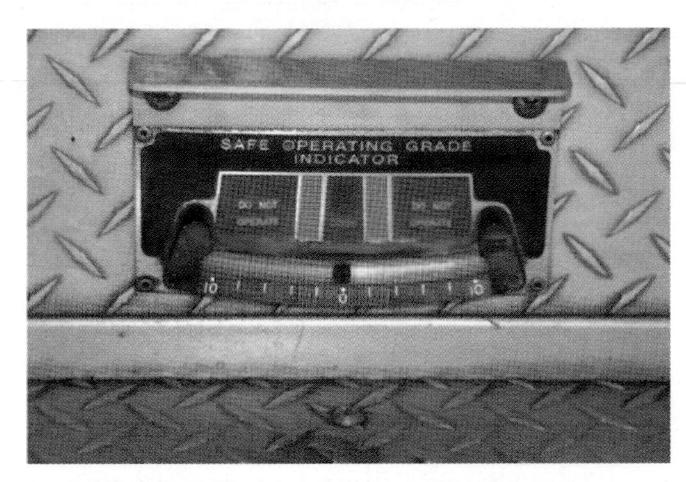

Figure 6.48 Most aerial apparatus are equipped with a leveling gauge that shows when the apparatus is level.

apparatus, the combination of hydraulic system holding valves and the interlock feature of the selector valve provide automatic locking capabilities. The interlock prevents the flow of hydraulic fluid into or out of the stabilization system while the aerial device is in use. The holding valves prevent the movement of fluid within the stabilization system. The holding valves are activated when the selector switch is moved to the aerial position. It may be necessary to set the holding valves by operating the stabilizer controls once in each direction ***after*** the selector valve has been moved to the aerial position.

Many newer apparatus are also equipped with an indicator light that shows when the interlocks are engaged (Figure 6.47). The apparatus may also be equipped with a site gauge level that also allows the driver/operator to easily determine if the apparatus has been raised evenly (Figure 6.48).

Aerial apparatus equipped with box-type stabilizers commonly have a series of holes drilled through the bottom post of the elevating jack (Figure 6.49). These holes are designed for the insertion of large steel pins that act as an additional safeguard against droppage of the jacks. Once the apparatus has been raised to the desired position, the pins should be inserted in the hole closest to the jack housing (Figure 6.50). Once all of this is done, the driver/operator is ready to mount the turntable and operate the aerial device.

Figure 6.49 Most box stabilizers have holes that allow safety pins to be inserted.

Raising the Stabilizers

When the use of the aerial device is complete and it is returned to its bed, the stabilization process is reversed in order to prepare the apparatus for departure from the scene. Once the aerial device is in its bed, the driver/operator should remove the safety pins from the jacks (if present) and return the selector valve to the stabilization position. The wheel chocks should also be moved slightly away from the tires at this time so that the apparatus is not set down on top of them. The driver/operator should also check to make sure that all personnel are clear of the apparatus before raising the stabilizers. When the pressure is released on the stabilization system, the apparatus often drops very quickly. Persons standing near the apparatus could be struck by an extending portion of the vehicle or the rising stabilizer; a tire or portion of the vehicle's body could drop on them. Once these steps have been taken, the stabilizers may be raised. On level ground, the order in which the stabilizers are raised is unimportant. On uneven terrain, the stabilizers should be raised in reverse order of lowering. Once the raising procedure is complete, return the selector valve to the neutral position, and disengage the PTO system. Do not forget to stow the stabilizer pads in their appropriate storage location.

Figure 6.50 Insert the pin in the closest hole once the stabilizers are deployed.

Manual Stabilizers

Some older pieces of aerial apparatus still in service may be equipped with manual stabilizers. Older midship and tractor-trailer aerials are the most likely to have manual stabilizers (Figure 6.51). Manual stabilizers are similar to the box-type hy-

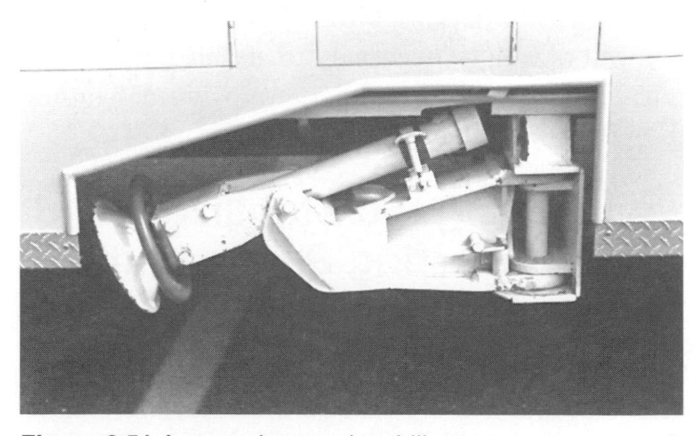

Figure 6.51 A stowed manual stabilizer.

draulic stabilizers described earlier in this chapter. They consist of an extension arm that has a screw jack attached to the end of it. The primary difference is that with the manual setup, the extension arm must be swung into position by hand; also the screw jack attached to the arm must be manually lowered.

The manual stabilizers are swung into position a little differently on midship and tractor-trailer apparatus. On midship aerials, the extension arm is stored perpendicular to the ground. To place this stabilizer in service, the driver/operator must first release the extension arm pin that holds it in the vertical position. Once the extension arm is lowered to the horizontal position, reinsert the pin through the extension beam and arm (Figure 6.52). The stabilizer pad should then be positioned under the jack, and the screw portion of the jack should be turned to lower the jack onto the plate (Figure 6.53). When the stabilizer boot rests firmly on the pad, repeat the process for the other side of the apparatus. *If the stabilizer lifts off the pad during aerial operations, do not adjust it.* This is normal load shifting.

On tractor-trailer aerials, the extension arm is stored in a recessed position beneath the gooseneck portion of the trailer (Figure 6.54). To put the extension arm in position, remove the bayonet pins holding the jack and the extension arm in place. Then swing the jack away from the vehicle and pull the sliding extension arm as far as it goes (Figure 6.55). Allow the jack to hang down to the ground and position the stabilizer pad beneath it. Lock the extension arm in place by replacing the pins. The jack is then lowered in the same manner as previously described for midship aerials. Stowing the stabilizers is accomplished by reversing this process.

Figure 6.52 Insert the pin between the extension arm and the beam.

Figure 6.53 Screw down the jack portion of the stabilizer.

Figure 6.54 Older tiller apparatus may have manual stabilizers.

Figure 6.55 Pull the sliding arm out as far as possible.

Stabilizing Tractor-Trailer Aerial Apparatus

While most of the principles discussed in this chapter apply to tractor-trailer aerial apparatus, one additional aspect must be applied to these special trucks. A critical amount of stability is given to tractor-trailer apparatus by the manner in which it is parked at the scene. By placing an angle between the tractor and trailer portions, the stability can be dramatically increased over the amount provided by the stabilizers (Figure 6.56). Maximum stability occurs when the angle of the tractor is 60° from the centerline of the trailer, and the aerial ladder is operated toward the outside of the angle. Angles less than 60° provide lesser stability, as do angles from 61° to 90°. The vehicle should never be positioned with the tractor and trailer at an angle greater than 90° as poor stability is produced and damage to the truck may occur. Many tractor-trailer vehicles are equipped with an alarm that sounds if an attempt is made to create an angle greater than 90°. Figure 6.57 shows several angles for positioning tractor-trailer apparatus.

The best way to put the vehicle into this angled position is to approach the intended objective until the turntable is right next to it. The wheels of the tractor should be cut sharply toward the center of the street (away from the objective) and the truck pulled slightly forward. Then the tiller wheels are cut sharply toward the building, and the truck is backed up slightly toward the building. This pushes the turntable toward the desired objective, and it produces the desired angle between the tractor and the trailer. Once in position, all brakes should be applied, including the one for the tractor axle, and further stabilization, as described earlier in this chapter, may be employed.

Figure 6.56 The tiller is parked at an angle to increase stability.

It should be noted that these techniques for jackknifing tillered aerial apparatus were more important on older models of apparatus than they are on more modern apparatus. Most modern manufacturers of tillered aerial apparatus equip the rigs with stabilization systems that are perfectly capable of securely stabilizing the apparatus, whether it is jackknifed or not (Figure 6.58). Check your manufacturer's recommendations for more specific guidelines.

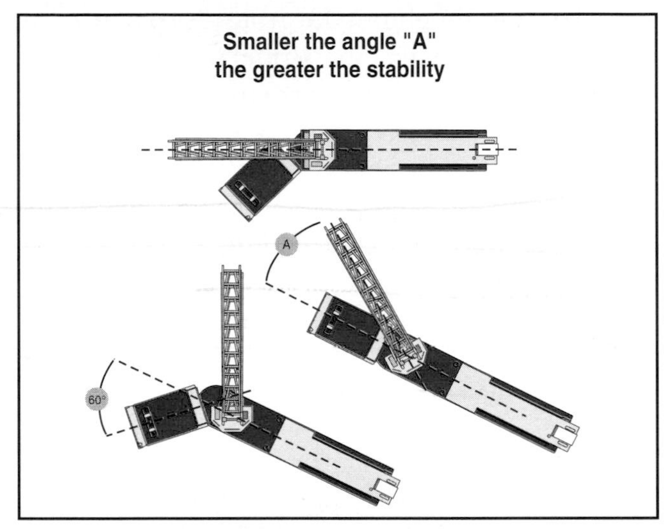

Figure 6.57 The appropriate angles for parking a tiller apparatus when the aerial device is deployed.

Figure 6.58 Modern tiller apparatus do not have to be jackknifed in order for the aerial ladder to be safely deployed.

Chapter 7

Operating Telescoping Aerial Equipment

Job Performance Requirements

This chapter provides information that will assist the reader in meeting the following job performance requirements from NFPA 1002, *Standard on Fire Apparatus Driver/Operator Professional Qualifications*, 1998 edition. Particular portions of the job performance requirements (JPRs) that are met in this chapter are noted in bold text.

4-2.3 **Maneuver and position the aerial device from each control station, given an incident location, a situation description, and an assignment, so that the aerial device is properly positioned to safely accomplish the assignment.**

(a) ***Requisite Knowledge*: Aerial device hydraulic systems, hydraulic pressure relief systems, gauges and controls, cable systems, communications systems, electrical systems, emergency operating systems, locking systems, manual rotation and lowering systems, stabilization systems, aerial device safety systems, system overrides, safe operational limitations of the given aerial device, safety procedures specific to the device, and operations near electric hazards and overhead obstructions.**

(b) ***Requisite Skills*: The ability to raise, rotate, extend, and position to a specified location and the ability to lock, unlock, retract, lower and bed the aerial device.**

4-2.4 **Lower an aerial device using the emergency operating system, given an aerial device, so that the aerial device is safely lowered to its bedded position.**

(a) ***Requisite Knowledge*: Aerial device hydraulic systems, hydraulic pressure relief systems, gauges and controls, cable systems, communications systems, electrical systems, emergency operating systems, locking systems, manual rotation and lowering systems, stabilization systems, aerial device safety systems, system overrides, safe operational limitations of the given aerial device, safety procedures specific to the device, and operations near electric hazards and overhead obstructions.**

(b) ***Requisite Skills*: The ability to rotate and position to center, unlock, retract, lower, and bed the aerial device using the emergency operating system.**

The purpose of this chapter is to familiarize the driver/operator with the operation of telescoping aerial devices. These devices include aerial ladders, aerial ladder platforms, telescoping elevating platforms, and telescoping water towers. Included are recommended procedures for placing the aerial device in position, raising and lowering the aerial device, and operating the aerial device under adverse conditions.

Aerial apparatus driver/operators and company officers must be thoroughly trained in operating the apparatus to which they are assigned. The general operating practices, limitations, and safety measures covered in this chapter provide a realistic approach to the operation of telescoping aerial apparatus. Safe, smooth, and efficient operation of these aerial devices is best ensured through a training program that includes practice, classroom study, and drills. These lessons should include use of technical data, preparation for use, general positioning of the apparatus, operating the normal and emergency controls, use with fire streams, and use during rescue.

All fire department aerial devices have operation levers or control devices that the driver/operator uses to make the aerial function. These controls may differ in certain respects, as do the controls of different automobiles. A driver/operator does not need to be a mechanical engineer or know the mechanical aspects of operation. However, the operator must be well trained in the location of each control, its function, how it operates, when it should be used, and the appropriate safety precautions. Driver/operators should also understand the function and operation of all stabilizing power and safety devices. In addition, the operator must know the limitations of the apparatus, proper spotting angles, maximum load limits, and safe operating operations. The importance of the operator being well versed in the layout and operation of the aerial device controls cannot be overemphasized. Early portions of any training program should be dedicated toward this goal. If the driver/operator may be expected to drive and operate more than one type of aerial apparatus, he should be trained on each of the different apparatus.

Proper positioning of the aerial apparatus was covered thoroughly in Chapter 5 of this manual. However, before any discussion of raising aerial devices, it is important to reemphasize several positioning points. The most important thing to remember is to make sure that the apparatus is positioned so that stress on the aerial device is minimized. Place the apparatus as close to the target as safely possible. Earlier-arriving engine companies or other apparatus should not have blocked the aerial apparatus from obtaining an optimum position. Make sure that the turntable is aligned with the target. Lastly, whenever possible, avoid positioning the apparatus in a location that will subject the aerial device to overhead obstructions, particularly power transmission lines.

NOTE: The following procedures for operating telescoping aerial devices assume that the apparatus has been properly positioned on a hard, level surface in negligible winds and nonfreezing conditions. Information on operating in adverse conditions is contained later in this chapter.

WARNING!

Prior to the application of the information contained in this chapter, it is important to properly stabilize the apparatus in accordance with manufacturer's recommendations and Chapter 6 of this manual. Failure to properly stabilize the apparatus before performing the applications covered in this chapter could result in tipping over the apparatus.

Raising the Aerial Device

To raise the aerial assembly means to elevate the entire assembly from its stored position toward a vertical position of use (Figure 7.1). The raising of the aerial device is actually a series of motions that include elevating, rotating, extending, and lowering the aerial device to its objective. Experienced aerial device operators may be able to perform some of these functions simultaneously; however, new operators should take one motion at a time.

The instructions in this section pertain solely to the actual deployment of the aerial device itself. Complete the following tasks before deploying the aerial device:

- Place the apparatus in the proper position for maximum efficiency of use. Properly apply all parking brakes (see Chapter 5).

Figure 7.1 The aerial device must be raised from its bed in order to be useful at the emergency scene. *Courtesy of Ron Jeffers.*

- Place the apparatus transmission into the proper gear, and operate the power take-off to provide power to the aerial device's hydraulic system (see Chapter 6).
- Fully deploy the stabilizers according to manufacturer's recommendations (see Chapter 6).
- Switch the selector valve from the stabilizer position to the aerial device position so that hydraulic power is provided for aerial operations (see Chapter 6) (Figure 7.2).

With the preceding tasks accomplished, the aerial device may be placed into operation. As described in the previous chapter for the stabilizer controls, the operation speed of the aerial device is directly dependent on the amount the control lever is pushed or pulled (Figures 7.3 a and b). Newer operators should avoid full-speed operations until they are extremely familiar with the operation of the aerial device. Full-speed operations, regardless of the skill of the operator, should be minimized due to the large amount of dynamic stress placed upon the aerial device when the motion is halted.

Figure 7.2 Move the selector valve to the aerial device position.

The following step-by-step procedure may be used to raise the aerial device to its working position. Some of these steps may be omitted depending on the exact type of aerial device being raised. However, the order for the remaining steps will remain the same. Also keep in mind that references to climbing the aerial device or working from the ladder are generally not applicable to elevating platform apparatus.

Step 1: *Release the hold-down locks.* Hold-down locks secure the aerial device in its cradle when the apparatus is in the road travel

Figure 7.3a The aerial device's speed is proportional to the amount the control lever is pushed.

Figure 7.3b The aerial device will move at maximum speed when the control lever is pushed as far as possible.

mode (Figure 7.4). They prevent bouncing of the aerial device that could lead to structural damage. On most modern aerial trucks, the hold-down locks automatically release when the power take-off is engaged, when the stabilizers are fully deployed, or when the selector valve is switched from the stabilizer to the aerial device position, depending on the manufacturer. On older model ladders, these locks may have to be manually released. This should be done according to the manufacturer's directions.

Figure 7.4 The hold-down locks prevent the aerial device from bouncing during road travel.

Step 2: *Move tiller operator's station.* This step is necessary only on tractor-tiller aerial ladders and generally only on the older ones. On these rigs, it is necessary to relocate all or a portion of the tiller operator's control station in order for the ladder to be raised. On some units, the whole tiller station is flipped to the side (Figure 7.5). On others, it is only necessary to remove the steering column (Figure 7.6).

Step 3a: *Attach ladder pipe and make hose connections.* If the aerial ladder is not equipped with a permanently attached waterway system and the use of an elevated master stream is required, it is necessary to install the portable ladder pipe and hoseline on the ladder before its elevation (Figure 7.7). This is the safest time to make these connections. For more information on this procedure, see Chapter 9 of this manual. If the aerial apparatus is equipped with a piped waterway system, the driver/operator may choose this time to connect supply lines to the apparatus (Figure 7.8). Piped waterway systems may be supplied from another apparatus equipped with a fire pump or through direct piping from a fire pump on the aerial apparatus (if so equipped).

Step 3b: *Allow personnel to board the elevating platform.* If the apparatus is an elevating platform, personnel who will be working from the platform may enter at this time (Figure 7.9). This eliminates the need for

Figure 7.5 Flip the tiller station to the side of the ladder.

Figure 7.6 Pull the steering column from its mounting.

Figure 7.7 Install the ladder pipe if an elevated master stream is going to be deployed. *Courtesy of Ron Jeffers.*

Figure 7.8 Connect the supply hose to the waterway inlet.

firefighters to climb a raised aerial device and then enter the platform. There are several ways that the platform may be entered. Firefighters may walk down the ladder to get to the platform, or they may use some other portion of the apparatus to reach the platform. Each occupant should connect to a manufacturer-approved portion of the railing with an approved safety harness. Once personnel are in the platform, it may be operated from either the ground control station or the platform control station. In some jurisdictions, this step may be skipped until the aerial device is put into operation. The platform is then moved to the ground so that firefighters may board without having to climb on the apparatus.

Some jurisdictions prefer that the firefighters in the platform have primary control of the aerial device. This is because they are in the best position to place the platform in the most desirable operational

Figure 7.9 Firefighters may board the platform before it is lifted from its bed.

location. However, even if the firefighters in the platform will be controlling the aerial device, an operator should be stationed at the lower control pedestal at all times. The operator at the lower control pedestal can override the firefighters in the platform if they are approaching a hazard that is unseen to them or in the event they become incapacitated.

Step 4: *Check the intended path of the aerial device for obstructions.* This consideration should have been covered during the positioning of the apparatus. However, occasionally obstacles not foreseen during positioning will come to light when the driver/operator assumes the aerial device operator's control position (at either the platform control station or the turntable control position). One last visual check should be made along the pending path of travel of the aerial device to make sure that no obstructions are present.

Figure 7.10 Elevate the aerial device.

Step 5: *Elevate the aerial device.* Operate the appropriate control to elevate the aerial device from its bedded position (Figure 7.10). The aerial function controls on most modern aerial devices are directly linked to the vehicle's engine throttle, so there will be a marked increase in engine speed (noise) when the control is operated. This noise should not be of concern to the operator. Lift the aerial device to an elevation that is slightly higher than that required to reach the intended target.

WARNING!

No personnel should be on the aerial *ladder* while the aerial device is being *elevated.* Personnel on the moving ladder dramatically increase the stress created on the aerial device. Personnel are also subject to injury caused by a loss of footing and the resultant fall.

Step 6: *Rotate the aerial device.* Rotate the aerial device until the tip of the device is in line with the intended target (Figure 7.11). Rotation should be done at a smooth, controlled speed. If high speeds are used during rotation, the aerial device should be gradually slowed before bringing it to a complete stop. Going directly from a high-speed rotation to a complete stop places an enormous amount of momentum-related stress on the aerial device and the rotational system.

WARNING!

No personnel should be on the aerial *ladder* while the aerial device is being *rotated.* Personnel on the moving ladder dramatically increase the stress that is created on the aerial device. Personnel are also subject to injury caused by a loss of footing and the resultant fall.

Figure 7.11 Rotate the aerial device until it is in line with the intended target.

Figure 7.12 Extend the aerial device until it is slightly above the target.

Step 7: *Extend the aerial device.* Extend the fly sections of the aerial device until the tip of the device is slightly above the intended target (Figure 7.12). New driver/operators may find it easier to sight along the top rails of the aerial device as it is being extended. This provides an extra margin of safety as the device is being extended. With experience, the driver/operator may choose to sight along the beams as his confidence in operation increases.

WARNING!

No personnel should be on the aerial *ladder* while the aerial device is being *extended.* Personnel on the moving ladder dramatically increase the stress created on the aerial device. Personnel are subject to injury caused by a loss of footing and the resultant fall. Personnel are also subject to crushing injuries should a limb be caught between the moving ladder rungs or cross braces.

Step 8: *Lower the aerial device to the objective.* Following extension, the aerial device may be lowered into the desired work position (Figure 7.13). Most manufacturers do not recommend that their aerial device be used in the supported position. Therefore, position the aerial device about 4 to 6 inches (100 mm to 150 mm) above the surface of the objective (roofline, windowsill, etc.) and allow to settle onto the surface from the weight of the firefighters climbing the aerial device (Figures 7.14 a and b). Take care not to damage any waterway system components when lowering the device. Make sure that the rung alignment indicator shows that the rungs on aerial ladders are in proper position for climbing.

Figure 7.13 Lower the aerial device into the objective.

Figure 7.14a The aerial ladder should be 4 to 6 inches (100 mm to 150 mm) above the target.

Figure 7.14b It is acceptable for the aerial device to settle onto the target as firefighters near the tip.

Step 9: *Activate all aerial device locks.* Once the aerial device is in position, activate all locks that are present. This includes hoisting cylinder (elevation) locks, rotational locks, and extension fly locks (Figures 7.15 a through c). On some models of aerial ladders, it may be necessary to slightly retract the ladder in order for the extension lock to properly seat and release tension from the cable or chain. The driver/operator should be able to visually confirm that these locks are seated. No personnel should be allowed on the ladder until the locks are engaged.

In particular, seating the extension locks is an important step in making the aerial ladder a rigid structure. This increases the overall strength of the device and greatly reduces the chance of an aerial device failure. No failure of an aerial ladder with properly seated extension locks, an elevation of at least 50°, and proper loading has ever been recorded.

Step 10: *Climb the aerial device.* Make sure that the rated load capacity of the aerial device is not exceeded. The waterway system or hoselines can be charged at this time as well. For more information on elevated master stream operations, see Chapter 9 of this manual.

Lowering the Aerial Device

The process of lowering the aerial device is the exact reverse of the process used to place it into position. The following section details the lowering procedure.

Step 1: *Remove personnel from the aerial ladder.* Do not attempt any part of the lowering process with personnel on the aerial lad-

Figure 7.15a The elevation lock.

Figure 7.15b The rotation lock.

der. If the device is equipped with an elevating platform, it is acceptable for personnel to remain on the platform during the lowering process.

> **WARNING!**
>
> **No personnel should be on the aerial *ladder* while the aerial device is being *retracted*. Personnel on the moving ladder dramatically increase the stress created on the aerial device. Personnel are subject to injury caused by a loss of footing and the resultant fall. Personnel are also subject to crushing injuries should a limb be caught between the moving ladder rungs or cross braces.**

Step 2: *Drain the waterway system.* It is best to drain the waterway system before any movement of the aerial device (Figures 7.16 a and b). This reduces the amount of stress on the aerial device as it is being lowered. It also prevents the buildup of excessive pressures within the piping system as the sections are retracted. Excess pressure can cause damage to the seals located between the sections. In order for the system to be properly drained, open both ends of the system (the nozzle and the intake valve). At least one end of the system should remain open during the lowering process to avoid air from becoming pressurized within the system.

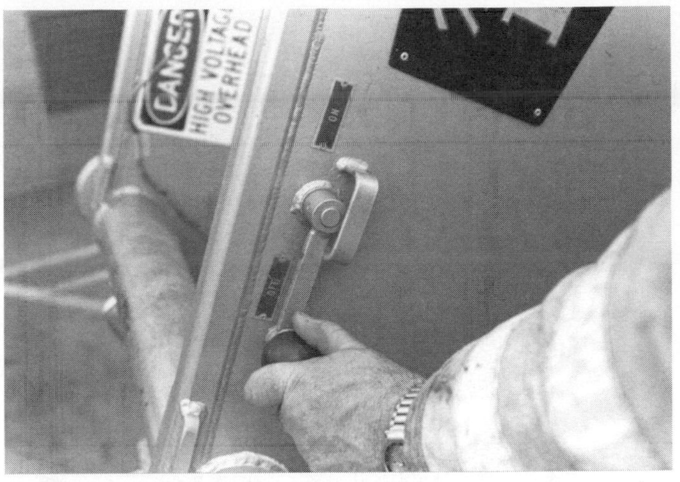

Figure 7.15c The extension lock.

Step 3: *Disengage the aerial device locks.* Disengage all locking devices so that the aerial device will be ready for lowering. These include hoisting cylinder (elevation) locks, rotational locks, and extension fly locks. It is again important to emphasize that no personnel should be allowed on the ladder after the locks are disengaged. It may be necessary to slightly extend some aerial ladders in order to unseat the extension locks. Once the lock is unseated and released, the aerial ladder may then be retracted.

Step 4: *Raise the aerial device away from its objective.* In order to avoid damage to the tip of the ladder or the platform, the aerial device should be raised slightly from its working position to ensure clearance (Figure 7.17).

Figure 7.16a Fixed waterway systems are equipped with drain controls.

Figure 7.16b Open the drain on the siamese appliance to drain the hose when a portable ladder pipe is used.

Figure 7.17 Raise the aerial device away from its objective.

Step 5: *Retract the aerial device.* Operate the controls necessary to fully retract all sections of the aerial device (Figure 7.18). In some cases, it may be necessary to use the "power assist" retraction control when the aerial device is at a low angle or ice has built up on the device. However, make sure to slow the retraction speed of the aerial device as it approaches the stops in the fully retracted position. Powering down against the stops at a high speed may cause structural damage to the device or the stops.

Step 6: *Check the intended path of the aerial device for obstructions.* Before rotating the aerial device, the operator should recheck the intended path of travel to make sure that it is free of obstructions.

Step 7: *Rotate the aerial device.* Rotate the aerial device until it is positioned directly above its travel cradle. Depending on the manufacturer of the apparatus, there are several ways of ensuring that the ladder is prop-

Figure 7.18 Fully retract the aerial device before rotating.

Figure 7.19 The turntable alignment indicators show when the device is positioned for stowing.

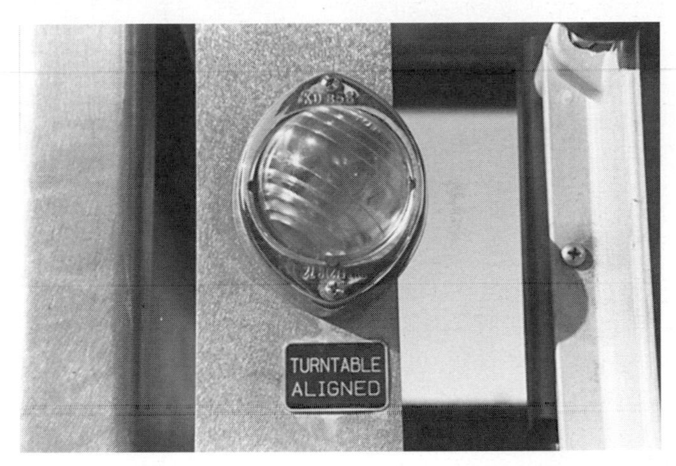

Figure 7.20 Some apparatus may have a turntable alignment light.

erly positioned for storage. Some manufacturers paint corresponding markers (usually arrows) on the base of the turntable and some portion of the apparatus adjacent to the turntable. When the two markers are lined up, the aerial is in position for stowing (Figure 7.19). Other manufacturers use sensors that are connected to a light on the control panel that illuminates when the aerial device is in position for stowing (Figure 7.20). On aerial ladders and water towers that are equipped with piped waterway systems, make sure that the nozzle is in the stowed position before proceeding to the next step.

Step 8: *Lower the aerial device.* With the aerial device lined up above its resting cradle, it may be lowered into the stored position. Slow the lowering speed as the aerial device approaches the resting cradle. Regardless of what the alignment indicators may show, make sure that the ladder is properly aligned within the arms of the cradle (Figure 7.21). Once proper alignment is confirmed, gently lower the device onto the rest. Some manufacturers recommend that once the device is on the rest, the lowering control should be operated with a quick downward motion to ensure that the device is firmly seated on the rest and that the hold-down locks will engage.

Step 9: *Allow personnel to exit the platform.* Any personnel located in the elevating platform may exit at this time.

Step 10: *Remove ladder pipe, hose, and associated equipment.* This applies only to aerial ladders not equipped with permanent waterway systems. Remove and properly stow all elevated master stream equipment before moving the apparatus.

Figure 7.21 Slowly lower the aerial device into its cradle.

Step 11: *Replace the tiller operator's station.* This step is only necessary on tractor-tiller aerial ladders, and generally only on the older ones. The tiller operator's controls must be put back into a state of readiness for road travel — a procedure that varies depending on the manufacturer of the apparatus.

Step 12: *Activate the hold-down locks.* On most modern apparatus, this automatically occurs when the selector valve is switched from the aerial device to the stabilizer position, or when the power take-off is disengaged, depending on the manufacturer. Older models may require manual activation of the hold-down locks. Consult the manufacturer's specific instructions for this procedure. However, in any case the driver/operator should visually confirm that the locks have activated.

When the aerial device has been lowered and stowed, the stabilizers may be raised, the power take-off disengaged, the fire pump disengaged (if so equipped and in use), and the truck readied for road travel.

◆ Operating Under Adverse Conditions

The previously described procedures for operating telescoping aerial devices assume that the apparatus is positioned properly on a hard, level surface, with negligible wind conditions and above-freezing temperatures. These are the safest conditions under which an aerial device may be operated. Unfortunately, it is a regular occurrence that one or more of these conditions may be working against the driver/operator of the aerial apparatus. In order to ensure the safety of the firefighters working on or around the aerial device and to ensure the physical soundness of the aerial device, the driver/operator must understand its limitations when operating under less than ideal conditions. The following sections contain information that will assist the driver/operator in understanding the special precautions that must be taken when operating the aerial device under adverse conditions.

Low-Angle Operations

Most aerial ladder failures or apparatus overturning incidents involve a combination of overloading and overextension of the ladder at low angles of elevation. Any aerial device is subject to the potential for a catastrophic failure when the ultimate capacity of one or more critical members or connections in the assembly is exceeded. These situations may be as a result of bending stresses, twisting stresses, or a combination of both.

The NFPA's design criteria for aerial ladders were upgraded in 1991. Aerial ladders built since 1991 are required to safely support a minimum load of 250 pounds (113.4 kg) at any position the ladder will reach. However, many ladders built before 1991 do not meet this requirement. Those aerial ladders have varying load and extension limitations. In some cases, it is operationally possible to place one of these "older design" aerial ladders into a position where the ladder was not designed to support *any* load whatsoever. When this situation occurs, the possibility of a catastrophic failure of the device is very real. The driver/operator must be absolutely sure of the limitations of the aerial device he is operating before deploying it under any "normal" or adverse conditions.

Several documented aerial ladder failures can also be traced to situations where a ladder pipe had been in operation with the aerial ladder placed at a low angle of elevation. The failures occurred shortly after the water to the ladder pipe was shut off. As long as water is flowing, the nozzle reaction from the ladder pipe helps to counteract the force of gravity and assists in supporting the weight of the ladder. However, when the water is shut off, the extra support is suddenly lost and the ladder may be too weak to support the weight of the ladder pipe and hose itself. This can result in a failure of the ladder. Some departments choose to supply aerial devices from two different pumpers so that if one pumper loses water, there is not a complete immediate loss of water to the aerial device. This reduces the chance of damage created by the sudden shut off of water.

Other aerial device failures have been noted when the device was deployed to perform special rescues at very low elevations. These include water rescues and below grade rescue incidents. Many of the older, light duty aerial ladders do not have the load capacity to be used under these conditions. Again, the driver/operator must know the exact capabilities of the device he is operating before attempting these maneuvers.

Operating on a Grade

As explained in Chapter 6 of this manual, operation of the aerial device when the apparatus is parked on a grade greatly compounds considerations concerning force on the aerial device system. Situations where the rungs of the aerial ladder are not parallel to the ground result in a twisting force on the entire aerial device. Aerial devices are not designed to routinely handle these twisting forces. The driver/operator must be conscious of these conditions and knowledgeable of the limitations of the particular aerial device being operated under these conditions.

Before the development of NFPA 1904, *Standard for Aerial Ladder and Elevating Platform Fire Apparatus* in 1991, no standards existed for the performance of aerial devices when operated on a grade. Thus, apparatus constructed prior to the adoption of this standard have a wide range of capabilities under sloped conditions. In general, most are capable of carrying their full-rated load capacities on grades of up to 6 percent (lateral or longitudinal); however, you should check the manufacturer's recommendations for the device you are operating. Past this point, the load on the aerial device may need to be reduced. Depending on the manufacturer, the maximum grade on which operations may be attempted will range from 9 to 14 percent.

WARNING!

The driver/operator must be knowledgeable of the grade requirements for the aerial device being operated. Failure to understand these restrictions may result in aerial device failure and resulting injury or death to personnel working on or around the apparatus.

When adopted in 1991, NFPA 1904 required new aerial apparatus to be capable of carrying their full-rated capacities on grades of up to 6 percent in either direction. Capabilities above and beyond this 6 percent figure varied depending on the manufacturer of the apparatus. However, when the requirements for aerial apparatus were rolled back into NFPA 1901, *Standard for Automotive Fire Apparatus* in 1996, the stability requirements for operating on a grade were changed again. Apparatus constructed since the adoption of that standard must be capable of sustaining a static load that is one and one-third times its rated tip load capacity when the apparatus is on a 5° (about 11%) slope that is downwards in the direction most likely to cause overturning.

The keys to operating on a grade are optimum positioning, proper stabilization of the apparatus, and knowing load restrictions for the given grade. For more information on optimum positioning and proper stabilization on grades, refer to Chapter 6 of this manual.

High Wind Conditions

Wind can adversely affect the operating capabilities of the aerial device. Wind gusts impose significant dynamic loads to the device, and sustained winds of sufficient velocity can cause deformation or twisting of ladder sections (Figure 7.22). This presents a serious hazard to personnel working on the ladder or platform. Most aerial device manufacturers specify wind speed limitations within which their aerial devices may be safely operated. These will vary among manufacturers; however, most allow full operation in winds of up to 35 or 40 miles per hour (56 km/h to 64 km/h).

The driver/operator should be able to roughly determine the wind speeds at the location of aerial device operation. Obviously, the most reliable methods of obtaining this information involve the use of calibrated wind-measuring equipment. Some fire departments have this equipment mounted on their apparatus. Other jurisdictions rely on having this information relayed from the dispatch center. If neither of these options is available, the driver/operator can roughly estimate wind speed using the information in Table 7.1.

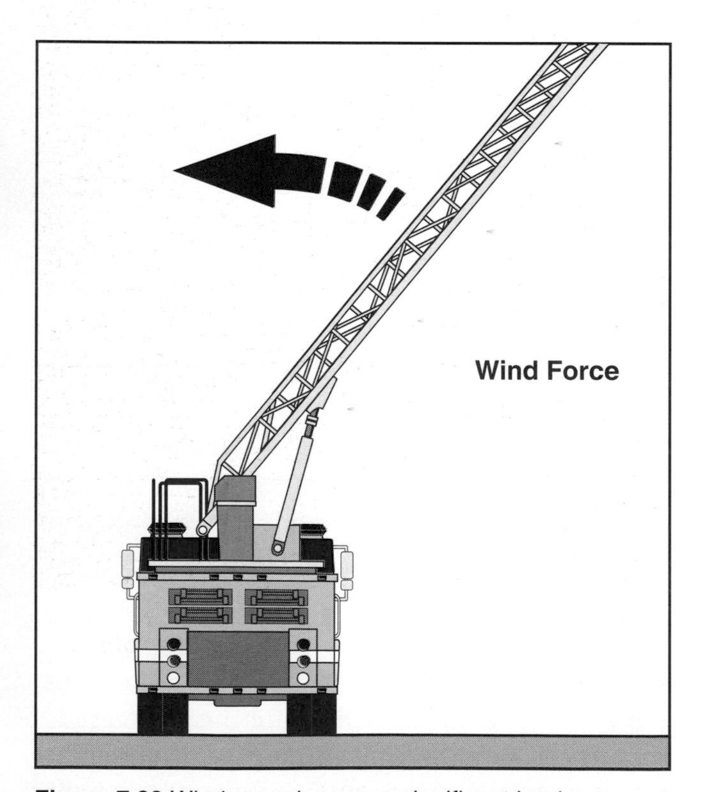

Figure 7.22 Winds may impose a significant load on a raised aerial device.

Table 7.1
Estimating Wind Speeds*

Estimated Wind Speed in MPH (KM/H)	Visible Conditions
15-24 (24-38)	Whitecaps visible on ponds and small lakes
25-31 (39-50)	Large tree branches in motion; whistling heard in utility wires
32-38 (51-61)	Whole trees in motion; inconvenience felt walking against wind
39-54 (62-87)	Twigs break off trees; walking against the wind is generally impeded

*Based on the Beaufort and Fujita Wind Scales

The manufacturers of several of the older, light-duty ladders (Seagrave, Pirsch, Maxim, American LaFrance) recommend the use of guy ropes in winds exceeding 35 miles per hour (56 km/h). Guy ropes should be attached to the top end of the fly section when it is necessary to extend the ladder 75 feet (23 m) or more. Tension should be maintained on the guy rope in the direction from which the wind is coming, especially if the wind is blowing sideways on the ladder (Figure 7.23). This may be eased off when the ladder is in position against the building, unless the wind is of exceptional velocity. (**NOTE:** Most aerial devices still in service today are not designed for use with guy ropes. Guy ropes should be used only if the manufacturer of the aerial device specifically approves them.)

Low Air Temperature Conditions

Low air temperatures present several special considerations for the aerial apparatus driver/operator. Low air temperatures result in increased viscosity of the hydraulic oil, slowing overall machine operation. The less obvious effects of cold weather are physical changes in the properties of

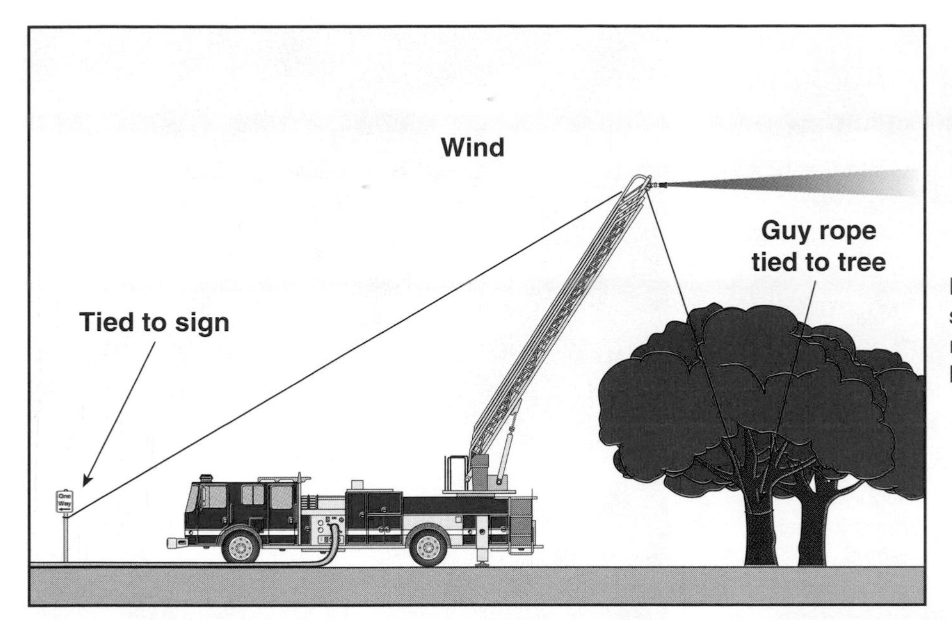

Figure 7.23 The manufacturers of some older aerial devices recommended guy ropes be used in high wind conditions.

the steel structural members of the device. When subjected to extreme cold temperatures, the device should be operated in a manner that minimizes or eliminates "shock loading" to its structural members.

Precipitation and/or airborne water droplets from elevated master stream operation, combined with low temperatures, will result in ice formation on the device. Due to its weight, ice significantly reduces rescue capability and adversely effects stability. The ice buildup on the aerial device can cause damage to the structural members or auxiliary systems when the device is moved from position. Grease applied to exposed sliding surfaces, such as slip tubes or rung rails, is effective to prevent immediate ice adhesion to the aerial device when ice formation begins. It also facilitates "shrugging off ice" by slowly extending and retracting the device. In some cases, the power assist control may be carefully activated to facilitate this movement. If the ice becomes so thick that it cannot be shrugged off, another method of deicing will have to be used. Common methods of removing ice include using high-pressure steam or liquid thawing agents similar to those used on aircraft (Figure 7.24). Some departments in climates subject to much freezing weather have special thawing apparatus that respond when conditions warrant (Figure 7.25). Any thawing agents that are used must be noncorrosive and otherwise harmless to the aerial device or general apparatus components. Check with the manufacturer of the aerial device to see what thawing agents are acceptable for your particular aerial device.

Figure 7.24 Use only manufacturer-approved methods for removing ice from the aerial device. *Courtesy of Ron Jeffers.*

Figure 7.25 Departments that frequently encounter freezing conditions may have special apparatus to assist with thawing operations. *Courtesy of Ron Jeffers.*

Exposure to Fire Products

Because the aerial device is often operated out and over a fire area, the device occasionally comes in contact with superheated air or, on rare occasions, is subject to direct flame contact (Figure 7.26). While it is preferred that these situations be avoided, they do, in fact, occur. The driver/operator should be continually aware of the conditions surrounding the aerial device and should reposition the device when the danger of exposure to extreme heat is present. If the aerial device cannot be moved immediately, protective water hose streams should be played on the device to keep it cool. When using water streams, the operator should be careful not to knock down any personnel operating on the ladder. Also, use extreme care if a master stream device is being used to cool the aerial device. A straight or solid stream could impose damaging forces on the aerial device. Many elevating platforms are equipped with spray-curtain nozzles beneath the platform to provide a protective water curtain in the event of a high heat situation (Figures 7.27 a and b).

Heat damage to the aerial device may or may not be visible. Obvious signs of damage include discoloration, disfiguration, deformed welds, or improper operation of any portion of the aerial device. When heat damage is noted, or even suspected, the aerial device should be removed from service immediately. It should be subjected to a thorough testing procedure, per NFPA 1914, before being placed back into service.

Figure 7.26 The aerial device may be subjected to high levels of heat. *Courtesy of Chris Mickal.*

Figure 7.27a A protective water curtain nozzle located beneath an elevating platform.

Figure 7.27b The shower nozzle discharges a wide cone of water to absorb heat approaching the elevating platform.

Aerial Device Mechanical or Power Failure

Occasionally, the adverse condition will be caused by the apparatus itself. While proper maintenance and use will minimize the chance of a mechanical failure occurring, in some cases they are unavoidable. The driver/operator must be constantly on the lookout for any of the following signs of mechanical trouble:

- Leaking fuel
- Leaking hydraulic fluid
- Leaking water from the engine cooling system
- Leaking motor oil
- Overheating of any mechanical components
- Unusual noises or vibrations
- Drifting of the aerial device when raised

If any of these indicators are present, personnel should be removed from the aerial device, and the device should be bedded and taken out of service immediately. Each manufacturer will have specific information in its operator's manual pertaining to the procedures for lowering aerial apparatus following a mechanical or power failure. Failure of the main hydraulic system will typically require the auxiliary hydraulic pump to be used. As mentioned previously in this manual, this pump should only be operated at one-minute intervals, followed by a minute of rest. This prevents overheating of the auxiliary pump. Before placing the unit back in service, the aerial device should be repaired by a qualified technician and thoroughly tested in accordance with NFPA 1914.

Safe Operating Practices for Telescoping Aerial Devices

Safe operations are best ensured by an adequate training program that requires frequent practice. Practice once a week is strongly recommended. Such practice should include using technical data, spotting the apparatus, operating controls, and using ladder pipes and aerial ladders. In addition to the information covered earlier in this chapter, driver/operators of aerial apparatus should recognize the following rules of thumb:

- All ladders are stronger when the load is applied perpendicular to the rungs than when applied laterally (Figure 7.28).
- The lower the angle of the ladder from horizontal, the less load it will safely carry.
- A shock load, such as a person jumping onto the ladder, imposes stresses several times greater than those involved when the load is gradually applied.
- If unable to extend the ladder over the exact front or rear of a straight-chassis aerial apparatus, try to keep it as close to these positions as possible.
- When placing the ladder, always ease it gently toward the objective. When moving the ladder in any manner, it must be clear of the objective to prevent scraping.
- The ladder locks should be engaged and the hydraulic lock valves closed before loading the ladder.

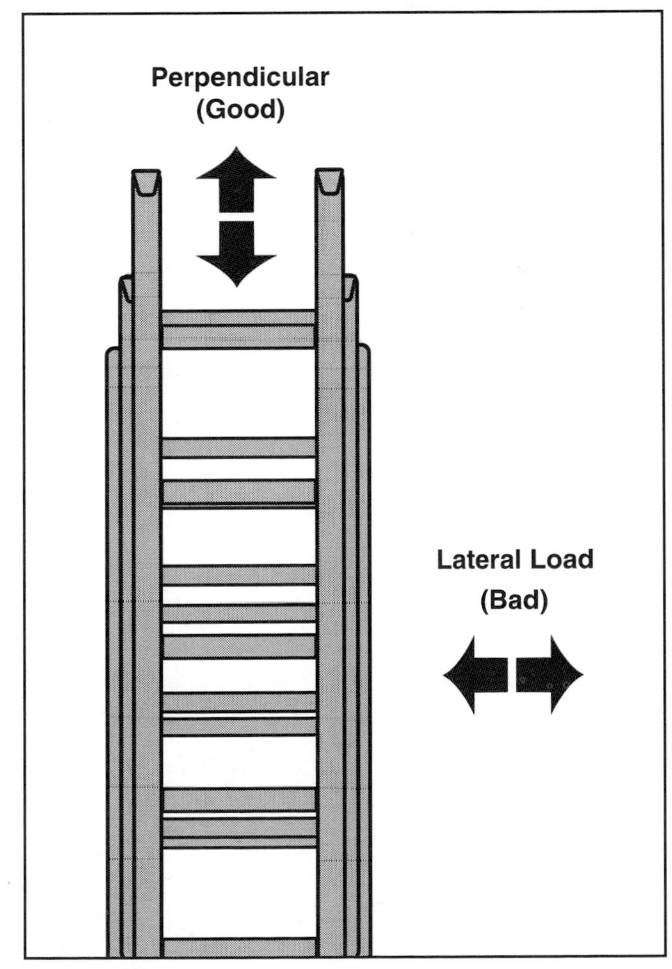

Figure 7.28 Aerial devices are designed to handle perpendicular loads.

- The aerial ladder should not be overloaded. Follow manufacturer's recommendations.
- It is important not to exceed the rated weight capacity of the platform. The rated capacity of the platform is the safe carrying capacity in ***any*** position. Depending upon elevation and extension, additional loads may be safely carried.
- The weight of equipment mounted in the platform after delivery should be subtracted from the rated capacity of the platform (Figure 7.29). Shock loads impose great stresses on the aerial device. Do not jump into the platform from above.
- You should be aware that strong winds will affect the load capacity and the stability of the aerial.
- The aerial device should not be used to lift items heavier than its rated platform capacity.
- If you are unsure about the safe operating principles and limitations of your specific make of aerial apparatus, contact the manufacturer for detailed information.
- Never extend or retract an aerial ladder with firefighters on the ladder.
- Always monitor for overhead obstructions, particularly overhead wires.

If you are involved in an incident where a failure of the aerial device occurs, gather as much information about the operating conditions as possible. This will greatly aid any professional reviews of the incident to determine the exact cause that led to the failure.

Figure 7.29 Some departments choose to mount commonly used equipment at the tip of the aerial device.

Chapter 8

Operating Articulating Aerial Equipment

Job Performance Requirements

This chapter provides information that will assist the reader in meeting the following job performance requirements from NFPA 1002, *Standard on Fire Apparatus Driver/Operator Professional Qualifications*, 1998 edition. Particular portions of the job performance requirements (JPRs) that are met in this chapter are noted in bold text.

4-2.3 Maneuver and position the aerial device from each control station, given an incident location, a situation description, and an assignment, so that the aerial device is properly positioned to safely accomplish the assignment.

(a) *Requisite Knowledge*: Aerial device hydraulic systems, hydraulic pressure relief systems, gauges and controls, cable systems, communications systems, electrical systems, emergency operating systems, locking systems, manual rotation and lowering systems, stabilization systems, aerial device safety systems, system overrides, safe operational limitations of the given aerial device, safety procedures specific to the device, and operations near electric hazards and overhead obstructions.

(b) *Requisite Skills*: The ability to raise, rotate, extend, and position to a specified location and the ability to lock, unlock, retract, lower and bed the aerial device.

4-2.4 Lower an aerial device using the emergency operating system, given an aerial device, so that the aerial device is safely lowered to its bedded position.

(a) *Requisite Knowledge*: Aerial device hydraulic systems, hydraulic pressure relief systems, gauges and controls, cable systems, communications systems, electrical systems, emergency operating systems, locking systems, manual rotation and lowering systems, stabilization systems, aerial device safety systems, system overrides, safe operational limitations of the given aerial device, safety procedures specific to the device, and operations near electric hazards and overhead obstructions.

(b) *Requisite Skills*: The ability to rotate and position to center, unlock, retract, lower, and bed the aerial device using the emergency operating system.

Aerial apparatus operators and company officers must be thoroughly trained in operating the apparatus to which they are assigned. The general operating practices, limitations, and safety measures covered in this chapter will provide a realistic approach to the operation of articulating aerial apparatus. Safe, smooth, and efficient operation of aerial devices is best ensured through an adequate training program. Practice, classroom study, and drills are strongly recommended. These study sessions should include using technical data, preparing for use, general positioning of apparatus, operating controls, using with fire streams, and using during rescue.

All fire department aerial devices have certain operation levers or control devices that a trained operator can use to make the aerial function. These controls may differ in certain respects, as do the controls of different automobiles. The operator must be well trained in the location of each control, its function, how it operates, when it should be used, and appropriate safety precautions. Aerial operators should also understand the function and operation of all stabilizing power and safety devices. In addition, the operator must know the limitations of the apparatus, proper positioning angles, maximum load limits, and safe operating practices. The importance of having the operator well versed in the layout and operation of the aerial device controls cannot be overemphasized. Early portions of any training program should be dedicated toward this goal. If the driver/operator may be expected to drive and operate more than one type of aerial apparatus, he should be trained on each of the different apparatus.

There are two types of articulating aerial apparatus: water towers and aerial platforms. Both types have two or more sections called booms. The water tower is equipped with only a master stream nozzle at the end of the upper boom (Figure 8.1). It is not intended to lift firefighters. The aerial platform is equipped with a platform and a master stream nozzle at the end of the upper boom (Figure 8.2). The booms are hydraulically operated by controls located below the turntable and in the platform, if so equipped (Figures 8.3 a and b). Included in this chapter are recommended procedures for placing the aerial device in position, raising and lowering the device, and operating the aerial device under adverse conditions. The procedures are similar for raising articulating aerial devices that have platforms and those that are equipped with only a master stream device. Although this chapter is written basically for aerial devices equipped with elevating platforms, most of the operational information (exclusive of references to platforms) is applicable to the operation of those equipped only with an elevated master stream device.

It should also be noted that some manufacturers offer aerial devices that both articulate and telescope (Figure 8.4). Elements from both this chapter

Figure 8.1 An articulating water tower. *Courtesy of Chris Mickal.*

Figure 8.2 An articulating aerial platform.

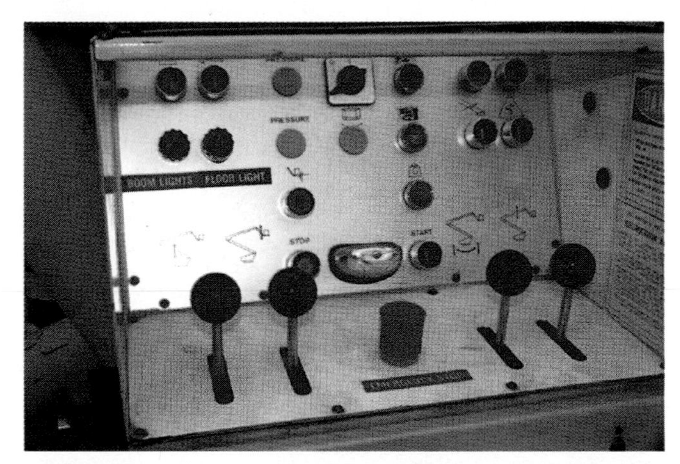

Figure 8.3a The lower control pedestal.

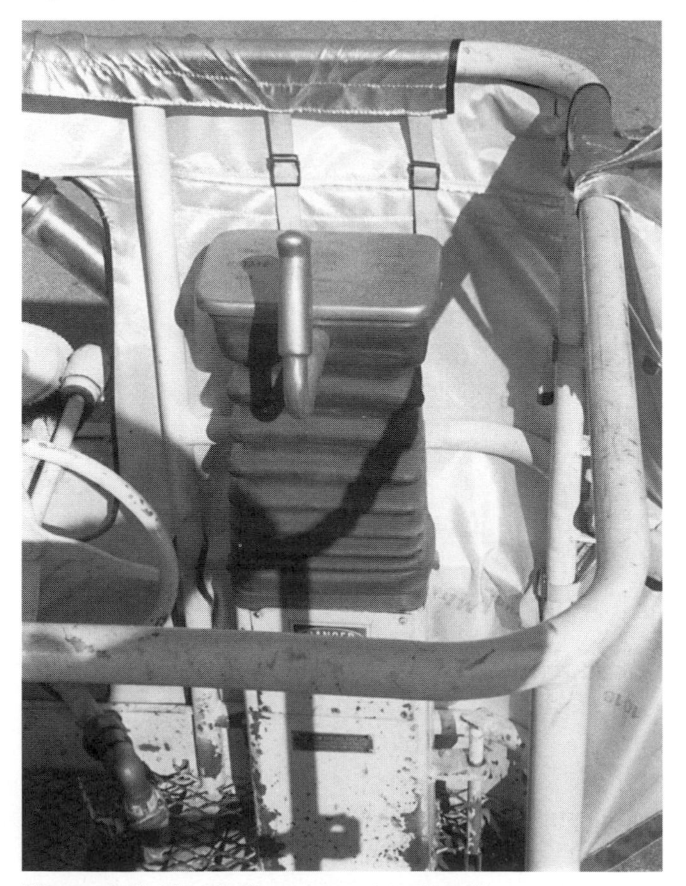

Figure 8.3b The platform control pedestal.

Figure 8.4 Some models of articulating aerial devices have sections that telescope within the articulating sections.

and the previous one on operating telescoping aerial devices apply to these hybrids. Because the operation of combination devices is extremely complex, this information should be obtained directly from the operator's guide provided by the manufacturer of the aerial device.

Proper positioning of the aerial apparatus was covered thoroughly in Chapter 5 of this manual; however, it is important to reemphasize several positioning points before discussing raising aerial devices. The most important is to make sure that the apparatus is positioned so that the stress on the aerial device is minimized as much as possible. Earlier-arriving engine companies or other apparatus should not have blocked the aerial apparatus from obtaining an optimum position. Place the apparatus as close to the target as safely possible. Make sure that the turntable is aligned with the target. Whenever possible, try to avoid positioning the apparatus in a location that will subject the aerial device to overhead obstructions, particularly power transmission lines. Operators of articulating

Figure 8.5 Care should be taken to not park the apparatus in a location that will make it difficult to raise the aerial device. *Courtesy of Chris Mickal.*

devices must be concerned about the elbow joint between the two boom sections, as well as the working end of the device. The knuckle is of particular concern when operating at low elevations. It is in this position that the knuckle may be well behind the position of the operator.

Another unique concern particular to positioning articulating aerial apparatus is that other apparatus must respect the operating space required by the articulating aerial device and position accordingly. If other apparatus, particularly telescoping aerial apparatus, position too close to the articulating aerial apparatus, its aerial device operating space may be impeded to the point of making it inoperable (Figure 8.5). This requires good judgment on the part of all driver/operators and incident commanders who position apparatus, not just the one driving the articulating aerial apparatus.

(**NOTE:** The following procedures for operating articulating aerial devices assume that the apparatus has been properly positioned on a hard, level surface, with negligible wind and nonfreezing conditions. Information on operating in adverse conditions is addressed later in this chapter.)

WARNING!

Before applying the information contained in this chapter, properly stabilize the apparatus in accordance with the manufacturer's recommendations and Chapter 6 of this manual. Failure to properly stabilize the apparatus before performing the applications covered in this chapter could result in a tip-over of the apparatus.

Raising the Aerial Device

To raise the aerial assembly means to elevate the entire assembly from its stored position toward a vertical position of use. Raising the aerial device is actually a series of motions that include elevating, rotating, extending, and lowering the aerial device to its objective. Experienced aerial device operators may be able to perform some of these functions simultaneously; however, new operators should take one motion at a time.

The instructions in this section pertain solely to the actual deployment of the aerial device itself. Prior to the deployment of the aerial device, the following tasks should have already been completed:

- Place the apparatus in the proper position for maximum efficiency of use. Properly apply all parking brakes (see Chapter 5).
- The apparatus transmission should be placed into the proper gear, and the power take-off should be operated to provide power to the aerial device and the stabilizer hydraulic systems (see Chapter 6).
- The stabilizers must be fully deployed according to manufacturer's recommendations (see Chapter 6). If the aerial device controls are located at ground level, the operator's platform should be pulled out so that no further operations are conducted with the driver/operator standing on the ground.
- The selector valve should be switched from the stabilizer position to the aerial device position so that hydraulic power is provided for aerial operations (see Chapter 6).

With the preceding tasks accomplished, the aerial device may be placed into operation. As with the stabilizers described in Chapter 6, the operational speed of the aerial device is directly dependent on the amount the control lever is pushed or pulled (Figures 8.6 a and b). Newer operators should avoid full-speed operations until they are extremely familiar with the operation of the aerial device. Full-speed operations, regardless of the skill of the operator, should be minimized due to the large amount of dynamic stress placed upon the aerial device when the motion is halted.

Figure 8.6a The speed in which the aerial device moves is proportional to the amount the control lever is moved.

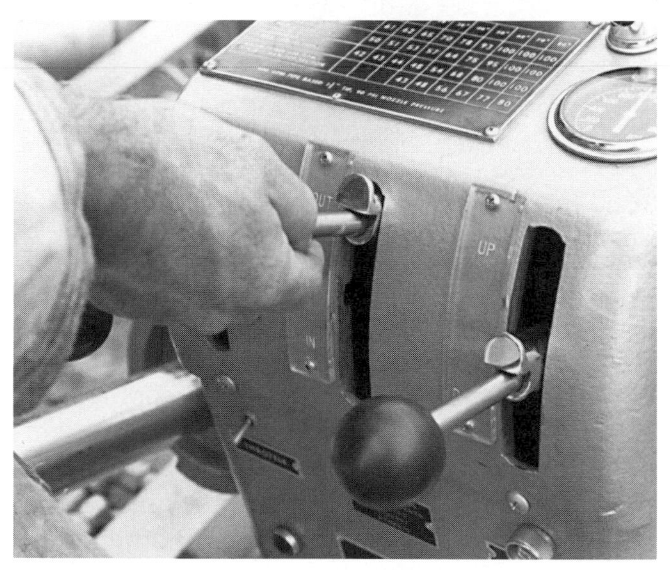

Figure 8.6b The aerial device will move at maximum speed if the lever is moved to the end of its travel distance.

The following step-by-step procedure may be used to raise the aerial device to its working position:

Step 1: *Release the hold-down locks.* Hold-down locks secure the aerial device in its cradle when the apparatus is in travel mode (Figure 8.7). These locks prevent bouncing of the aerial device that could lead to structural damage. Some articulating platforms are also equipped with basket stabilizers that support the platform when the apparatus is in travel mode (Figure 8.8). Disengage these locks and stabilizers before operating the aerial device. The most common of the articulating aerial platforms, manufactured by the Snorkel Company, has both upper boom hold-down locks and basket stabilizers. Manually release both by pulling on the lever that is a part of each (Figure 8.9). When the upper boom hold-down is released, it falls to the side and out of the way (Figure 8.10). When the basket stabilizer arms are released, they must be flipped up and stowed on the platform (Figure 8.11).

Figure 8.9 The driver/operator must release the boom hold-down locks before deploying the aerial device.

Figure 8.7 The hold-down locks prevent the aerial device from bouncing during road travel.

Figure 8.10 The upper boom hold-down device falls out of the way once it is opened.

Figure 8.8 Platform stabilizers hold the platform in place during road travel.

Figure 8.11 Stow the platform stabilizer in its intended storage position.

Step 2: *Make hose connections.* Connect the intake fire hose(s) to the appropriate fire pump or waterway intake connection(s) (Figure 8.12). Determination of which intake connection to use depends on whether the apparatus fire pump or an external fire pump will be used to supply the master stream device. Connections for taking water from an external fire pump are usually found on the rear step area of the apparatus.

Step 3: *Check the intended path of the aerial device for obstructions.* This consideration should have been covered during the positioning of the apparatus. Occasionally, obstacles not foreseen during positioning will come to light when the driver/operator assumes the aerial device operator's control position (at either the platform control station or the turntable control station). Make one last visual check along the intended path of travel to make sure that no obstructions are present. Platform control station operators must be extra careful because protruding portions of the upper and lower booms are behind the operator and not readily in view (Figure 8.13). This is particularly true when the platform is being operated at low elevations.

Step 4: *Energize the platform controls.* Many articulating elevating platforms have a switch on the ground control station panel that is used to activate or deactivate the platform control panel (Figure 8.14). If personnel will operate the aerial device from the platform, the control should be activated at this time. Some departments require these controls to be activated so that the firefighters in the basket have control in the event a safety concern becomes apparent.

Some jurisdictions prefer that the firefighters in the platform have primary

Figure 8.12 Connect the supply hose to the waterway system inlet.

Figure 8.13 Keep in mind that the booms and knuckle are behind the platform whenever it is in motion.

control of the aerial device. This is because they are in the best position to place the platform in the most desirable operational location. However, even if the firefighters in the platform will be controlling the aerial device, an operator should be stationed at the lower control pedestal at all times. The operator at the lower control pedestal can override the firefighters in the platform if they are approaching a hazard that is unseen to them or in the event they become incapacitated.

Step 5: *Elevate the lower boom, and move the platform to ground level.* Operate the appropriate control to elevate the lower boom section so that the platform may be moved to ground level, allowing personnel to board (Figure 8.15). Use caution to ensure that no personnel are beneath the platform when this process is initiated. Once personnel are in the platform, it may be operated from either the ground control station or the platform control station.

Figure 8.14 Most articulating aerial platforms have a switch located on the lower control pedestal that activates the platform control station.

Figure 8.15 The platform may be placed on the ground for personnel to board.

WARNING!

Whenever operating the aerial device from the ground control station, the driver/operator must be standing on the anti-electrocution platform (Figure 8.16). Failure to do so could result in electrocution should the aerial device come into contact with energized electrical equipment.

Step 6: *Place the aerial device in the desired position.* Once personnel have mounted the platform, the aerial device may be moved into position for the desired function (Figure 8.17). Unlike telescoping aerial devices, there is no set procedure for elevating, ro-

Figure 8.16 When operating the aerial device from the lower control station, the driver/operator should be standing on the apparatus to avoid electrocution.

Figure 8.17 An articulating aerial device in a deployed position.

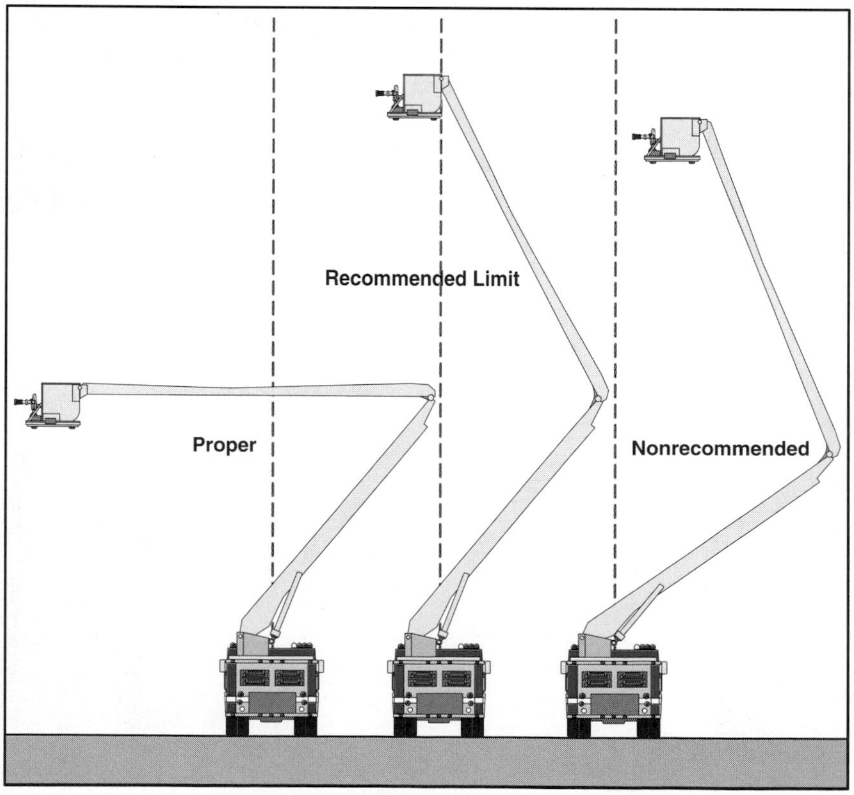

tating, extending, and lowering articulating aerial devices. Training and practice allow the operator to develop the skills and judgment necessary to place the aerial device in the desired position. Articulating platforms should always be operated forward of the turntable (Figure 8.18). When performing rescue operations, try to avoid placing the platform in a position that allows panicked victims to jump into the platform before it is properly placed. Raise the device so that the platform is slightly away from and directly above the victim(s). Then, lower it down to them.

Step 7: *Charge the waterway system, and operate the elevated master stream, if desired.* If the articulating device is to be used for deploying an elevated master stream, position the device first and then operate the stream. Follow manufacturer's recommendations for movement of the aerial device while operating the elevated master stream.

Step 8: *Engage cylinder and/or turntable locks if applicable.* If the aerial device is equipped with cylinder and/or turntable locks and the aerial device is going to stay in a set position for an extended period of time, activate the locks (Figure 8.19). Only do this if recommended by the manufacturer of the aerial device. Remember to release the locks if it becomes necessary to reposition the aerial device.

Recommended Limit

Proper

Nonrecommended

Figure 8.18 The platform must always be operated ahead of the turntable.

Figure 8.19 The cylinder locks may be engaged if the aerial device is going to be stationary for an extended period of time.

Lowering the Aerial Device

The process of lowering the aerial device is the exact reverse of the process used to place it in position for operation. The following steps describe the procedure for lowering the aerial device:

Step 1: *Disengage cylinder and/or turntable locks if applicable.* If the aerial device is equipped with locks and they have been engaged during the operation, disengage them before further lowering operations.

Step 2: *Drain the waterway system.* Drain the waterway system before lowering the aerial device (Figure 8.20). Unlike telescoping aerial devices, there is little chance of damaging the waterway system while lowering an articulating device. However, a drained waterway reduces the amount of unnecessary stress placed on the device.

Step 3: *Check the intended path of the aerial device for obstructions.* Before rotating the aerial device, the operator should recheck the intended path of travel to make sure that it is free of obstructions.

Figure 8.20 Drain the waterway system before lowering the aerial device.

Step 4: *Rotate the aerial device until the booms are in line with the boom support travel cradle.* If the aerial device was raised to a building or other object, move the platform safely away from the object and then rotate the device until it is aligned with the boom support travel cradle. Depending on the manufacturer of the apparatus, there are several ways to ensure that the aerial device is properly positioned for storage. Some manufacturers paint corresponding markers (usually arrows) on the base of the turntable and some portion of the apparatus adjacent to the turntable (Figure 8.21). When the two markers are lined up, the aerial is in position for stowing. Other manufacturers use sensors connected to a light on the control panel that illuminates when the aerial device is in position for stowing (Figure 8.22).

Figure 8.21 The turntable alignment arrows show when the aerial device is properly positioned for stowing.

Figure 8.22 Some manufacturers use a light to indicate turntable alignment.

> **WARNING!**
>
> **When operating an articulating aerial device, watch for the two primary points of contact: the platform end of the device and the hinge (knuckle) between the upper and lower booms (Figure 8.23). Oftentimes, the hinge is behind the operator when placing the platform or nozzle into position. Failure to watch both locations on the aerial device may result in electric lines or other dangerous objects being struck by the hinge.**

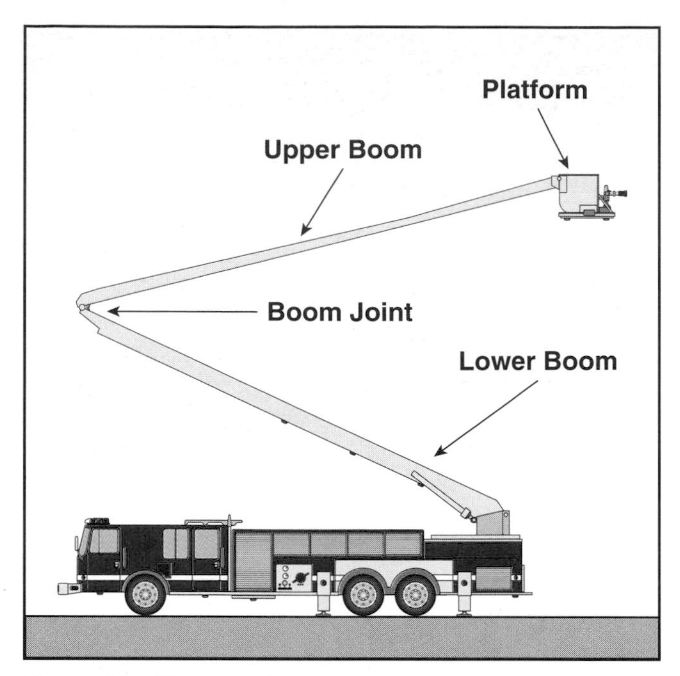

Figure 8.23 The driver/operator must be conscious of the potential points of contact on an articulating aerial device.

Step 5: *Lower the platform to the ground to allow personnel to exit.* This will not be necessary if the apparatus is designed to allow personnel to exit the platform from its stowed position.

Step 6: *Check lower boom hold-down hooks to make sure that they are open and ready to receive the aerial device.* If these hooks are not open, follow the manufacturer's recommendations for opening them (Figure 8.24). On most articulating devices, this is accomplished by extending the stabilizers slightly, causing the hooks to automatically open.

Figure 8.24 Make sure the lower boom hold-down locks are open and ready to receive the aerial device.

Step 7: *Lower the aerial device onto the boom support travel cradle.* Slowly lower the booms into the cradle (Figure 8.25). During this process, the booms should be kept as close to each other as practicably possible.

Step 8: *Hook and latch the upper boom hold-down locks and the platform stabilizer arms.* This step is necessary only for those devices manufactured by the Snorkel Company. Both the upper boom hold-down locks and the platform stabilizers must be manually attached to the aerial device (Figures 8.26 a and b).

When the aerial device has been lowered and stowed, the stabilizers may be raised, the power take-off disengaged, the fire pump disengaged (if so equipped and in use), and the truck readied for road travel.

Figure 8.25 Slowly lower the boom into the cradle.

Figure 8.26a Reconnect the upper boom hold-down lock.

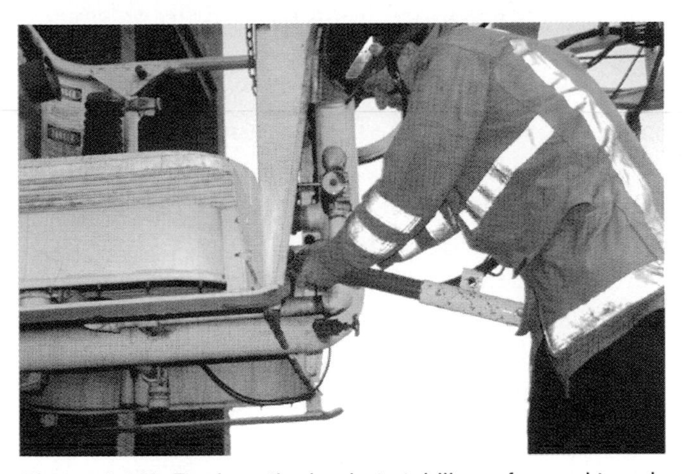

Figure 8.26b Replace the basket stabilizers for road travel.

Operating Under Adverse Conditions

The previously described procedures for operating articulating aerial devices assume that the apparatus is positioned properly on a hard, level surface, with negligible wind conditions and above-freezing temperatures. Those are the safest conditions under which an aerial device may be operated. Unfortunately, often one or more of these conditions may be working against the driver/operator of the aerial apparatus. In order to ensure the safety of the firefighters working on or around the aerial device and the physical soundness of the aerial device, the driver/operator must understand its limitations when operating under less than ideal conditions. The following sections contain information that will assist the driver/operator in understanding the special precautions that must be taken when operating the aerial device under adverse conditions.

Operating on a Grade

As explained in Chapter 6 of this manual, operation of the aerial device when the apparatus is parked on a grade greatly compounds the considerations concerning force on the aerial device system. Situations in which the floor of the platform and/or the knuckles between the boom are not parallel to the ground result in a twisting force on the entire aerial device. Aerial devices are not designed to routinely handle these twisting forces. The driver/operator must be conscious of these conditions and knowledgeable of the limitations of the particular aerial device being operated under these conditions.

Before the development of NFPA 1904, *Standard for Aerial Ladder and Elevating Platform Fire Apparatus* in 1991, no standards existed for the performance of aerial devices when operated on a grade. Thus, apparatus constructed before the adoption of this standard have a wide range of capabilities under sloped conditions. In general, most are capable of carrying their full rated load capacities on grades of up to 6 percent (lateral or longitudinal). Past this point, the load on the aerial device may need to be reduced. Depending on the manufacturer, the maximum grade on which operations may be attempted will range from 9 to 14 percent.

WARNING!

The driver/operator must be knowledgeable of the grade requirements for the aerial device being operated. Failure to understand these restrictions may result in aerial device failure and resulting injury or death to personnel working on or around the apparatus.

When adopted in 1991, NFPA 1904 required that new aerial apparatus be capable of carrying their full-rated capacities on grades of up to 6 percent in either direction. Capabilities above and beyond this 6 percent figure varied depending on the manufacturer of the apparatus. When the requirements for aerial apparatus were rolled back into NFPA 1901, *Standard for Automotive Fire Apparatus* in 1996, the stability requirements for operating on a grade were changed again. Apparatus constructed since the adoption of that standard must be capable of sustaining a static load that is one and one-third times its rated tip load capacity when the apparatus is on a 5° (about 11%) slope that is downwards in the direction most likely to cause overturning.

The keys to operating on a grade are optimum positioning, proper stabilization of the apparatus, and knowing load restrictions for the given grade. For more information on optimum positioning and proper stabilization on grades, refer to Chapter 6 of this manual.

High Wind Conditions

Wind can adversely affect the operating capabilities of the aerial device. Wind gusts impose significant dynamic loads to the device, and sustained winds of sufficient velocity can cause deformation or twisting of the booms of the aerial device (Figure 8.27). This presents a serious hazard to personnel working on the platform. Most aerial device manufacturers specify wind speed limitations within which their aerial devices may be safely operated. These will vary among manufacturers; however, most allow operation in winds of up to 35 or 40 miles per hour (56 km/h or 64 km/h). Consult the manufacturer of your particular aerial device for recommendations on operating in high wind conditions.

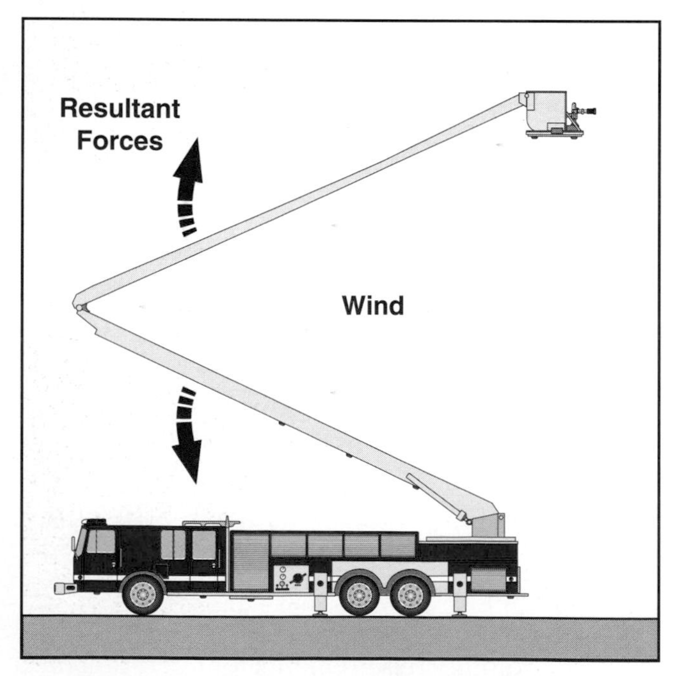

Figure 8.27 Wind can create a significant load on the aerial device.

The driver/operator should be able to roughly determine the wind speeds at the location of aerial device operation. Obviously, the most reliable methods of obtaining this information involve the use of calibrated wind-measuring equipment. Some fire departments have this equipment mounted on their apparatus. Other jurisdictions rely on having this information relayed from the dispatch center. If neither of these options is available, the driver/operator can roughly estimate wind speed using the information in Table 8.1.

Low Air Temperature Conditions

Low air temperature conditions present several special considerations for the aerial apparatus driver/operator. Low air temperatures result in increased viscosity of the hydraulic oil, slowing overall machine operation. The less obvious effects of cold weather are physical changes in the properties of the steel structural members of the device. When subjected to extreme cold (sub-zero) temperatures, the device should be operated in a manner that will minimize or eliminate "shock loading" to its structural members.

Table 8.1
Estimating Wind Speeds*

Estimated Wind Speed in MPH (KM/H)	Visible Conditions
15-24 (24-38)	Whitecaps visible on ponds and small lakes
25-31 (39-50)	Large tree branches in motion; whistling heard in utility wires
32-38 (51-61)	Whole trees in motion; inconvenience felt walking against wind
39-54 (62-87)	Twigs break off trees; walking against the wind is generally impeded

*Based on the Beaufort and Fujita Wind Scales

Precipitation and/or airborne water droplets from the water tower operation combined with low temperatures will result in ice formation on the device (Figure 8.28). Due to its weight, ice significantly reduces rescue capability and adversely effects stability. The ice buildup on the aerial device can cause damage to the structural members or auxiliary systems when the device is moved from position. Grease applied to exposed sliding surfaces, such as slip tubes, is effective in preventing immediate adhesion to the aerial device when ice formation begins. It also facilitates "shrugging off" ice by slowly extending and retracting the device. If the ice becomes so thick that it cannot be shrugged off, another method of deicing will have to be used. Common methods of removing ice include high-pressure steam or a liquid thawing agent similar to those used on aircraft (Figure 8.29). Some departments in climates subject to much freezing weather have special thawing apparatus that respond when conditions warrant (Figure 8.30). Any thawing agents used must be noncorrosive and otherwise harmless to the aerial device or general apparatus components. Check with the manufacturer to see what thawing agents are acceptable for your particular aerial device.

Figure 8.28 Ice that builds up on the aerial device can create excessive loads on the device. *Courtesy of Tom McCarthy, Chicago Fire Department.*

Figure 8.29 Use manufacturer-approved means for thawing the aerial device. *Courtesy of Tom McCarthy, Chicago Fire Department.*

Exposure to Fire Products

Because the aerial device is often operated out and over a fire area, the device occasionally comes in contact with superheated air or, on rare occasions, is subject to direct flame contact (Figure 8.31). The

Figure 8.30 Departments in regions subject to frequent freezing conditions may have special apparatus to assist with thawing operations. *Courtesy of Ron Jeffers.*

Figure 8.31 Be alert for heat damage when operating at large fires. *Courtesy of Harvey Eisner.*

driver/operator should be continually aware of the conditions surrounding the aerial device and should reposition the device when the danger of exposure to extreme heat is present. If the aerial device cannot be moved immediately, water streams should be played on the device to keep it cool. When using water streams, the operator must be careful not to knock down personnel operating on the platform. Also, use extreme care if a master stream device is being used to cool the aerial device. A straight or solid stream could impose damaging forces on the aerial device. Many elevating platforms are equipped with spray-curtain nozzles beneath the platform to provide a protective water curtain in the event of a high heat situation (Figures 8.32 a and b). The spray-curtain nozzles are most commonly operated by applying pressure to a foot pedal on the floor of the platform (Figure 8.33).

Heat damage to the aerial device may or may not be visible. Obvious signs of damage include discoloration, disfiguration, deformed welds, or improper operation of any portion of the aerial device. When heat damage is noted, or even suspected, remove the aerial device from service immediately. It should be subjected to a thorough testing procedure, per NFPA 1914, before being placed back into service.

Figure 8.32a A typical shower-curtain nozzle beneath an elevating platform.

Figure 8.32b The water fog will absorb heat that is being directed toward the bottom of the platform.

Figure 8.33 This shower curtain nozzle is controlled by a gate valve.

Aerial Device Mechanical or Power Failure

Occasionally, the adverse condition will be caused by the apparatus itself. While proper maintenance and usage minimizes the chance of mechanical failure, in some cases mechanical failure is unavoidable. The driver/operator must be constantly on the lookout for any of the following signs of mechanical trouble:

- Leaking fuel
- Leaking hydraulic fluid
- Leaking water from the engine cooling system
- Leaking motor oil
- Overheating of any mechanical components
- Unusual noises or vibrations

If any of these indicators are present, personnel should be removed from the aerial device, and the device should be bedded and taken out of service immediately. Each manufacturer will have specific information in its operator's manual pertaining to the procedures for lowering aerial apparatus following a mechanical or power failure. Failure of the main hydraulic system will typically require the auxiliary hydraulic pump to be used. As mentioned previously in this manual, this pump should only be operated at one-minute intervals, followed by a minute of rest. This will prevent overheating of the auxiliary pump. Before placing the unit back in service, the aerial device should be repaired by a qualified technician and thoroughly tested in accordance with NFPA 1914.

Safe Operating Practices for Articulating Aerial Devices

All manufacturers of articulating aerial apparatus include precautionary measures and recommendations for safe operation in their operator's handbook. Many of these recommendations apply to a specific functional part and may be applicable to only one model or brand of apparatus. The following general safety features may be applied as rules of thumb to all articulating aerial devices. However, the list is not all-inclusive and does not give complete safety rules for all articulating aerial devices. Consult the manufacturer's documentation for more specific safety recommendations.

- Most aerial apparatus are designed to be used on firm, level ground. Attempts to operate them on slopes exceeding manufacturer's recommendations is not recommended.
- During master stream operations, the maximum load capacity of the personnel platform decreases in direct proportion to the nozzle reaction force in pounds.
- Apparatus brakes and stabilizers must be securely set before operation.
- Caution must be exercised when operating near power lines, even if the upper boom and personnel platform are insulated. Remember, any item that leads from the platform to the ground — such as electrical cords, hydraulic or water lines, or communication headsets — conducts electricity and can render protective insulation ineffective.
- Care must be taken during master stream operations to prevent water hammer when nozzles are closed. Avoid sudden movement of the fire stream from side to side because such movement puts excessive lateral stress on the booms.
- The aerial operator on the ground should always be standing on the operator's platform when operating the aerial device.
- When securing the apparatus for road travel, the operator must make sure that the booms are locked in the transport position and the stabilizers are fully retracted.
- To avoid unsafe positioning of the tip of the aerial device platform, the operator should always rotate the turntable until facing the working area before moving forward.
- If you are unsure about the safe operating principles and limitations of your specific make of aerial apparatus, contact the manufacturer for detailed information.

If you are involved in an incident where a failure of the aerial device occurs, gather as much information about the operating conditions as possible. This will greatly aid any professional reviews of the incident to determine the exact cause that led to the failure.

Chapter 9

Aerial Apparatus Strategies and Tactics

Job Performance Requirements

This chapter provides information that will assist the reader in meeting the following job performance requirements from NFPA 1002, *Standard on Fire Apparatus Driver/Operator Professional Qualifications*, 1998 edition. Particular portions of the job performance requirements (JPRs) that are met in this chapter are noted in bold text.

4-2.3 Maneuver and position the aerial device from each control station, given an incident location, a situation description, and an assignment, so that the aerial device is properly positioned to safely accomplish the assignment.

(a) *Requisite Knowledge*: Aerial device hydraulic systems, hydraulic pressure relief systems, gauges and controls, cable systems, communications systems, electrical systems, emergency operating systems, locking systems, manual rotation and lowering systems, stabilization systems, aerial device safety systems, system overrides, safe operational limitations of the given aerial device, safety procedures specific to the device, and operations near electric hazards and overhead obstructions.

(b) *Requisite Skills*: The ability to raise, rotate, extend, and position to a specified location and the ability to lock, unlock, retract, lower and bed the aerial device.

4-2.5 Deploy and operate an elevated master stream, given a master stream device and a desired flow, so that the stream is effective and the device is operated safely.

(a) *Requisite Knowledge*: Nozzle reaction, range of operation, and weight limitations.

(b) *Requisite Skills*: The ability to connect a water supply to a master stream device and control an elevated nozzle manually or remotely.

As mentioned in earlier chapters, the uses of aerial apparatus are many. Some of the most common uses include the following:

- Rescue
- Exposure protection
- Ventilation
- Elevated fire attack

These uses are termed *operational strategies*. The operational strategies chosen depend on prefire plans, prearrival information, and initial size-up of the emergency. To successfully achieve any of the strategies, firefighters must perform certain tactics. These tactics will be covered in this chapter. Each strategy will be examined separately, and specific tactics to achieve each strategy will be given.

This chapter focuses solely on the aerial device's role in each of these strategies and the accompanying tactics. Obviously, there are many other actions that must be taken by truck company personnel to accomplish these functions at an emergency scene. In many cases, the driver/operator may be physically involved in chores other than simply operating the aerial device. However, it is not the purpose of this manual to be a truck company operations manual. For more information on strategy and tactics, see the Fire Protection Publications manual *Fire Fighting Strategy and Tactics* by Harry Carter, Ph.D.

Rescue

Rescue is always the first priority on the fireground. When an interior stairway is available, it is the preferred route for evacuation for victims located above ground level. Reliable, exterior fire escapes would be the second choice if present. If fire conditions do not allow victims to be brought down the stairs or fire escapes it will be necessary to use ground ladders or an aerial device to get them down, though using an aerial device for this purpose should always be a last resort. If it is necessary to employ an aerial device to rescue victims, the tactics in this section can be used to achieve faster, safer, and more efficient rescue operations. For information on rescuing victims using ground ladders, see the IFSTA **Fire Service Ground Ladders** or **Essentials of Fire Fighting** manual.

Regardless of whether ground ladders or aerial devices are being used for rescuing victims, the firefighters involved in the rescue operation must always be aware of the weight limitations of the ladder or aerial device. Overloading the aerial device in a rescue situation could lead to a sudden failure of the device, resulting in injury or death to all parties involved.

Priority Considerations

In situations that require using aerial apparatus for rescue, the main objective is to reach as many victims or points of egress as possible with a minimum number of aerial movements. Remove victims in the following order of priority:

1. Most severely threatened by current fire conditions
2. Largest number or groups of people
3. Remainder of people in the fire area
4. People in exposed areas

It is obvious that those who are in the greatest amount of real or perceived danger should be given the highest priority. Determining which victims are in the most danger is a judgment call that must be made by the incident commander, truck company officer, or the aerial apparatus driver/operator. Typically, those occupants located on or immediately above the fire floor will be in greatest danger (Figure 9.1). Visible fire conditions will be a strong indication of which victims are in the worst situation. Additional preference should be given to those individuals who are in a panicked state and appear ready to jump if they do not see help arriving soon. This could even include panicked victims who are not in the immediate area of the fire but are so upset that they may be ready to jump.

The second priority to be considered involves multiple victims who may be located in different parts of the fire building. When two or more groups of victims appear to be in the same amount of danger, the larger of the two groups should have the aerial device extended to them first (Figure 9.2). In a worst-case scenario, it is best to rescue the largest number of people possible, given the time available. If both an aerial ladder and an elevating platform are present in this situation, the aerial ladder should be raised to the largest number of victims. This will ensure that the greatest number of victims can exit the building in a reasonable amount of time. The elevating platform may not have room for all victims to be removed in one trip.

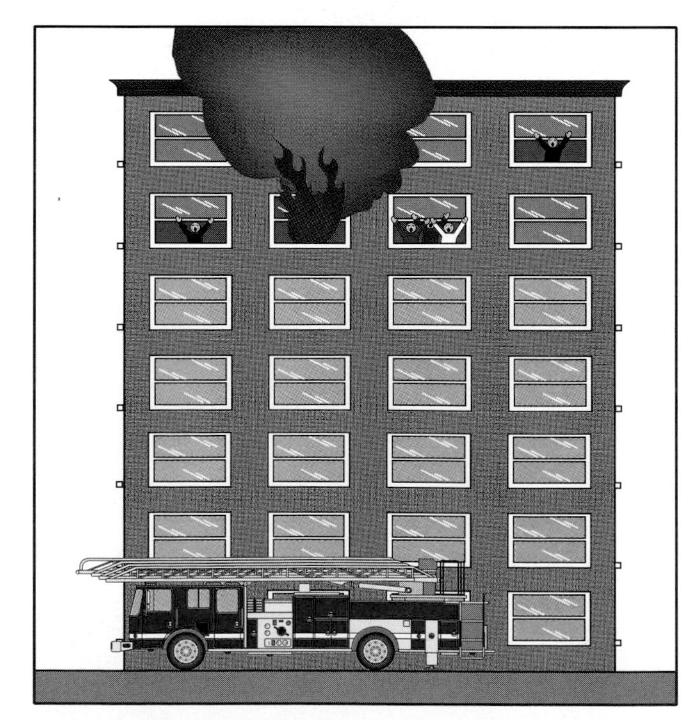

Figure 9.1 Occupants who are on the fire floor are in the greatest amount of danger.

The third priority involves the remainder of people in the fire area. Remaining groups of victims should be removed in descending order of numbers. During this process, the driver/operator and/or officer should continue to monitor fire conditions for changes that might increase the danger presented to any of the waiting victims.

The fourth priority is the people in the exposed area. Once all those who appear to be in imminent danger have been removed, evacuate other victims who are unable to escape the fire building or exposures by standard means.

In the case of victims who fall into both the third and fourth priority categories, some judgment should be used to determine whether or not the victims are in enough danger to warrant evacuation by aerial device. In many cases, these victims can simply be reassured that they are not in imminent danger and left in place until fire conditions improve. At that time, they may then be evacuated using interior stairways or fire escapes. The risk of injuring a victim during a ladder evacuation should be avoided if it is not truly necessary.

Be alert for changing fire conditions during rescue operations. In some cases, it is possible that victims who did not appear to be a high rescue priority when operations began could suddenly become the top priority due to fire growth and spread. In many cases, when rescue is the highest priority on the fireground, fire attack may not be getting the normal amount of attention. Thus, the fire may be continuing to grow and spread at will.

Raising the Aerial Device to a Victim

In all situations, the best position is one in which the extended and rotated aerial device is perpendicular to the objective (Figure 9.3). In the perpendicular position, twisting stresses imposed on the aerial are reduced. If an aerial ladder is designed to be used in the unsupported position, place the tip 4 to 6 inches (100 mm to 150 mm) above the target (Figure 9.4). This allows the ladder to contact the building as people climb or descend without placing unusual stress on the device. If the aerial ladder is designed to be used in the supported position, perpendicular positioning allows for maximum load carrying capabilities when both base rails are evenly in contact with the supporting object (Figure 9.5).

In many cases, optimum positioning is not possible, and the aerial ladder must be extended toward the objective at an angle. In these situations,

Figure 9.2 Raise the aerial device to the group of victims first. Remove the single victim after the group has been evacuated.

Figure 9.3 Try to keep the aerial device perpendicular to the victim.

extend the ladders so that the beam on the building side of the ladder is above and over the objective (Figure 9.6). When victims crawl onto the ladder, their weight will cause the ladder to settle, allowing the beam to rest on the objective. If the beam is initially rested on the objective, the weight of victims will cause the opposite beam to bend downward, which could eventually twist the ladder.

When maneuvering any aerial device to reach a victim trapped in an elevated position, initially aim the aerial device above the victim and then lower it to meet the victim (Figures 9.7 a and b). If the aerial device is raised beneath a panic-stricken victim, the victim may jump for the tip of the ladder or the platform. This could injure the victim or a firefighter or could damage the aerial device.

Once an aerial ladder is in place for a rescue, seating the extension locks is an important step in making the aerial ladder a rigid structure. This increases the overall strength of the ladder and greatly reduces the chance of ladder failure. There is no recorded failure of an aerial ladder that had the extension locks properly seated, an elevation of at least 50°, and proper loading.

Always be alert for dangling or falling debris when positioning the aerial device for a rescue. Fire or otherwise damaged buildings may have unstable debris present that could fall onto the aerial device (Figure 9.8). This may result in injury to the people on the device and/or damage to the device. When moving the aerial device into position, use care not to knock debris from the structure.

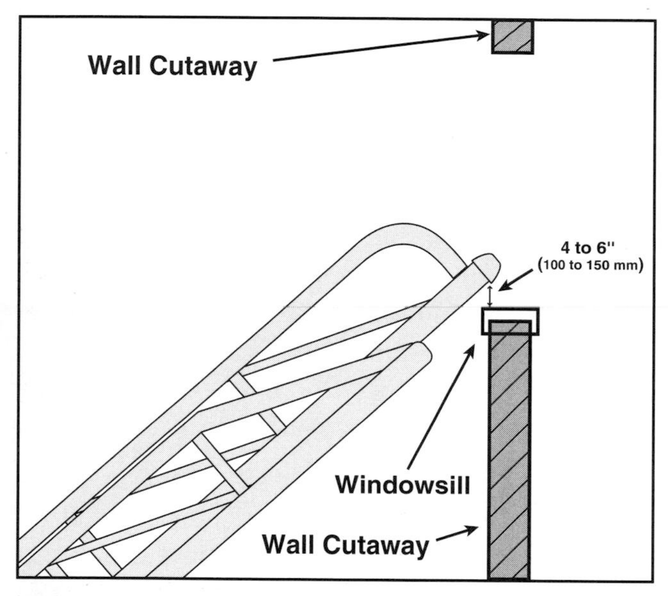

Figure 9.4 The unsupported aerial device tip is left slightly above its target.

Figure 9.5 The supported aerial device tip rests on its target.

Figure 9.6 When the ladder is extended to the roof at an angle, the beam closest to the building should be slightly above the roofline.

Figure 9.7a Raise the ladder to a point above the victim to prevent him from attempting to jump onto the ladder.

Figure 9.7b Lower the ladder into position to rescue the victim.

Figure 9.8 Be careful not to place the aerial device in a position that might subject it to debris falling from the building.

Positioning the Aerial Device for a Rescue from a Window

Placement of the aerial device tip is an important consideration when victims are to be removed from the interior of a building through a window. The tip must be positioned so that the area of the window opening is not diminished, yet safe access to the aerial device is provided for both victims and firefighters.

When using an aerial ladder, place the tip with the first rung even with the sill (Figure 9.9). In this position, the ladder will not block much of the window opening. This positioning also allows firefighters to place victims directly on the ladder without undue lifting.

When using aerial platforms, most jurisdictions prefer that the platform be placed so that the top rail of the platform is even with the windowsill (Figure 9.10). This way, victims can crawl or be

Figure 9.9 Place the first rung of the ladder even with the windowsill when performing a rescue.

Figure 9.10 The top rail of the platform is placed even with the windowsill when performing a rescue.

Figure 9.11 Depending on the angle at which the platform is approaching the window, it may be possible to make a rescue through the gate on the elevating platform.

placed directly into the bucket. An alternative to this method is to position the platform so that the access gate through the platform railing is adjacent to the window opening (Figure 9.11). In this case, open the gate to allow the victim entry to the platform. This method may not be possible depending on the design of the aerial device.

Positioning the Aerial Device for a Rescue from a Roof

When removing trapped victims from a roof (or other similar, open, flat objectives), place the tip of the aerial ladder so that at least 6 feet (2 m) extends above the edge of the roof (Figure 9.12). Place platforms so that the bottom of the platform is just above and over the edge of the roof (Figure 9.13). This allows victims to have a handhold while climbing onto the ladder or to crawl directly into the platform through the access gate.

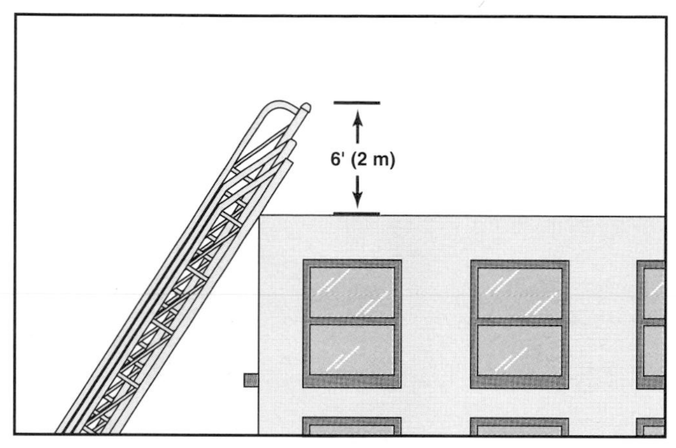

Figure 9.12 The tip of the aerial ladder should be at least 6 feet (2 m) above the roof line.

Figure 9.13 The floor of the elevating platform should be at roof level.

Placement of the aerial device for rescue from balconies with railings or roofs with parapets should follow these same guidelines. However, when rescuing victims off parapet roofs, it may be necessary to use roof ladders to get from the roof to the aerial device (Figure 9.14). Again, the aerial device should only rest against the target if supported operations are approved by the manufacturer of the aerial device. Use caution when laddering parapets because some of them tend to weaken over time. The weight of a ladder against the parapet may cause it to topple. In these cases, it may be safer to rest the top of the roof ladder against the platform or the tip of the aerial ladder.

When dealing with parapets, articulating aerial devices are advantageous over telescoping devices. In some cases, the articulating device may be able to reach over the parapet and set the platform down at roof level. Three-section (two knuckle) articulating devices are especially effective at this maneuver.

Removing Victims from Elevated Positions Using Aerial Devices

Once the aerial device is properly positioned, victims may be removed from the structure. The techniques used will vary depending on the type of aerial device and the age and condition of the victims being removed. The following sections highlight the procedures used to move victims down aerial ladders, elevated platforms, and via a Stokes stretcher using either aerial device.

Moving Victims Down Aerial Ladders

After properly positioning the aerial ladder, at least one and preferably two firefighters should be assigned to assist the victims in boarding and descending the ladder. When two firefighters are available, one should assist the victim from inside the dwelling, while the other remains on the tip of the ladder (Figure 9.15). If only one firefighter is available, the situation will dictate whether it is better for the firefighter to be on the ladder or inside the structure. Capable adults should be guided down the ladder under their own power. A firefighter should always lead the adult victim down the ladder (Figure 9.16). Caution should be taken to make

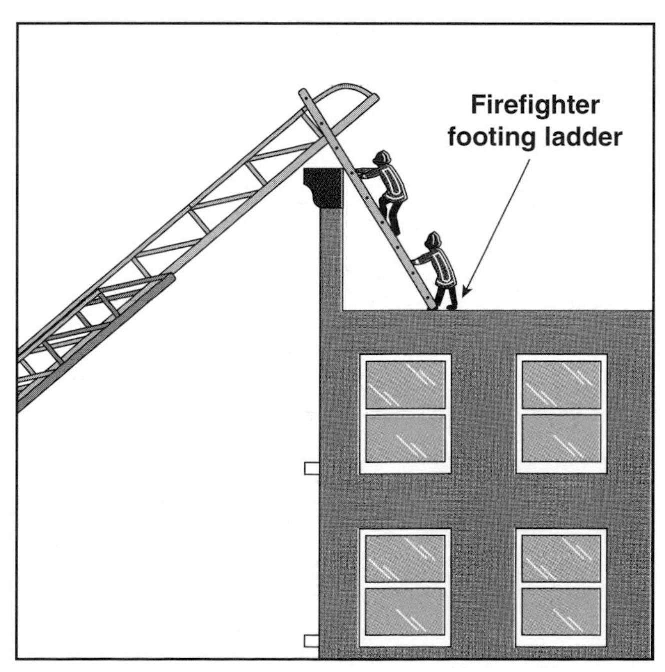

Figure 9.14 It may be necessary to use a ground ladder to traverse a tall parapet.

Figure 9.15 The firefighter inside the building assists the victim onto the ladder while the firefighter on the ladder makes sure the victim does not fall during the process.

Figure 9.16 The firefighter on the ladder should lead the victim down the ladder.

sure that multiple victims are adequately spaced down the length of the ladder according to the manufacturer's recommendation to avoid overloading it.

Small children and adults incapable of climbing the ladder by themselves will have to be carried. Remember, this is a last resort tactic and should only be done when all other means of egress have been ruled out. Infants and small children may be cradled in one arm while the other arm is used to assist with balancing while descending the ladder (Figure 9.17). Larger victims will require the firefighter to use an over-the-shoulder method of carrying. When using this method, at least two and preferably three firefighters will be needed. One firefighter should be on or inside the structure to help load the victim onto the shoulder of the firefighter who will carry the victim down the ladder (Figure 9.18). The third firefighter should provide backup for the second firefighter as the victim is moved down the ladder. In order to provide the best possible balance, position the victim face down across the firefighter's shoulder. The victim's waist should be on the firefighter's shoulder, and the victim's knees should be in the middle of the firefighter's chest so that the firefighter can hold the victim in position (Figure 9.19). Once the victim has been positioned, the firefighter should back down the ladder. The assisting firefighter should remain in close contact with the carrying firefighter as the descent is being made (Figure 9.20). An alter-

Figure 9.17 Cradle infants (bundled in a blanket) or small children in your arm as you descend the ladder.

Figure 9.18 Firefighters inside the building should assist in loading the unconscious victim onto the ladder.

Figure 9.19 The victim's waist should be on top of the rescuer's shoulders.

Figure 9.20 If it can be done without overloading the ladder, one firefighter should back up the rescuer as the ladder is descended.

Figure 9.21 The victim may be slid down the ladder handrails.

native method for carrying an unconscious victim down an aerial ladder is to place the victim perpendicular to the ladder, across both rails (Figure 9.21). The firefighter performing the rescue cradles the victim across both arms and places each hand on the handrail for support. The victim is then simply slid down the ladder as the firefighter descends.

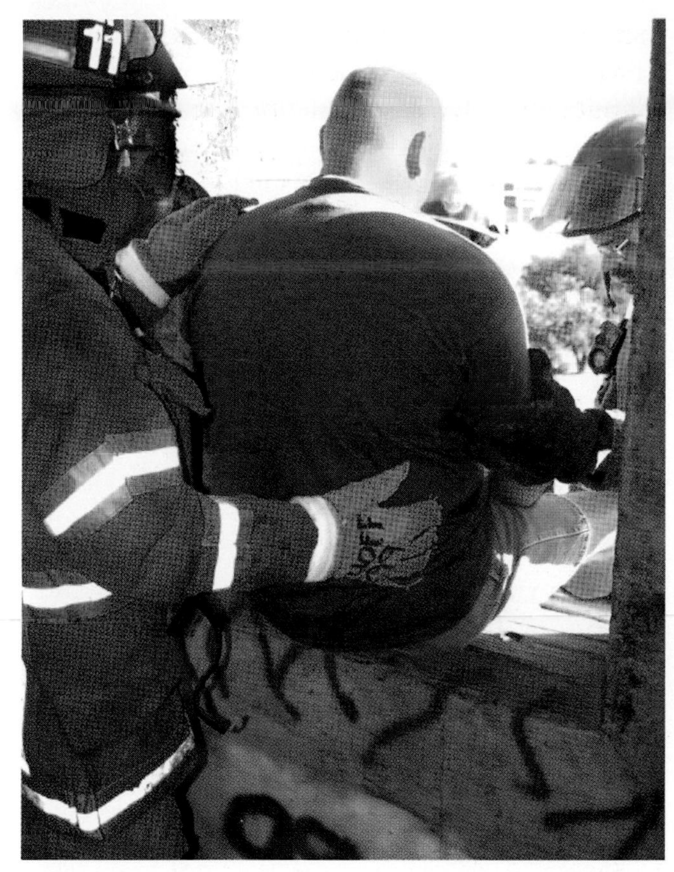

Figure 9.22 Make sure the victim does not jump into the platform.

Moving Victims in Aerial Platforms

Using aerial platforms to move victims is considerably easier than backing them down aerial ladders. However, it is somewhat slower and not suited to mass evacuations from a single point. The removal of conscious victims requires at least one and preferably two firefighters. With the top railing positioned even with the windowsill or balcony railing, victims should be assisted down into the platform (Figure 9.22). Do not allow victims to jump into the platform as this creates a severe and potentially damaging dynamic load on the aerial device. If local standard operating procedures dictate that the victims enter the platform through the railing access gates, position the platform accordingly and assist the victim(s) through the opening.

The number of passengers who may be lowered into the platform will be determined by the load capacity of the aerial device and space in the platform. Whenever possible, this load rating should not be exceeded. One firefighter lowers the aerial device to the ground to unload the victims. The second firefighter, if available, remains at the loading point to calm and reassure the remaining victims while the aerial device is being unloaded and returned to the rescue point (Figure 9.23).

Figure 9.23 Leave a firefighter at the rescue point until the aerial device can return to retrieve the remaining victim(s).

The removal of unconscious victims requires a minimum of two firefighters. At least one firefighter will have to lift the victim from the structure to another firefighter waiting in the platform (Figure 9.24). If the victim is not on a litter, set or lay the victim on the floor of the platform. If the victim is in a litter, preferably a basket-type litter, the litter may be laid on the floor of the platform. If the litter will not fit on the floor of the platform, it should be laid across the top railings and lashed in place (Figure 9.25).

Lowering a Stokes Litter Using an Aerial Ladder

If victims are unconscious or injured and unable to move themselves, it may be necessary to evacuate them down an aerial ladder via a basket litter. As with other aerial evacuations previously discussed, this tactic should be a last resort. It is much easier and safer to carry a litter down a stairwell, fire escape, or move it to a safe place of refuge until conditions improve. These should be the first choices whenever possible. When it is not possible to carry the litter down a stairwell, there are two methods of lowering a stokes stretcher via the aerial ladder. The easiest method is to slide the litter down the ladder. Two methods may be used to slide the litter down the ladder. The simplest method works only if the stokes litter will fit between the aerial ladder rails. In this case, the litter is simply placed on the rungs of the ladder. At least one and preferably two firefighters should guide the litter down the ladder, taking care not to exceed the ladder's rated capacity. A rope, controlled by one or more firefighters near the tip of the ladder, may be used to help control the descent of the litter (Figure 9.26).

If the litter does not fit between the rails of the ladder, place it on top of the handrails, perpendicular to the ladder. If this method is used, use a guide rope from the topside to help control the descent of

Figure 9.24 It will be necessary for firefighters in the structure to pass the victim to the firefighters in the platform.

Figure 9.25 The litter may be lashed to the rails of the platform.

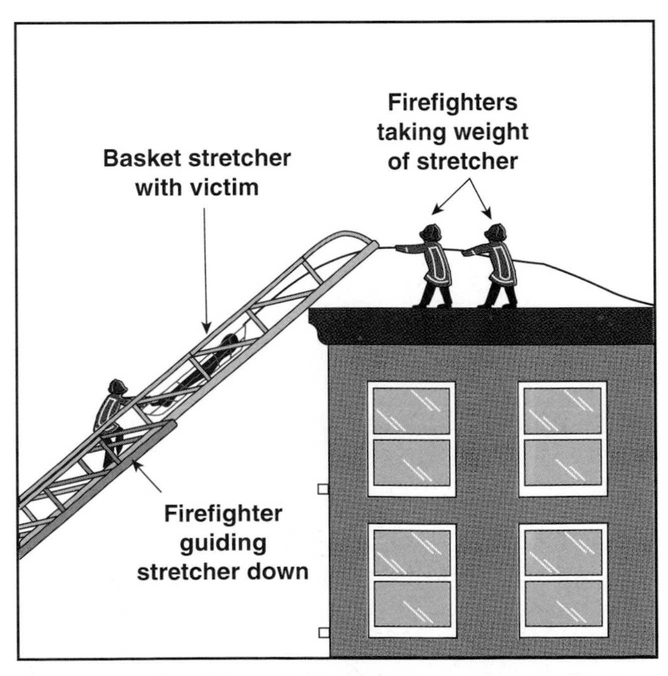

Figure 9.26 On wide aerial ladders, it may be possible to slide the litter down the ladder between the handrails.

the litter. Some departments also prefer lashing two long-handled tools, such as pike poles, to the bottom of the litter to assist its sliding ability (Figure 9.27). A firefighter should back down the ladder with the litter to make sure that it does not slip off the rails. As the litter nears the bottom of the ladder, at least two firefighters on the aerial device turntable should assist in removing the litter from the ladder.

In the past, some fire departments have used aerial ladders as part of a rope lowering system for removing victims in basket litters from elevated positions (Figure 9.28). Under the best of circumstances, this is a dangerous evolution. In some cases, the strain placed on the aerial ladder exceeds the manufacturer's recommended load limitations. As well, these evolutions have the potential to place strains on the rope that exceed their breaking strength. Before ever attempting such an evolution, make sure that the manufacturer of the aerial ladder, rope, and rope hardware approve of the use. With the rapid increases in rope rescue technology in the past decade, the need to use aerial ladders as part of the lowering system has decreased significantly. For more information on rope lowering systems, see the IFSTA **Fire Service Rescue** manual.

Using Aerial Devices for Water Rescues

In the past, aerial apparatus have often been called upon to assist personnel who are faced with water rescue situations. The two most common scenarios include vehicles that have driven into high water across roadways and removing people who have become trapped on a tree or similar object in a rain-swollen moving body of water. Numerous successful rescues using aerial apparatus in these operations can be documented. However, in general, it is recommended that fire departments discourage the use of aerial apparatus (particularly light- and medium-duty aerial devices) for these operations.

The first safety issue that affects aerial apparatus at water rescue incidents is often a lack of a suitable location to adequately stabilize the apparatus. In order to reach the victim, it is often necessary to spot the apparatus only a few feet (meters) from the water's edge. In many cases, this is muddy or otherwise unstable ground that will not safely support the weight of the apparatus. This is especially critical when you consider that the aerial device is most

Figure 9.27 The basket litter may be slid down the ladder rails.

Figure 9.28 In the past some departments have used aerial ladders as part of a rope lowering system.

likely to be deployed to near its full extension and at a very low angle of elevation. This is the most unstable scenario for aerial device deployment.

Another safety issue that makes the use of aerial apparatus dangerous for water rescues is the potential for the aerial device to make contact with swiftly moving water. Moving bodies of water generate tremendous amounts of force on objects that enter them. As mentioned earlier in this manual, aerial devices are typically not designed to handle a considerable amount of lateral stress. Should an extended aerial device come into contact with swiftly moving water, the force on the ladder could result in serious damage to the device (Figure 9.29). At the very least, the device could be bent or twisted in a manner that could result in people on the ladder being cast into the water. If a firefighter was connected to the aerial device via a safety harness, it could be extremely difficult for him to free himself once the aerial device entered the water. The potential to incur a fatal drowning injury in this circumstance is very real.

Lastly, on aerial devices that have a low tip-load rating when fully extended to a horizontal objective, it is difficult to perform such a rescue without seriously overloading the aerial device. For example, if the aerial device has a 250 pound (113.4 kg) tip load, it is very likely that the weight of one firefighter and one victim will well exceed this limit. This again creates the potential for damage to, or failure of, the aerial device.

Figure 9.29 The aerial device can be damaged if it is lowered into moving water.

There are a number of safer methods for performing water rescues than by using aerial apparatus. For more information on these methods, consult the IFSTA **Fire Service Rescue** manual or *Swiftwater Rescue* by Slim Ray.

Aircraft Rescue

Aerial apparatus will have a limited role in rescue operations involving aircraft rescue and fire fighting (ARFF) incidents. As mentioned in the discussion in Chapter 5 on positioning aerial apparatus at ARFF incidents, aerial apparatus will serve as little more than support apparatus at most aircraft crash scenes. There is a more realistic possibility of using the aerial device for rescue purposes at no-crash fire scenarios involving large-frame aircraft. However, even at these incidents, use of the aerial device would be well down on the scale of desirable rescue methods.

All larger commercial passenger aircraft and some military aircraft are equipped with escape chutes or slides at the exit door openings (Figure 9.30). When deployed, these slides either inflate automatically or can be manually activated by a crew member or passenger. These slides normally provide a much faster means of evacuation than do steps or ladders. If these chutes or slides are provided and are in use when ARFF units arrive, they should not be disturbed unless they have been damaged by use or are threatened by fire exposure.

The second best option for rescuing victims from an aircraft is to place a set of portable stairs at the door of the aircraft. These stairs are typically

Figure 9.30 Inflatable escape chutes are the most common method of escape for occupants of large frame aircraft. *Courtesy of Dallas/Fort Worth International Airport.*

mounted on some type of truck or similar vehicle that can be driven into place. These stairs will also provide a more rapid method of escape than would a fire department aerial device.

If the first two options are not available, the aerial device may be used to perform rescues (Figure 9.31). The positioning of the aerial device in an aircraft exit opening should follow the principles that were described for performing window rescues from a building (Figure 9.32). If the device is being raised to the wing of a large aircraft, follow the procedures that were described for rescues from a roof (Figure 9.33). Use extreme caution when mounting and dismounting an aerial device from the wing of an aircraft. The surface of a wing is slippery under perfect conditions. It can be treacherous when water and foam have collected on it.

Figure 9.31 Many airport fire departments have aerial apparatus to assist with incident operations. *Courtesy of Joel Woods, Maryland Fire & Rescue Institute.*

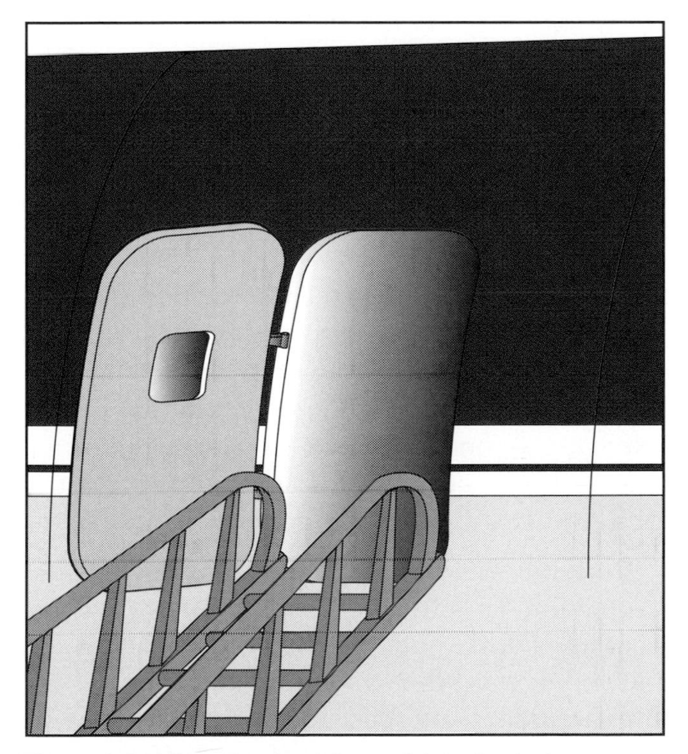

Figure 9.32 Place the tip of the aerial device in the doorway for rescue purposes.

In some cases, an aerial apparatus may be needed when a small aircraft has crashed and has come to rest in trees or other elevated locations. In these situations, the aerial device should be used to gain access to the aircraft and to remove the victim(s). The aerial device should not be used to stabilize the aircraft or lower it from its resting position. Make sure that the apparatus is parked and that the aerial device is operated in such a manner that if the aircraft becomes dislodged, it will not strike the apparatus. If the aircraft has come to rest in or near electric lines, you should refrain from deploying the aerial device to the aircraft until the appropriate utility company can verify that the lines have been de-energized.

For more information on rescuing victims from aircraft crashes and fires, see IFSTA's **Aircraft Rescue and Fire Fighting** manual.

Below Grade Operations

There are some instances where it may be desired to deploy the tip of the aerial device to a location that is below the level on which the apparatus is

Figure 9.33 The aerial device may be raised to the wing of the aircraft to provide access or egress.

parked. The aerial device may also be used to support other operations that are being conducted below grade, without actually having to place the tip of the device that low. These are more commonly referred to as below grade operations. Before attempting this type of operation with any aerial apparatus, consult the manufacturer's operating guide to ascertain whether or not the device can be safely deployed to a negative elevation.

WARNING!

If the aerial device manufacturer does not recommend deploying its aerial device below grade, do not attempt this type of maneuver. Aerial device failure or apparatus tip over could occur.

The most common scenario where aerial device below-grade operations are desired is special rescue. These situations include victims that are trapped in deep holes or other subterranean features or victims of automobile accidents that have gone over an embankment. Most aerial devices are not designed to have their working tip actually deployed to these locations. However, sometimes it is desired to use the aerial device to lift the victim from below grade. Again, the driver/operator must be familiar with the manufacturer's load and usage limitations. In most cases, aerial ladders are not designed to lift these types of weight loads. Aerial platforms may be more suited to these functions. Some aerial platforms are equipped with winches on the platforms that allow lifting operations to be safely performed, as long as weight restrictions are not exceeded.

One type of aerial device that is particularly well-suited for below grade rescue operations is the three-boom, articulating aerial platform. This device has the ability to reach out, down, and under the surface on which the apparatus is parked. For example, this type of device could be used to rescue someone who is trapped underneath a bridge that the apparatus is actually parked on. However, even these devices have their limitations and manufacturer's use recommendations must be followed explicitly.

Some aerial devices may also be used to deploy master streams to below grade locations. When the water system is charged, make sure to closely follow any range of motion restrictions imposed by the manufacturer.

Exposure Protection

The second priority for aerial device operations is exposure protection. An *exposure* is defined as a structure or separate part of the fireground to which the fire could spread. Fire spreads to exposures by either radiated or convected heat. *Radiated heat* is heat movement in all directions in the form of energy waves (Figure 9.34). *Convection* is heat travel through a heat-circulating medium, the most common being air (Figure 9.35). Direct flame contact is considered to be a form of convection.

Conditions that may affect exposure hazards include the following:

- Weather
- Building construction
- Spacing between the fire building and the exposure

Both present and recent weather conditions affect a fire's ability to spread to exposures. Obviously, rain reduces the potential for fire spread. Even if it is not actually raining at the time of the fire, the increased moisture content in exposed combustible material prolongs its resistance to ignition. Rain also reduces the chance of natural cover catching fire and spreading to surrounding structures.

On its own, rain is not a reliable method of ensuring adequate exposure protection. In situations where the exposure is severely threatened, it is probable that the amount of rain falling on the exposure is not sufficient to prevent ignition. Standard tactical means of providing exposure protection will have to be employed in these cases.

Certainly wind plays a large part in exposure protection. The effects of radiated and convected heat transfer are enhanced by wind. If the wind is blowing toward the exposure, the chance of the fire spreading is increased. If the wind is blowing away from the exposure, the chance of fire spread is reduced, although not eliminated. In severe

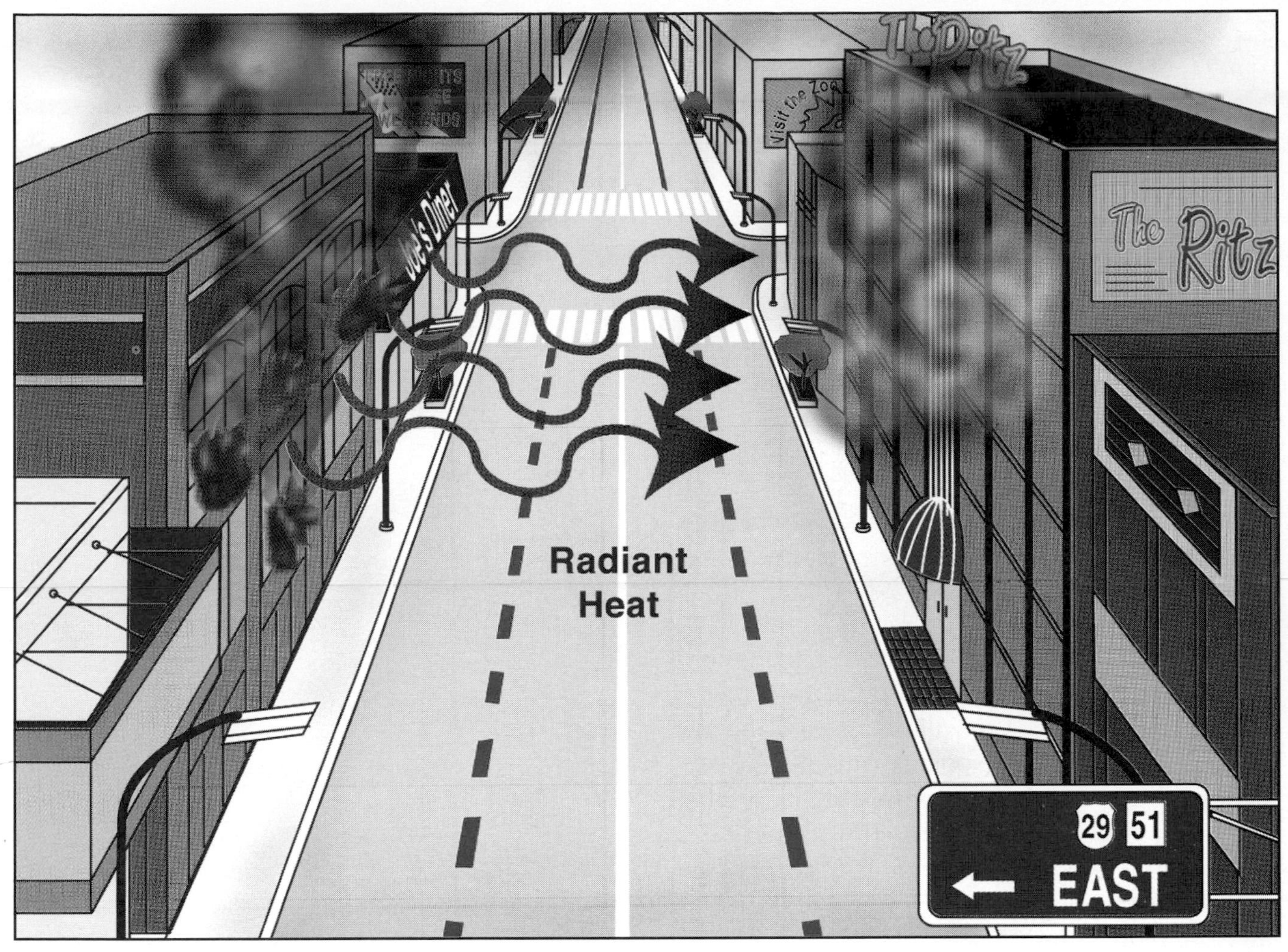

Figure 9.34 Radiated heat is the biggest factor in exposure protection.

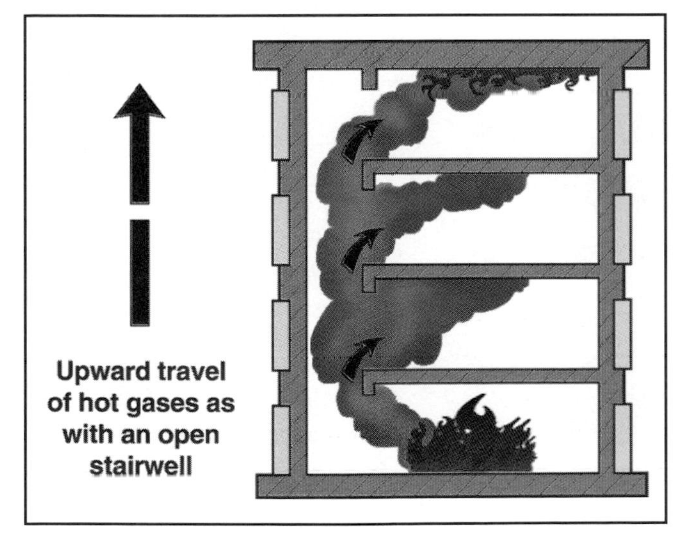

Figure 9.35 Convected heat can affect exposure protection to some extent.

fire situations, the fire itself will create very strong winds.

The type of construction and the proximity of the exposure to the fire building also affect the chance of the exposure catching fire. For example, exposures with combustible exterior finishes have an increased risk of secondary fire involvement. Some combustible construction elements include wood or plastic siding, wooden window frames, and interior combustible items (such as curtains or blinds) close to large windows or glass areas (Figure 9.36). Obviously, the closer the exposure is to the fire building, the more likely it will catch fire under exposure conditions. Proximity and construction materials should be considered during response and size-up of the incident.

Also keep in mind that the apparatus itself must be considered an exposure hazard. In some cases, it may even be the most valuable exposure at the scene. Do not place the apparatus in a position of needing the same level of protection that other structures close to the fire building need. After all, you can move the apparatus; that is generally not possible with the other exposures you are protecting.

Figure 9.36 Note the vertical blinds behind the glass beginning to burn.

Because radiated heat transmits through water, simply directing fire streams into the air between the fire and the exposure will not appreciably reduce the chance of the exposure catching fire. Exposure protection is best accomplished by placing water directly on the exposed structure (Figure 9.37). Water placed directly on the exposure absorbs the heat energy developed on the surface of the exposure and reduces its ability to burn.

Figure 9.37 Water must be placed directly on an exposure to ensure proper protection. *Courtesy of Ron Jeffers.*

Protection from convected fire spread can be accomplished by placing large fog streams into the thermal column (Figure 9.38). This may reduce the chances of burning embers traveling through the air and igniting other structures. When resources are limited, it may be necessary to use one master stream to both fight the fire and protect the exposures. This can be done by positioning the apparatus so that the master stream can be moved between the fire and the exposure (Figure 9.39). In these situations, care should be taken not to neglect the exposure by devoting too much attention to the fire itself. Some elevating platform apparatus are equipped with two nozzles on the platform. This allows one stream to be played on the fire while the other protects the exposure (Figure 9.40).

Figure 9.38 A fog stream placed into the thermal column may help reduce the amount of convected heat that is directed toward exposures. *Courtesy of Ron Jeffers.*

Positioning aerial (or other) apparatus to provide exposure protection provides some interesting challenges that require a delicate balance be considered. Position apparatus to allow for effective and safe stream placement. Position apparatus for maximum coverage of the exposure and, if desired, coverage of the fire building as well. However, if the fire is large enough to create exposure problems for other buildings in the area, the apparatus assigned to protect the exposed buildings

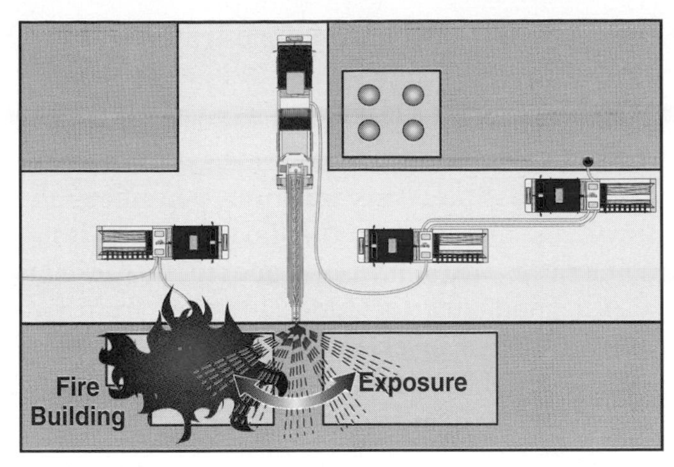

Figure 9.39 Alternate the stream between the exposure and the fire building.

also become exposures. While it is desired to position the apparatus to provide maximum protection to the exposed buildings, do not park it in a location that will require drastic measures to protect the apparatus itself (Figure 9.41). Always leave a straight, forward exit route out of the area in the event that fire conditions worsen and exposure protection operations must be abandoned.

In exposure protection operations, wide sweeping nozzle patterns are generally used to cover the involved area. When using sweep patterns, take care not to rotate the stream beyond the safe limits imposed by the manufacturer of the aerial device. Also, do not operate elevated streams in man-made or natural ventilation openings at any time. Take care to avoid breaking windows. Broken windows may result in extensive water damage and could also enhance the fire's ability to spread inside the exposure.

The exact type of stream to use will depend on the conditions present at each fire. When possible, a fog pattern will allow for maximum coverage of the exposure. However, fog patterns may not always work due to a variety of conditions, including:

- High wind conditions
- Excessive distance between the nozzle and the exposure
- Extensive heat from the fire evaporating the stream prior to it reaching the intended target

In these cases, it will be necessary to use a straight or solid stream. Straight and solid streams

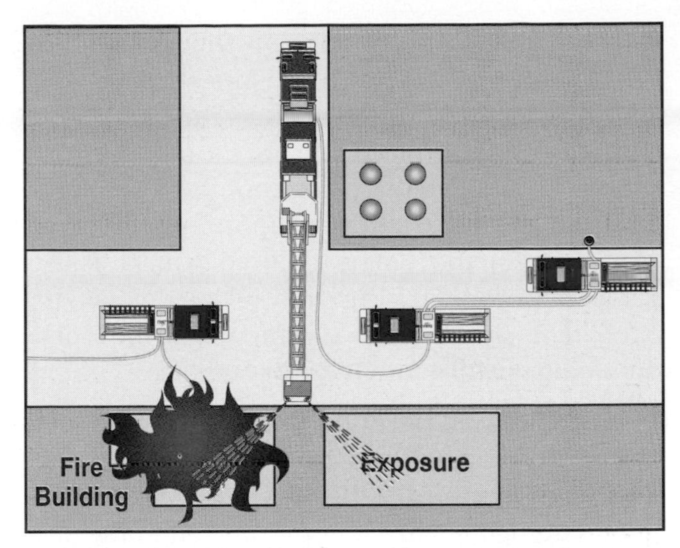

Figure 9.40 Elevating platforms with two master stream nozzles may protect the exposure and attack the fire at the same time.

Figure 9.41 The need to spray water on the apparatus to keep it cool is indicative of poor positioning and the need to immediately relocate the apparatus. *Courtesy of Ron Jeffers.*

must be constantly maneuvered to cover the entire face of the exposure. Because of their power, exercise particular care with these streams to avoid hitting and breaking windows.

◆ Ventilation

The third priority for aerial device operations is to assist in ventilating buildings. Aerials are often used to place firefighters on the roof or to provide access to upper-story windows. Recognizing the importance of ventilation in the overall combat plan for fire operations, it is important to note that ventilation operations should not leave rescue operations deficient of apparatus or manpower.

When putting firefighters on the roof with an aerial ladder, extend the fly sections so that a minimum of 6 feet (2 m) protrudes over the edge (Figure 9.42). This makes the location of the ladder plainly visible to firefighters on the roof, even in smoky conditions. This could be important should the firefighters on the roof be required to make a hasty retreat. Before exiting the aerial device, the firefighters should check the integrity of the roof by sounding it with a forcible entry tool.

When using an elevating platform, the bottom of the platform should be positioned even with or extended slightly over the edge (Figure 9.43). This allows firefighters to easily exit and enter the platform. When operating on parapet roofs, it may be necessary to use a small roof ladder as a means to get from the aerial device to the roof. Use caution when laddering parapets, as some of them tend to weaken over time. The weight of a ladder against the parapet may cause it to topple. It may be safer to rest the top of the roof ladder against the platform or the tip of the aerial ladder.

Figure 9.42 The tip of the ladder should be 6 feet (2 m) above the roofline when firefighters are ascending for ventilation purposes.

Position the aerial device as close to the work area as possible. This position minimizes the travel distance back to the device should it become necessary to hastily evacuate the roof. It is also a good idea to ladder the roof from two different sides to permit an alternate route of escape should the crew be cut off from the first aerial device (Figure 9.44).

If firefighters cannot safely operate on the roof, or if other conditions warrant a different approach, horizontal or cross ventilation can be accomplished with the assistance of aerial devices. A firefighter on the tip of a ladder or in the platform can break upper-story windows with a pike pole or similar tool. When operating in this manner, the firefighter doing the ventilation should be secured to the ladder or platform with a safety belt. Position the aerial device slightly above the window and slightly to the upwind side (Figure 9.45). If a wind is blowing across the face of the building, take out windows on the downwind portion first, using proper ventilation practices. By taking a position slightly above the window, the firefighter is not subject to fall-

Figure 9.43 The firefighters going to the roof to vent will maneuver the platform so that its floor is at roof level.

ing glass. Ensure that firefighters below the window(s) being broken stay clear to avoid falling glass and debris (Figure 9.46). For more information on ventilation techniques, see the IFSTA **Fire Service Ventilation** or **Essentials of Fire Fighting** manuals.

> **WARNING!**
> All firefighters operating on an aerial device should be attached to the aerial device by an appropriate safety harness at all times.

Elevated Fire Attack

The final priority for aerial device operations is the use of elevated master streams for fire attack (Figure 9.47). Large-scale, defensive fire fighting opera-

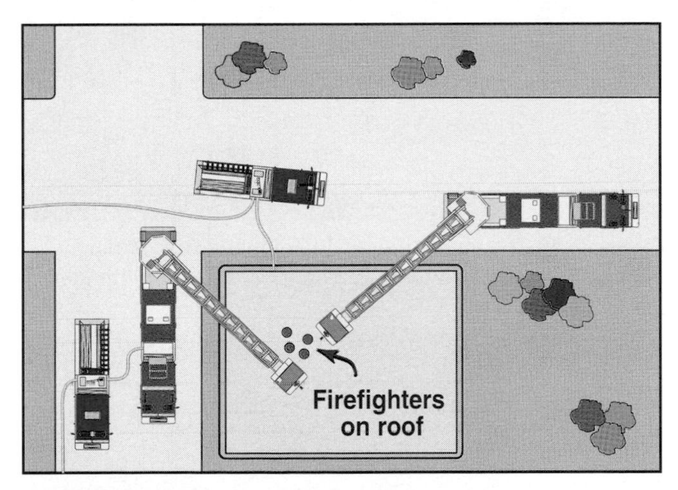

Figure 9.44 When possible, ladder the roof from more than one side during venting operations.

Figure 9.45 The proper positioning for venting a window from the ladder.

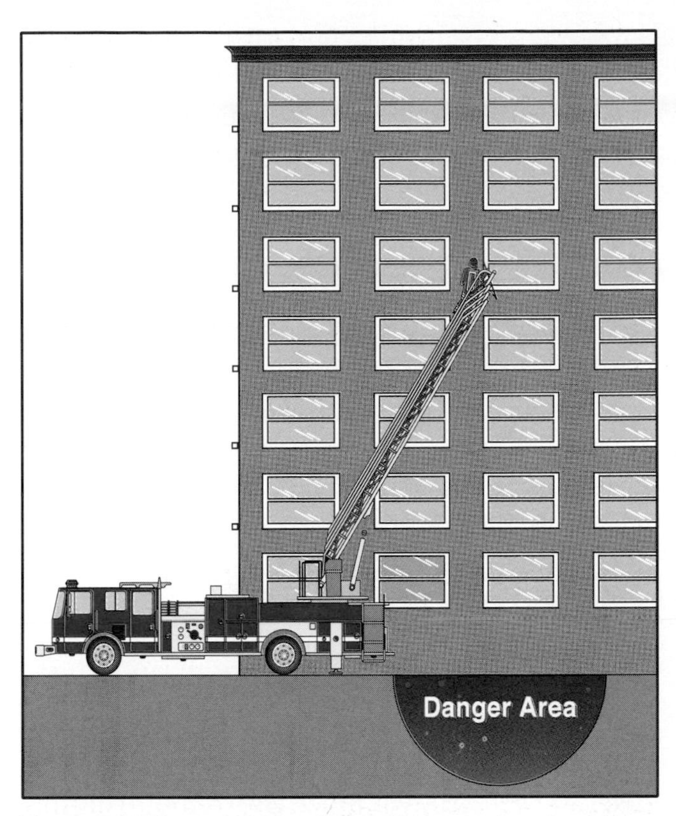

Figure 9.46 The area below the venting operation should be kept clear to avoid injuries from falling glass and other debris.

Figure 9.47 Defensive operations are the most common use for elevated master streams. *Courtesy of Chris Mickal.*

tions often require the use of elevated master streams. The following sections describe the basics of elevated master stream operations and proper stream placement. For more information on calculating the hydraulic requirements for elevated master streams, see the IFSTA **Pumping Apparatus Driver/Operator Handbook.**

Operating Elevated Master Streams

The types of elevated water delivery systems that are used on aerial apparatus were covered in detail in Chapter 2 of this manual. For operational purposes, elevated water delivery systems can be divided into three categories:

- Aerial ladders with piped waterways and water towers
- Aerial ladders with detachable waterways
- Elevating platforms

Aerial ladders with piped waterways and water towers may both be equipped with remote controls that allow the fire stream to be manipulated from the aerial platform control pedestal, the pump panel of the apparatus, or from a remote control box. Typically the stream pattern and direction can be changed using these controls (Figures 9.48 a through c). It is always most desirable to operate the nozzle with the remote controls. This negates the need to place a firefighter in a hazardous position on the tip of the ladder.

One advantage provided by some water tower devices is the addition of a video camera on the tip of the aerial device (Figure 9.49). A corresponding video monitor is then also placed close the aerial device controls (Figure 9.50). This allows the aerial

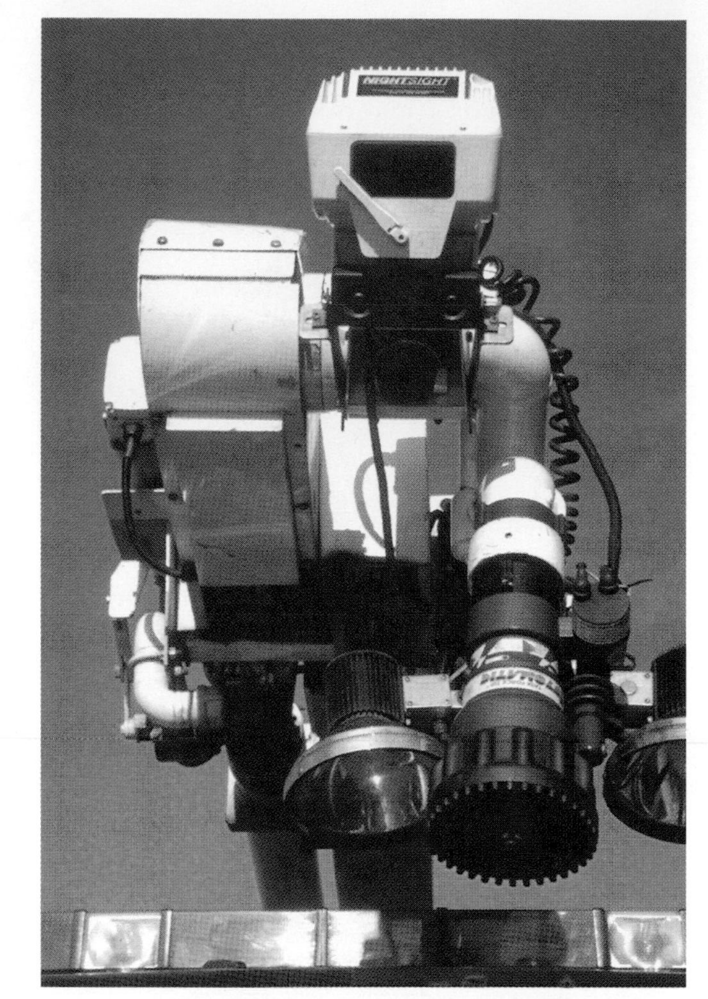

Figure 9.49 This aerial device is equipped with both a standard video and a thermal imaging camera at the tip.

Figure 9.48a Remote nozzle controls on the lower control pedestal.

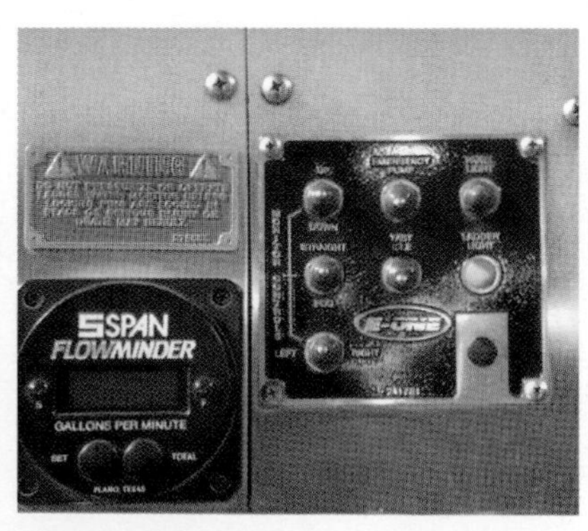

Figure 9.48b Nozzle controls may be located on the pump panel of a quint.

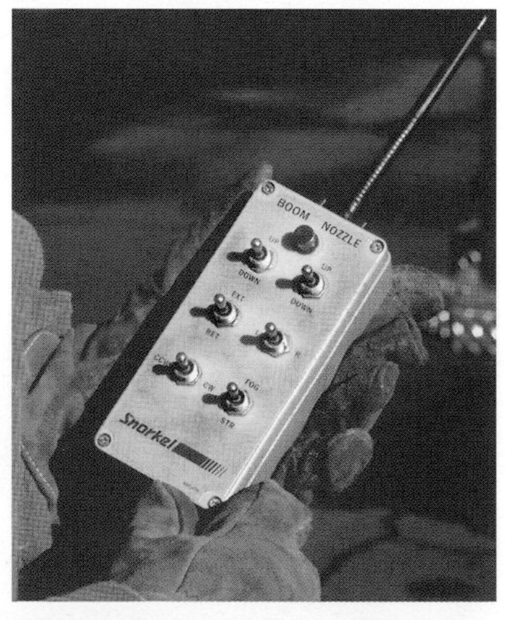

Figure 9.48c The nozzle controls are truly "remote." This hand-held unit allows the aerial device or the nozzle to be operated from hundreds of feet (meters) away from the apparatus.

device operator to view the deployment of the device from the tip and accurately direct the fire stream to the target. Aerial ladders and elevating platform apparatus may also be outfitted with these cameras and monitors. In some cases, the camera used on these devices may have thermal imaging capabilities that allows the operator to see the intended target through the smoke.

Aerial ladders with detachable waterways present more of an operational challenge than do those with piped waterways. Historically, firefighters have been placed at the tip of the ladder to manually operate the nozzle. However, this practice places the firefighter in danger from a number of standpoints, including:

- The firefighter is exposed to the heat and other products of combustion (Figure 9.51). Should a rapid change in fire conditions occur, the firefighter may be forced to quickly descend the ladder through these conditions.
- These types of ladders typically do not have aerial device operational controls at the tip. Thus the firefighter is unable to move himself out of danger should conditions worsen and the operator at the pedestal controls is distracted or not readily available.
- It is generally not considered safe to move aerial ladders with detachable waterways a great amount while the hose and nozzle are charged. The aerial ladder is not designed for these types of stresses.
- Should the nozzle come loose or the hose supplying the nozzle burst, the firefighter could be struck and injured or thrown from the ladder.

Figure 9.50 The video monitor is located adjacent to the pump panel and aerial device controls.

IFSTA recommends that detachable ladder pipes be operated from the turntable or ground level using ropes that are attached to the nozzle (Figure 9.52). These ropes allow the nozzle to be raised and lowered as desired. It will still be necessary for a firefighter to climb the ladder and manually change the stream pattern if desired. If this is necessary, the firefighter (wearing full personal protective clothing and an SCBA) should climb to the tip of the ladder, attach himself to the ladder using an ap-

Figure 9.51 Firefighters who are at the tip of the aerial device during fire stream operations may be subjected to dangerous conditions.

Figure 9.52 It is safer to operate the elevated master stream from the ground, even if ropes are needed to accomplish this. *Courtesy of Ron Jeffers.*

proved safety harness, change the stream, and then immediately descend the ladder. An even better choice would be to shut down the water flow before the firefighter ascends the ladder.

Some other general safety principles that should be followed when using detachable ladder pipes are as follows:

- Never should more than one person be on the top section of the aerial ladder when changing the stream pattern. ***Do not move the ladder while the firefighter is at the tip.***
- Avoid sudden movement or surges when using the ladder pipe. Slowly elevate or lower the ladder pipe barrel and gradually increase the nozzle pressure. Sudden pressure changes produce a surge that results in shock loading and violent ladder sway, with possible adverse results.
- Always locate the feeder hose in the center of the ladder (Figure 9.53). Never hang charged hose over the side of the ladder.
- When possible, supply the ladder pipes from two different water supply sources. This prevents damage to the ladder should one water supply source suddenly be shut off.

Figure 9.53 The supply hose should be in the center of the ladder.

- Make sure that turntable movement is smooth and slow when rotating the turntable to direct the ladder pipe stream in different directions.
- Use the 75-80-80 rule of thumb for quick ladder pipe use: 75-degree elevation, 80 percent extended length, and 80 psi (560 kPa) nozzle pressure for a 1½-inch (38 mm) tip (Figure 9.54).
- Use ladder pipes perpendicular to the rungs on older units, with a maximum lateral movement of 15 degrees to either side (Figure 9.55). Many new models offer 180° sweep capability.

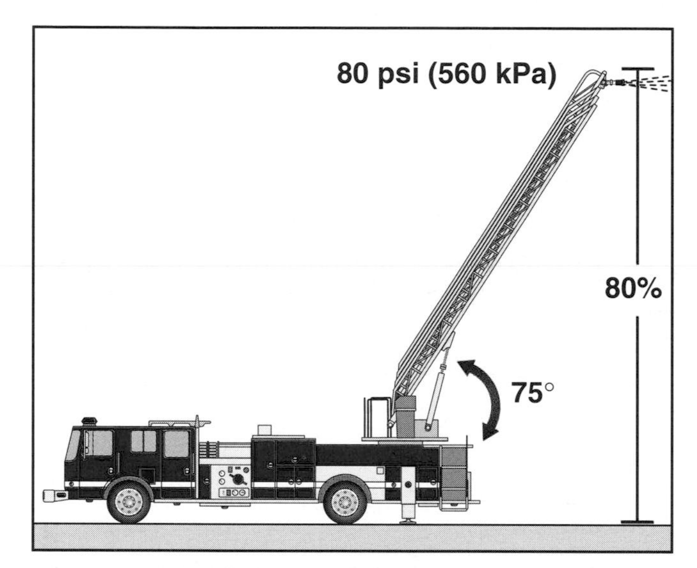

Figure 9.54 The ladder should be elevated to 75°, extended no more than 80 percent of its length, and the nozzle pressure for a solid stream nozzle should be no more than 80 psi (560 kPa).

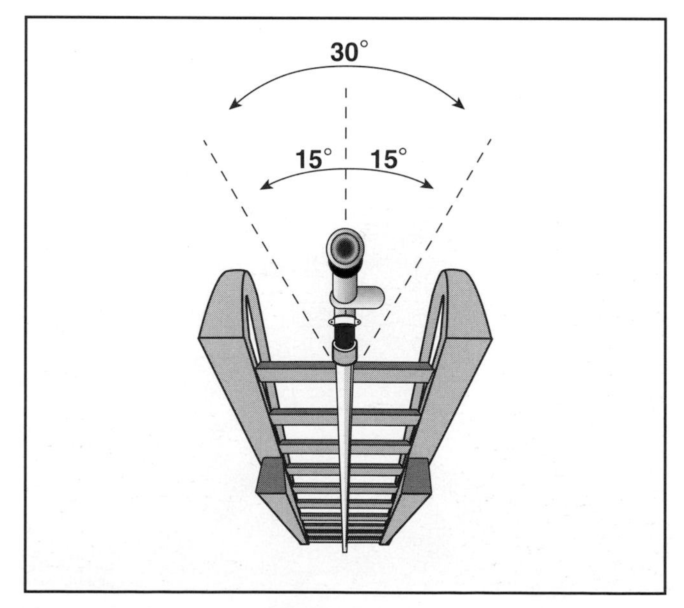

Figure 9.55 The ladder pipe's horizontal sweep should be limited to no more than 15° in either direction from center.

- Do not use a 2½-inch (65 mm) handline and nozzle strapped to the ladder as a ladder pipe.
- Always follow any specific operating guidelines that are provided by the manufacturer of the aerial device.

If it is desired to place one or more firefighters at the position where the elevated master stream is going to be deployed, it is preferable that some type of elevating platform apparatus be used. These devices provide a safer standing position for firefighters operating the stream. Most elevating platforms also have piped breathing air systems that allow the firefighter(s) to breathe clean air for extended periods of time (Figure 9.56). They also have a protective water curtain nozzle beneath the platform that can be used in high heat situations. Lastly, elevating platforms have their own control pedestals that allow the firefighters at the tip to move the device should they need to do so for safety (or operational) purposes.

Fog vs. Solid or Straight Streams

Both the straight tip and the fog nozzle have their respective place in aerial operations. If at all possible, both types of nozzles should be carried on the apparatus and be available for use from the aerial.

Fog nozzles provide wider coverage than straight tip nozzles, and they are able to break up the water to effect better steam conversion. The drawback to fog nozzles is that they do not provide the reach and penetration often needed to reach the seat of the fire (Figure 9.57). They are also adversely effected by atmospheric or fire-created wind. In a severe fire situation, the water from a fog nozzle may be converted to steam before reaching the seat of the fire. The best applications for fog streams are close-up blitz attacks and exposure protection. For most defensive or otherwise long-range operations, set the fog stream nozzle to the straight stream setting. This will closely emulate the performance of a solid stream nozzle.

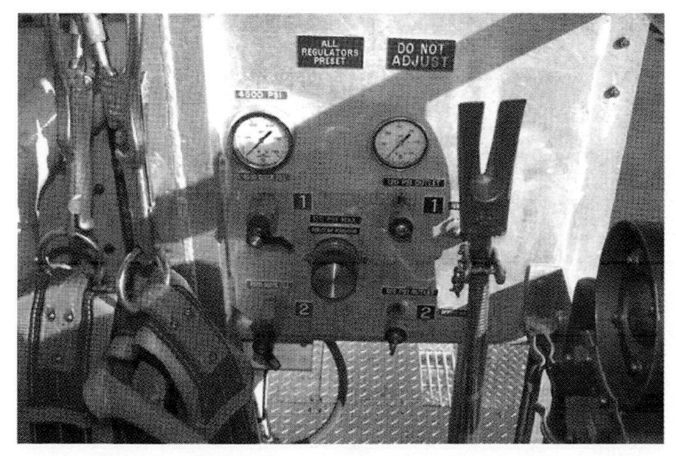

Figure 9.56 Most elevating platform apparatus have fixed breathing air systems.

Solid stream nozzles provide excellent penetration into fire areas (Figure 9.58). If an outside el-

Figure 9.57 In many cases, the water from a fog stream will evaporate before reaching the seat of the fire. *Courtesy of Chris Mickal.*

Figure 9.58 Solid streams are more likely to hit the seat of the fire.

evated stream is to be directed into the interior of a structure, these nozzles can provide the power needed to reach the seat of the fire. Solid streams are not as affected by wind conditions or rapid steam conversion as are fog streams. Solid streams are also able to reach long distances when the apparatus cannot be positioned close to the building. The best applications for solid streams are reaching fires deeply seated within a building or defensively attacking large, well-developed fires from a long distance.

Regardless of the type of nozzle chosen for the operation, it is imperative that the nozzle be provided its rated flow and pressure. All nozzles lose some effectiveness when they are not discharging their rated flow. For instance, a 750 gpm (3 000 L/min) fog nozzle discharging 750 gpm (3 000 L/min) will perform more effectively than a 1,000 gpm (4 000 L/min) nozzle flowing 750 gpm (3 000 L/min). For more information on effective fire streams, see the IFSTA **Pumping Apparatus Driver/Operator Handbook**.

Blitz Attack/Defensive Operations

Elevated streams can be used in both blitz attack and defensive operations. However, at no time should both operations be conducted simultaneously. Elevated master streams used while firefighters or victims are inside the building increase the possibility of injury. These injuries would be a result of the heat inversion caused by the disturbance created by the master stream device or by being struck by the stream. The added weight of water in a weakened building may also increase the collapse potential.

Regardless of whether a blitz or defensive attack is being undertaken, the driver/operator should always be alert for falling debris that may cause injury to personnel on the device or damage to the aerial device itself. When deciding where to deploy the aerial device, take into consideration avoiding falling debris.

Blitz Attacks

In offensive fire fighting operations, the strategy is to attack the fire on the inside and immediately stop the progress of the fire in the area of origin. Making this type attack from the exterior with a large caliber fire stream is often referred to as a *blitz attack*. A blitz attack can be made with an elevated master stream if conditions are right. The aerial apparatus must be able to position fairly close to the building, and the seat of the fire must be located where an outside stream can reach it. Blitz attacks are often conducted when the fire is on upper stories and while firefighters are preparing to enter the building. Again, ensure that no firefighters or victims are in the building when outside streams are directed into the building.

The best way to effect a blitz attack with an elevated master stream is to try to deflect water off the ceiling of the fire room. This can be accomplished by positioning the aerial device so that the nozzle is about even with the bottom of the window. This allows water to enter the area at an angle of about 30° (Figure 9.59). The water will hit the ceiling and the highest parts of the fire area where the most heat is concentrated, allowing steam conversion to take place (Figure 9.60). If too large an entry angle is used, water will hit the ceiling ahead of the fire area and cascade to the floor.

(**CAUTION:** The water adds weight to the structure at the rate of 1 ton per minute for every 250 gpm (1 000 kg for every 1 000 L/min) and could contribute to the collapse of an already fire-weakened structure. When possible, find avenues of escape from the building for water that is being introduced during fire fighting operations.)

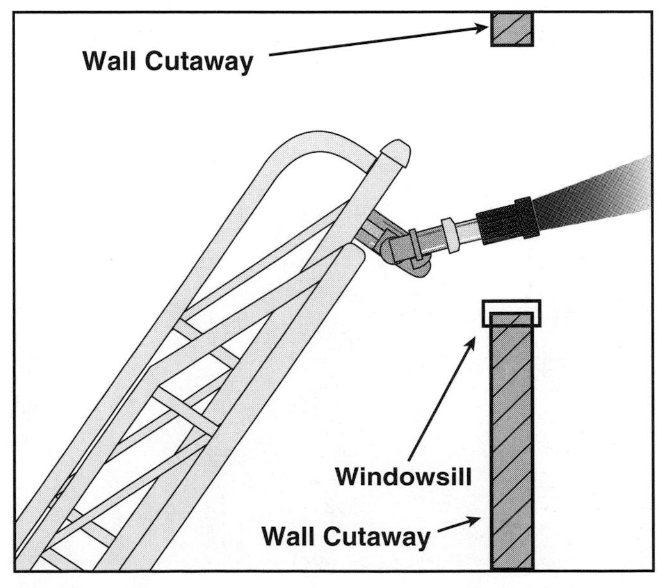

Figure 9.59 Water should enter the window at an angle of about 30°.

When elevated master streams are used in blitz attacks, they should only be flowed long enough to blacken out the fire. To avoid undue water damage, advance handlines into the fire area for final extinguishment and overhaul operations. Exercise caution when entering a structure that has been subjected to a long master stream attack.

Defensive Operations

A more common use of elevated streams is in defensive operations. A defensive attack is an exterior attack with emphasis on exposure protection (Figure 9.61). The objective of the defensive attack is to contain the fire within a specified area. Indicators of a defensive attack are:

- Fire/building conditions that prohibit safely advancing handlines into the building
- The incident commander giving up part or all of the building to the fire
- Large amounts of water needed to extinguish the fire

Fire conditions at the incident will dictate what type of stream should be used. Most commonly a solid or straight stream will be required to reach the seat of the fire (Figure 9.62). This is because the intense heat of the fire forces the apparatus to be positioned farther from the fire, and the fire may tend to convert a fog stream to steam before reaching the seat of the fire.

When trying to stop the spread of fire through a single building, care should be taken not to push the fire to uninvolved areas. The power of elevated streams is such that it is possible to change the direction of fire spread and even reverse the rise of

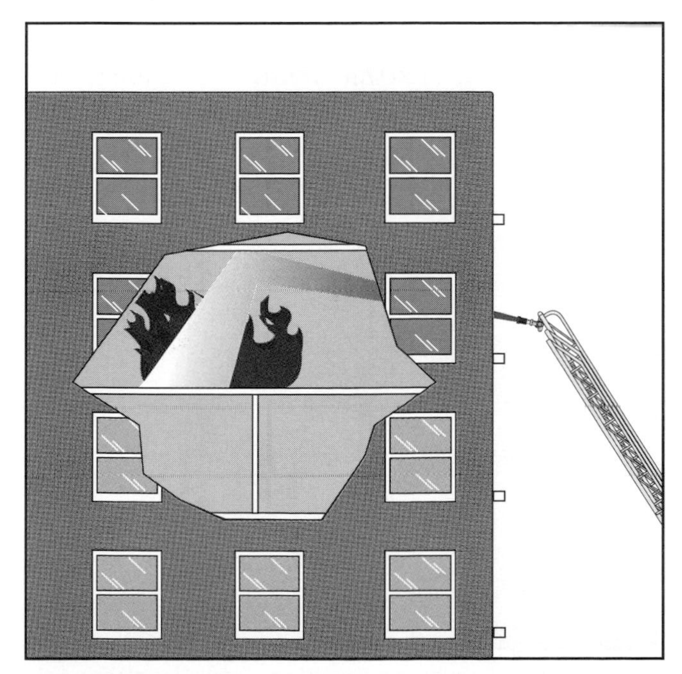

Figure 9.60 The fire stream should deflect off the ceiling above the seat of the fire.

Figure 9.61 A typical defensive fire attack.

Figure 9.62 Solid or straight streams are the preferred stream for defensive fire attacks. *Courtesy of Ron Jeffers.*

products of combustion in the thermal column. If possible, position the apparatus on the unburned side of the building. This allows the master stream to push the fire away from the unburned portions.

It is not uncommon for more than one elevated master stream to be in use at large, defensive fire fighting operations. When this is the case, driver/operators or firefighters who are operating the streams should exercise caution so that they do not strike another aerial device with their own stream. This could injure firefighters working at the tip of the device and could also cause damage to the other aerial device.

Ventilation Disturbance

When using elevated master streams, give continual attention to their effect on overall fire conditions. The force of the stream coupled with the air that is ingested with the stream can redirect smoke, heat, and fire gases throughout a building. Incident commanders should be particularly observant that ventilating operations — whether they be conducted by venting through a hole cut in the roof or cross ventilating through windows and doors — are not nullified by directing elevated master streams into these ventilation openings (Figure 9.63).

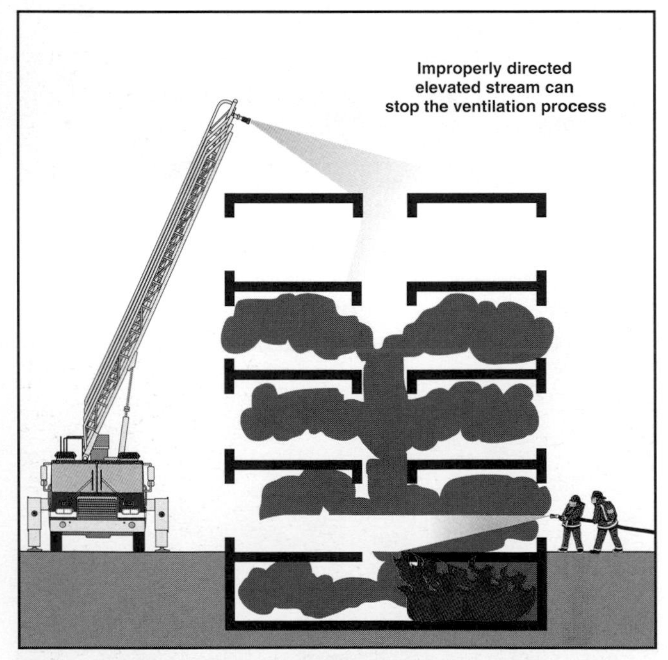

Figure 9.63 Never direct an elevated master stream into the ventilation opening if an interior attack is still in progress.

Improvised Standpipe Operations

Using the aerial as an exterior standpipe to upper floors is a tactic that can eliminate difficult hose lays up interior stairwells. These operations may be necessary when there is an interior fire on an upper floor and the building does not have an operable standpipe system. Using the aerial as an exterior standpipe is also an effective tactic for fires in parking garages, on building roofs, bridges, overpasses, and in buildings under construction where the standpipe system may end several floors below the fire. Interior handlines can be provided by removing the nozzle and attaching hoses to the outlet. This is a much easier method than dragging hose up several flights of stairs. It is also more efficient since friction loss is reduced greatly.

Many aerial platforms provide outlets for handlines on the platform itself (Figure 9.64). With these outlets, handlines can be preconnected and kept in the platform, saving valuable time in setting up for exterior standpipe operations (Figure 9.65). Remember, however, that adding preconnected lines to the platform reduces the amount of additional weight that may be added to the platform.

Foam Streams

Aerial apparatus can be used to apply large quantities of foam to flammable and combustible liquid fires. This may include fires in bulk fuel storage tanks, aircraft, industrial facilities, or transportation incidents. When planning to use an aerial device to deploy an elevated master stream, there are

Figure 9.64 Some elevating platforms have outlets for attaching handlines.

some basic principles of foam use that must be followed:

- It is generally not practical to discharge aerated foam through an elevated master stream. Aerated foam has a very high air content within the stream. This makes the stream susceptible to breakup caused by ambient winds or the fire's thermal column. In either case, it may result in the bulk of the foam failing to hit the seat of the fire.

- Most aerial devices will be used to discharge nonaerated foam onto these types of fires. Nonaerated foam often will provide a faster knockdown than will aerated foam; however, it has less of an ability to form a thick blanket over the fuel once the fire is extinguished. Nonaerated foam streams have much the same physical characteristics as plain water streams. Some aeration of the foam solution will occur as it travels through the air between the nozzle and the fire. When planning to use an aerial device to deploy a nonaerated foam stream, keep two important factors in mind:

  1. You must use a fog nozzle to discharge the stream (Figure 9.66). The nozzle may be set on any pattern. Solid stream nozzles will not allow the foam to proportion properly nor will they help aeration to occur after the foam has been discharged.

  2. You must use either aqueous film forming foam (AFFF) or film forming fluoroprotein (FFFP) concentrates when applying nonaerated foam streams. The other types of concentrate will not produce effective foam unless an aerating nozzle is used.

When the driver/operator is required to operate an elevated foam master stream on a fire, it is important to use the correct application techniques. If incorrect techniques are used, such as plunging the foam into a liquid fuel, the effectiveness of the foam is reduced. The techniques for applying foam to a liquid fuel fire include the roll-on method, bank-down method, and rain-down method.

Roll-On Method

The *roll-on method* directs the foam stream on the ground near the front edge of a burning liquid pool (Figure 9.67). The foam then rolls across the surface of the fuel. The driver/operator continues to apply foam until it spreads across the entire surface of the fuel and the fire is extinguished. It may be necessary to move the stream back and forth to different positions along the edge of a liquid spill to cover the entire pool. This method is used only on a pool of liquid fuel (either ignited or unignited) on the open ground. The aerial apparatus driver/operator should be aware of this technique, but it will probably be the least commonly used of the three techniques described here.

Bank-Down Method

The *bank-down method* may be employed when an elevated object is near or within the area of a burning pool of liquid or an unignited liquid spill. The object may be a wall, tank shell, or similar structure (Figure 9.68). The foam stream is directed off the object, allowing the foam to run down onto the surface of the fuel. As with the roll-on method, it

Figure 9.65 This department has a preconnected handline on the elevating platform.

Figure 9.66 A fog nozzle will be required to discharge foam from an aerial device.

Figure 9.67 The roll-on method for applying foam.

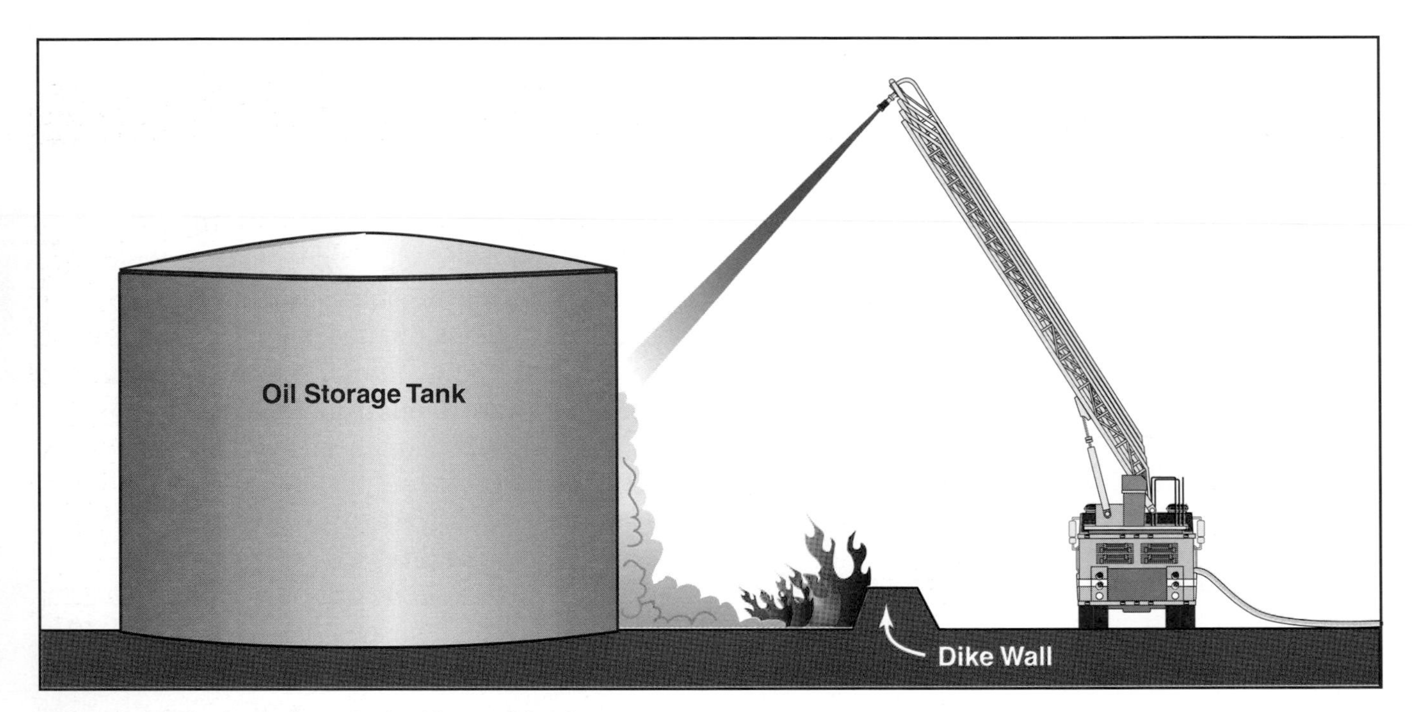

Figure 9.68 The bank-down method for applying foam.

may be necessary to direct the stream off various points around the fuel area to achieve total coverage and extinguishment of the fuel. This method is used primarily in dike fires and fires involving spills around damaged or overturned transport vehicles.

Rain-Down Method

The *rain-down method* is used when the other two methods are not feasible because of either the size of the spill area (either ignited or unignited) or the lack of an object from which to bank the foam. It is also the primary manual application technique used on aboveground storage tank fires. This method directs the stream into the air above the fire or spill and allows the foam to float gently down onto the surface of the fuel (Figure 9.69). On small fires, the driver/operator sweeps the stream back and forth over the entire surface of the fuel until the fuel is completely covered and the fire is extinguished. On large fires, it may be more effective for the driver/operator to direct the stream at one location to allow the foam to take effect there and then work its way out from that point (Figure 9.70). This is commonly referred to as establishing a foam "footprint."

For more detailed information on foam fire fighting, see Chapter 15 of the IFSTA **Pumping Apparatus Driver/Operator Handbook** or IFSTA's **Principles of Foam Fire Fighting** manual.

Figure 9.69 The rain-down method for applying foam.

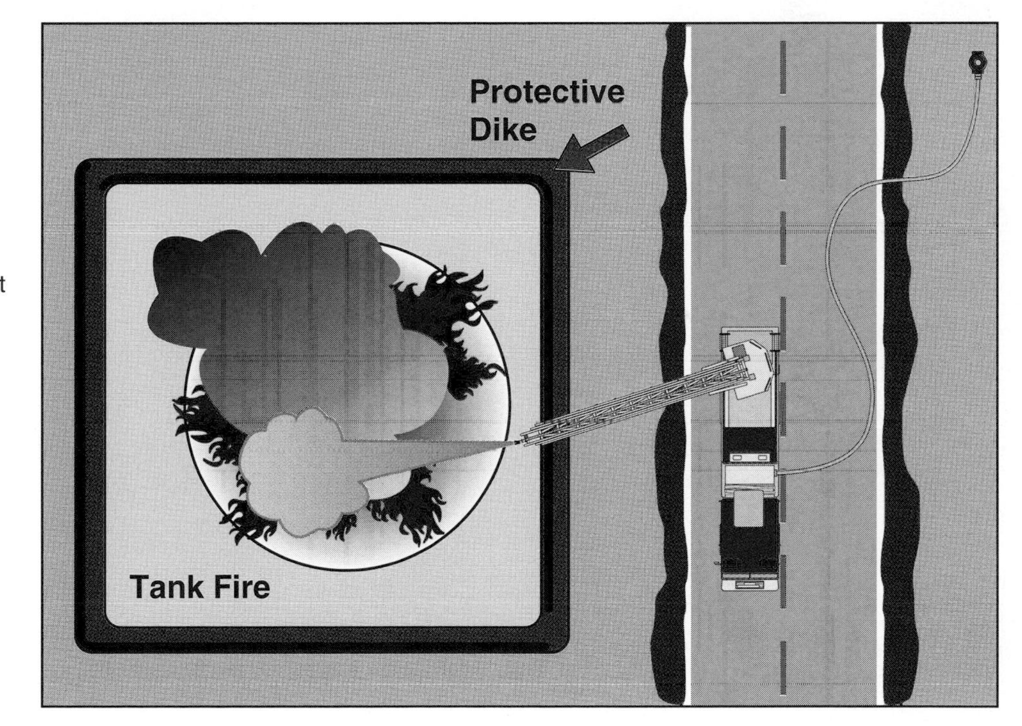

Figure 9.70 Establish a footprint of foam when using the rain-down method.

Chapter 10

Aerial Apparatus Testing

Job Performance Requirements

This chapter provides information that will assist the reader in meeting the following job performance requirements from NFPA 1002, *Standard on Fire Apparatus Driver/Operator Professional Qualifications,* 1998 edition. Particular portions of the job performance requirements (JPRs) that are met in this chapter are noted in bold text.

2-2.2 **Document the routine tests, inspections, and servicing functions, given maintenance and inspection forms, so that all items are checked for proper operation and deficiencies are reported.**

(a) ***Requisite Knowledge:*** Departmental requirements for documenting maintenance performed, **understanding the importance of accurate record keeping**.

(b) ***Requisite Skills:*** **The ability to use tools and equipment and complete all related departmental forms.**

4-4.1 **Perform the routine tests, inspections, and servicing functions specified in the following list in addition to those specified in the list in 2-2.1, given a fire department aerial apparatus, so that the operational readiness of the aerial apparatus is verified.**

- **Cable systems (if applicable)**
- **Aerial device hydraulic systems**
- **Slides and rollers**
- **Stabilizing systems**
- **Aerial device safety systems**
- **Breathing air systems**
- **Communication systems**

(a) *Requisite Knowledge*: Manufacturer specifications and requirements, policies, and procedures of the jurisdiction.

(b) *Requisite Skills*: The ability to use hand tools, recognize system problems, and correct any deficiency noted according to policies and procedures.

Fire apparatus is tested immediately after its construction—before the purchaser accepts it—to ensure that it performs in the manner for which it was designed. Once it is placed into service, it is tested at least yearly to ensure that it will continue to perform properly under emergency conditions. An organized system of apparatus testing, plus regular maintenance, is the best assurance that apparatus will perform within its design limitations. Furthermore, the insurance industry requires that apparatus be tested in order for the community to receive full insurance rating credit. This, in turn, affects the insurance rates for customers in the jurisdiction.

Apparatus tests can be grouped into two basic categories: preservice tests and service tests. All tests should conform to NFPA 1914, *Standard for Testing Fire Department Aerial Devices,* and NFPA 1901, *Standard for Automotive Fire Apparatus.* Preservice tests are conducted before the apparatus is placed into service. Usually, the driver/operator is not involved in preservice tests. However, the driver/operator must have a basic understanding of the preservice tests in order to appreciate and understand the service tests. *Preservice tests* include manufacturer's tests, aerial device certification tests, and acceptance tests.

Service tests are conducted on at least a yearly basis while the apparatus is in service. The driver/operator will often be required to perform these tests or at least assist technicians who are doing them. This chapter concentrates on both types of tests. *Service testing* includes inspecting and testing components such as the stabilizers, torque box, aerial device, and water systems.

This chapter begins with an explanation of the types of personnel who are typically involved with aerial apparatus testing. It continues with descriptions of the various basic types of test methods that are used on aerial devices. Finally, the chapter concludes with detailed descriptions of preservice and service testing procedures. (**NOTE:** This chapter contains only that information that is pertinent to the driver/operator being able to assist in conducting aerial device testing and inspections. It is not a blueprint for a complete aerial testing procedure. For complete details on aerial device testing, consult NFPA 1914.)

Personnel for Aerial Testing

In order for aerial device testing to be successful, properly trained personnel must carry out the testing. Regardless of the test being conducted, only personnel qualified to operate that particular piece of equipment, and who are acceptable to the owner of the equipment, should be allowed to operate the apparatus or the aerial device.

Preservice tests are usually conducted at the manufacturer's facility. Depending on the type of tests being conducted, they are performed either by the manufacturer's personnel or a third-party testing organization. Manufacturer's tests are usually carried out by employees of the apparatus builder. Certification tests must be performed by third-party testing agency personnel who are subject to the mutual approval of the manufacturer and the purchaser. *Third-party testing agencies* are independent firms not associated with an aerial device manufacturer or fire department. These agencies specialize in performing nonbiased testing on fire department equipment such as ground ladders, aerial devices, fire pumps, and other important pieces of equipment. These tests give an unbiased opinion as to whether or not a piece of apparatus meets NFPA requirements and is worthy of being placed into service.

Most of the service inspections and tests outlined by NFPA 1914, which are covered in this chapter, are designed to be performed by qualified fire department personnel. If the fire department does not have personnel qualified to perform these tests, the manufacturer or a third-party testing agency can be hired to perform the testing. Typically, third-party involvement will be required only for the nondestructive testing that must be performed every five years. Any person who is actually performing nondestructive testing must be certified as an American Society for Nondestructive Testing (ASNT) Level II NDT Technician in the test method being performed. Any third party employed to do nondestructive testing must comply with the standards set forth in the American Society for Testing and Materials (ASTM) Standard E543, *Standard Practice for Evaluating Agencies that Perform Nondestructive Testing.*

Fire departments should plan on having personnel witness the testing. By doing so, they may identify areas of the inspection that they can coordinate into their own routine checks of the apparatus. Individual fire departments generally do not own nondestructive test equipment; however, awareness of the potential defective areas can help them visually detect the onset of problems that may occur between inspections.

Testing Methods

A variety of scientific test procedures are used during the performance of aerial apparatus preservice and service testing. They can be generally grouped into one of three categories:

- Load testing
- Nondestructive testing
- Operational testing

Each of these is explained in more detail in the following sections.

Load Testing

Load testing is used to determine whether or not an aerial device is capable of safely handling the amount of weight it is rated to carry (Figure 10.1). Load testing has been an accepted method for aerial device acceptance and service testing for many years. However, in recent years, advancements in technology have made nondestructive testing available to the fire service.

While load testing still holds a place in the overall testing program, it may not be a totally accurate prediction of an aerial device's performance capability in situations normally encountered on the fireground. Load testing is a good indication that a static (nonmoving) load can be supported at the tip of a fully extended aerial ladder or in the basket of an elevating platform. Common fireground occurrences, such as the following, are not accurately tested if only static load tests are conducted:

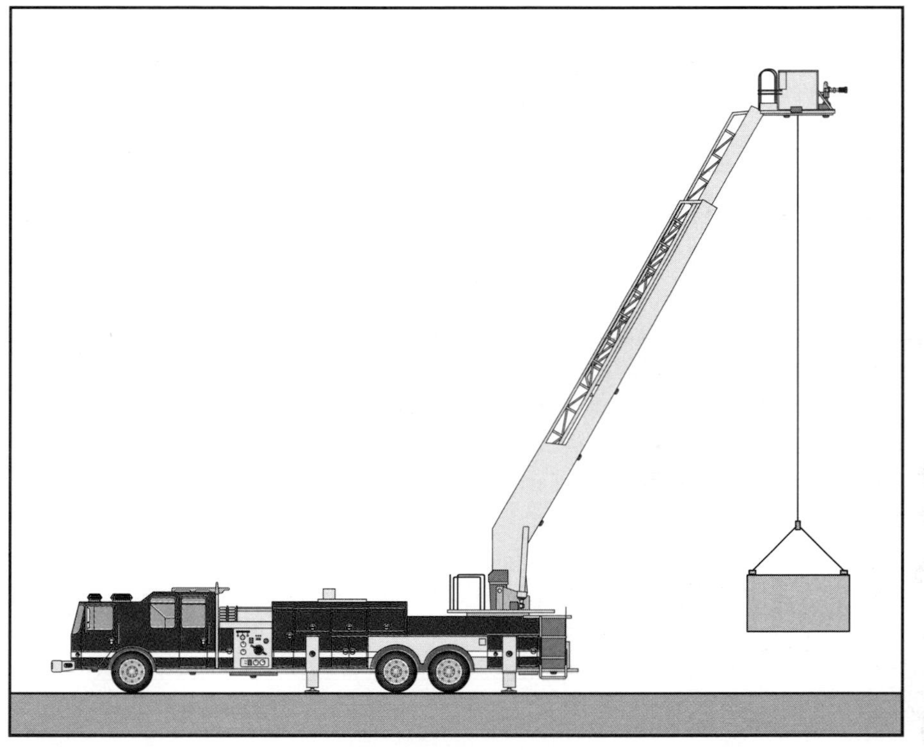

Figure 10.1 Load testing involves suspending an appropriate weight from the tip of the aerial device.

- The operation of an aerial master stream that causes stress on the aerial device due to nozzle reaction (Figure 10.2). The stress may be to either side, up or down, depending on the direction of water discharge. These stresses may cause a twisting action on the aerial device that is not considered in load testing.
- A firefighter stepping onto an aerial from a window or roof that causes an impact load on the aerial (Figure 10.3). A 200-pound (91 kg) firefighter jumping 3 feet (1 m) to an aerial can cause an impact load in excess of 700 pounds (317 kg).

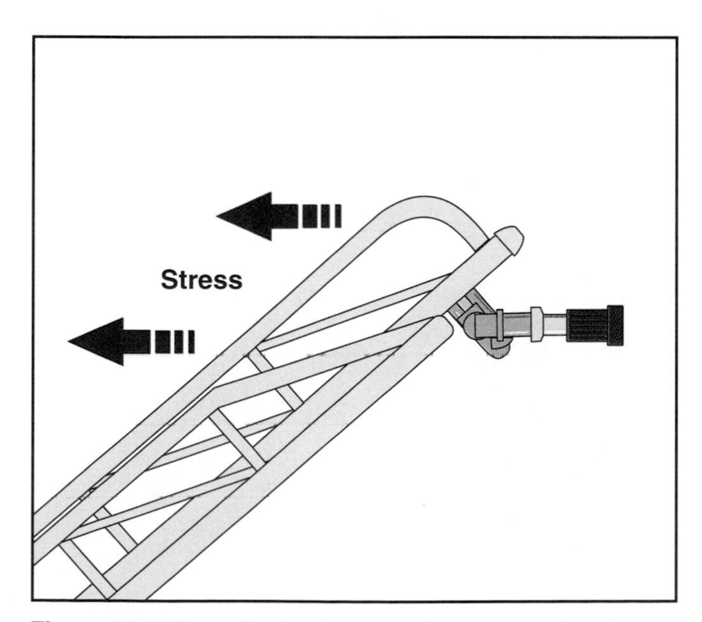

Figure 10.2 Water flowing from an elevated master stream nozzle imparts a reactive force on the aerial device.

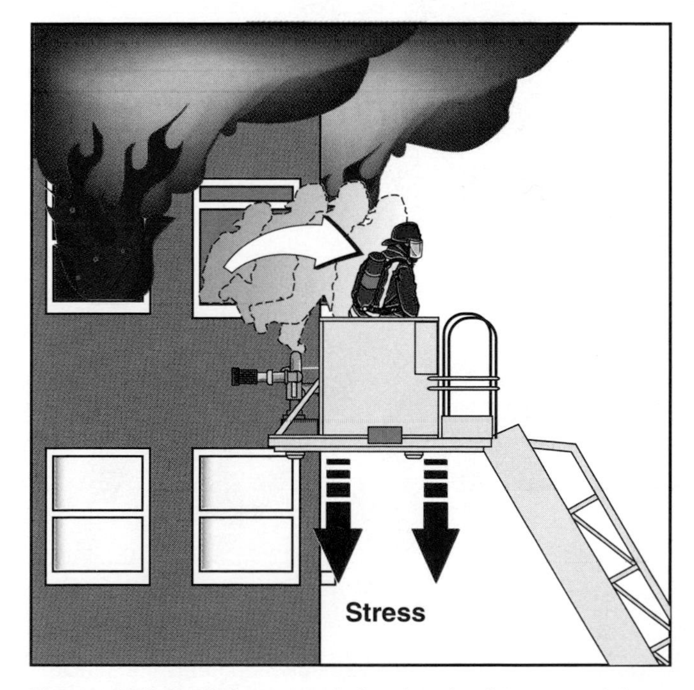

Figure 10.3 Firefighters or victims leaping into an aerial platform cause excessive stress on the aerial device.

Nondestructive Testing

Nondestructive testing involves any one of several testing methods used to inspect structural components (usually metal) without physically altering, placing under load or stress, or otherwise damaging the component. The term *nondestructive testing* (NDT) is used continuously throughout the rest of this chapter. Few fire departments have the equipment or expertise required to perform NDT; therefore, these tests are most commonly performed by independent testing agencies. Nondestructive tests can be performed on all aerial fire apparatus including water towers.

Nondestructive tests evaluate the materials used to construct the aerial device. Bolts can be examined for internal flaws without being removed. Cracks in welds can be detected and corrected before they fail. The NDT procedures used depend on the material used in the manufacture of the ladder. Common types of NDT that may be used on aerial devices include the following:

- *Magnetic particle inspection* — Magnetic particle inspection is performed on accessible welds and steel components. This test method involves magnetization of the area to be tested (Figure 10.4). Flaws on the surface create magnetic poles within the material, attracting iron powder to the defective area and identifying the extent of the flaw.
- *Conductivity readings* — Aluminum ladders can be nondestructively tested through the registering of hardness and by taking conductivity readings throughout the length of the ladder (Figure 10.5). A wide divergence in these readings indicates a change in the integrity of the material in that area.
- *Liquid penetrant testing* — A further examination of suspect areas involves dye penetrant testing. In this method, the surface of the test material is saturated with a dye or fluorescent penetrant, and a developer is applied. Dyes bleed visibly to

Figure 10.4 Magnetic particle inspection helps find cracks in the aerial device. *Courtesy of Underwriters Laboratories, Inc.*

Figure 10.5 Conductivity tests are performed on aluminum aerial devices. *Courtesy of Underwriters Laboratories, Inc.*

the surface indicating defects; fluorescents show the defect areas under ultraviolet light.

- *Ultrasonic inspection* — Ultrasonic inspection is performed on stabilizer pins and mounting bolts, rotation bearing bolts, raising and extension cylinder mounting pins and bolts, aerial hinge pins, and aerial support structure mounting bolts. This procedure involves the injection of high-frequency vibrations into the surface of the test component (Figure 10.6). These vibrations "bounce back" to their source from the opposite surface. If a flaw exists within the pin or bolt, the signals return in a different pattern revealing the location and extent of the flaw.
- *Radiographic testing* — Radiographic testing involves the use of X rays to check the integrity of welds that are used to construct the aerial device.
- *Hardness testing* — Hardness testing involves the use of one or more of a family of hardness tests to check the composition of metal components to determine if any significant changes have occurred. This is most commonly used on aluminum aerial devices and components.
- *Acoustic emission inspections* — Acoustic emission inspections use special equipment that identifies flaws in the metal of an aerial device by emitting sound waves into the device and reading them at a distant point.

Figure 10.6 Ultrasonic testing detects flaws in the aerial device by noticing changes in sound waves as they return to the sender. *Courtesy of Underwriters Laboratories, Inc.*

Operational Testing

Operational timing tests are performed to determine that the hydraulic system and structural mechanisms are performing in the proper manner and in accordance with the manufacturer's specifications. The specifics of operational testing are described individually in the appropriate sections of this chapter.

Preservice Tests

As mentioned previously, preservice tests are conducted before the apparatus is placed into service. The requirements for preservice tests are contained in NFPA 1901. Preservice tests include manufacturer's tests, aerial device certification tests, and acceptance tests. Generally, fire department personnel are not involved in performing manufacturer's or certification tests. Manufacturer's tests are performed by the manufacturer and certification tests are performed by a third-party testing agency such as Underwriters Laboratories (UL). Fire department personnel may be involved with acceptance testing. Acceptance testing is commonly performed after the apparatus has been delivered to the purchaser but before final acceptance has been made. The following sections outline the various tests that may be conducted before a piece of aerial apparatus is placed into service.

Manufacturer Testing

These tests are conducted by the apparatus manufacturer before delivery of the vehicle to the purchaser. NFPA 1901 does not require or specify particular manufacturer's tests for aerial devices or components. However, it does require that the apparatus successfully complete a road test. If the apparatus is equipped with a permanently mounted water delivery system or fire pump, hydrostatic testing is also required. Each of these is described below.

Road Test

NFPA 1901 requires the following minimum tests to be conducted on the fire apparatus after its construction is complete. The apparatus should be fully loaded in the same manner as it would be once in service. This includes making sure that the water and/or foam tanks are full and the weight of hose and equipment that will be carried on the apparatus are accounted for. The road tests should be conducted in a location and manner that will not violate any applicable traffic laws or motor vehicle codes. The test surface should be a flat, dry, paved road surface that is in good condition. At a minimum, the apparatus must meet the following criteria:

- The apparatus must accelerate to 35 mph (56 km/h) from a standing start within 25 seconds. This test must consist of two runs, in opposite directions, over the same surface.
- The apparatus must achieve a minimum top speed of 50 mph (80 km/h). This requirement may be dropped for specialized wildland apparatus not designed to operate on public roadways.
- The apparatus must come to a full stop from 20 mph (32 km/h) within 35 feet (10.7 m) (Figure 10.7).
- The auxiliary braking system must conform to the specifications listed by the braking system manufacturer.

Beyond these minimums, road tests are centered around the specific needs of each department. These needs are outlined in its apparatus bid specifications. For example, departments that protect hilly jurisdictions may have special requirements for apparatus acceleration, deceleration, and braking abilities on nonlevel surfaces. These special situations are the reason why many jurisdictions prefer to write performance-based specifications as opposed to engineering/equipment specifications. If the specifications are based on desired performance, the purchaser has a stronger position if the apparatus that is delivered does not meet the expectations.

Hydrostatic Test

Hydrostatic tests determine whether the permanent waterway system, pump, and pump piping can withstand pressures normally encountered during fire fighting operations. Aerial apparatus that are equipped with any of these systems must be hydrostatically tested.

Pumps are tested hydrostatically at 250 psi (1 725 kPa) for 3 minutes. During this test, close the tank fill line, tank-to-pump line, and by-pass line valves. Open and cap the discharge valves. Close and/or cap the intake valves. The test pressure should be maintained on the system for a minimum of three minutes without the failure of any component of the system.

If the apparatus is equipped with a permanently mounted waterway system, it must be hydrostatically tested at the maximum operating pressure required to achieve a flow of 1,000 gpm (3 785 L/min) at 100 psi (690 kPa) nozzle pressure when the aerial device is fully elevated and extended. The standard does not specify a duration for the test.

Certification Testing

NFPA 1901 requires extensive certification testing of aerial apparatus. This includes tests on both the

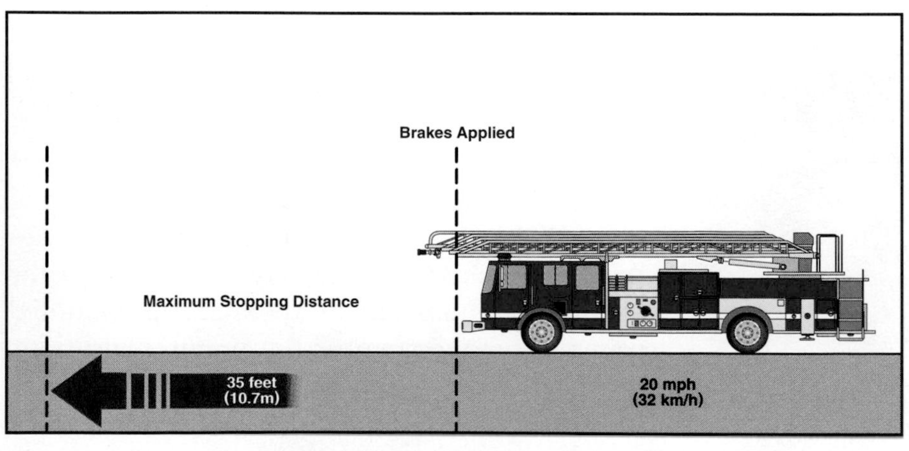

Figure 10.7 The apparatus must come to a complete stop within 35 feet (10.7 m) when traveling at 20 mph (32 km/h).

aerial device and the fire pump, if the apparatus is so equipped. These tests should be conducted at the manufacturer's facility, and the results should be certified by an independent testing agency acceptable to the purchaser.

Aerial Device Certification Testing

First, the aerial device must have complete testing, including nondestructive tests, as required in NFPA 1914. Following the required testing, the following additional tests are conducted.

Stability Testing

The stability of the apparatus should be tested to assure the purchaser that the apparatus will not be in danger of tipping over during normal operation. Stability is tested on both even and sloped surfaces. On an even surface, with the stabilizers deployed enough that the interlock system allows the aerial device to be operated, a weight that is one and one-half times the rated capacity is suspended from the tip of the aerial device when it is extended to its least stable position (Figure 10.8). Typically, the aerial device is in its least stable position when it is fully extended, parallel to the ground (horizontal), and at a 90° angle to the longitudinal centerline of the chassis. Under these conditions, the apparatus should show no signs of instability. It should be noted that the NFPA standard does highlight the fact that the lifting of a tire or stabilizer on the side opposite the deployed aerial device does not necessarily mean that the apparatus has become unstable (Figure 10.9).

If the manufacturer specifies a rated capacity while flowing water, then an amount equal to the water load, plus an amount equal to the worst-case nozzle reaction, shall be added to the test weights.

For elevating water towers, the even surface stability test should be conducted with a weight that is equal to one and one-half times the weight of water in the system and one and one-half times the worst case nozzle reaction when the water tower is in its position of least stability. If the water tower is equipped with a ladder that is intended for climbing, it must meet the aerial ladder requirements previously listed.

When testing for stability on sloped surfaces, the apparatus is parked on a 5° downward, sloped surface in the direction most likely to cause overturning (Figure 10.10). If the stabilizer system is capable of leveling the apparatus, this may be done. The test

Figure 10.9 A stabilizer lifting slightly off the ground does not necessarily mean that the apparatus has become unstable. *Courtesy of Boyd Cole.*

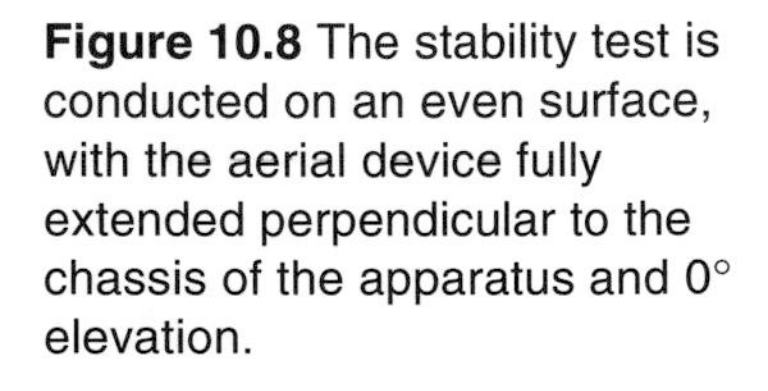

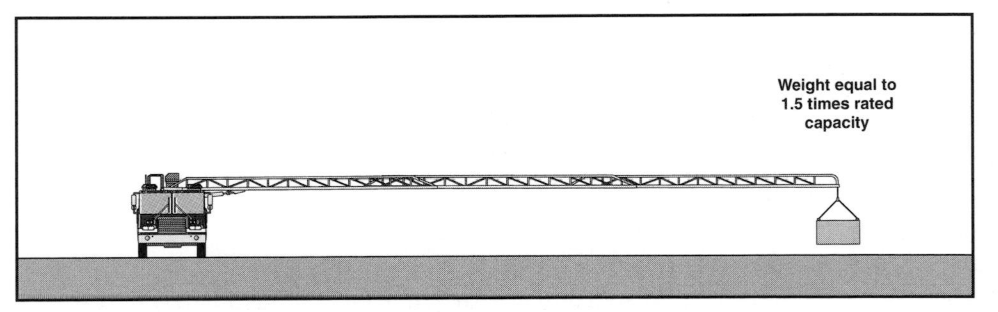

Figure 10.8 The stability test is conducted on an even surface, with the aerial device fully extended perpendicular to the chassis of the apparatus and 0° elevation.

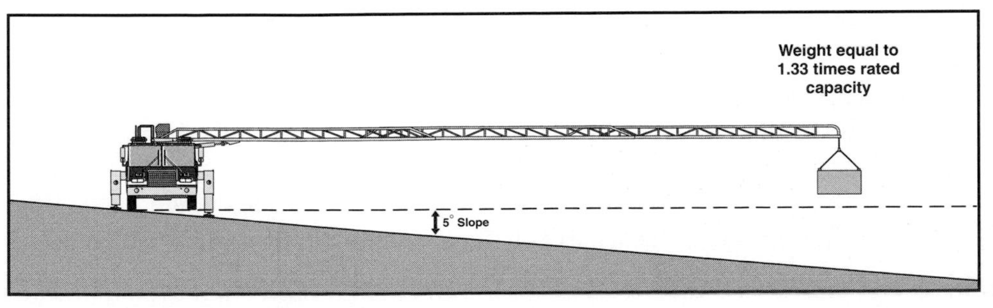

Figure 10.10 The uneven-surface stability test is conduct on a 5° slope.

procedure is the same from that point, except that the weight suspended from the end of the aerial device should be one and one-third times the rated capacity of the aerial device. The apparatus should show no signs of instability under these conditions.

Flow Testing

If the aerial device is equipped with a permanent water system, conduct a flow test to ensure that the system is capable of flowing at least 1,000 gpm (3 785 L/min) at 100 psi (690 kPa) nozzle pressure, with the aerial device at full extension and elevation. If the apparatus is equipped with its own fire pump, the onboard pump must be used to conduct the test. The intake pressure to the pump should not exceed 20 psi (138 kPa). On an aerial device where the water system has a maximum height of 110 feet (34 m) or less, conduct a flow test to ensure that the water system friction loss does not exceed 100 psi (690 kPa) between the piping swivel at the turntable and the monitor outlet. According to the NFPA standard, conduct this test with the aerial device at maximum extension and at any elevation. Third-party testing firms typically elevate the aerial device between 0° and 10° when performing this test.

Fire Pump Certification Testing

Apparatus equipped with fire pumps should be subjected to complete fire pump testing as described in NFPA 1901. These tests are the same as those covered in NFPA 1901. These tests are described in detail in Chapter 16 of the IFSTA **Pumping Apparatus Driver/Operator Handbook**.

Acceptance Tests

Acceptance tests are conducted to assure the purchaser that the apparatus meets bid specifications. These tests are most often performed after the apparatus is delivered to the purchaser. A representative of the manufacturer is present during testing. This ensures that the apparatus will perform in that jurisdiction, as specified. The types of tests and test criteria vary widely with local jurisdiction preference and conditions. The acceptance tests should include another pump test (if so equipped) and an aerial device operational test, even if certification tests were performed at the factory. Many instances have been documented where a pump or aerial device was certified at the factory, but did not perform as desired or rated once it was delivered. These tests do not have to mirror the certification tests but rather may follow the service test procedure listed later in this chapter. This should be proof enough that the certification test was accurate.

If the apparatus fails to perform according to the requirements detailed in the bid specifications, it should be rejected. On occasion, the manufacturer or salesperson may attempt to offer the purchaser alternative "credits," such as chrome wheels or gold-leaf lettering, to compensate for an apparatus that fails to meet all of the performance specifications. It is generally an unwise decision to accept this trade-off. The most important factor in making this purchase is getting an apparatus that performs as needed during an emergency.

If the apparatus is equipped with a fire pump, an important acceptance testing issue arises when the jurisdiction purchasing the apparatus is located at an altitude in excess of 2,000 feet (610 m) above sea level. In these cases, the pumping engine overload test must be performed during acceptance testing. This ensures that the engine develops the necessary power to operate in the jurisdiction it serves.

Service Testing Aerial Ladders

Service tests ensure that aerial ladders and other aerial devices offer a minimum degree of safety under continued use. They are more extensive than the inspection procedures covered in Chapter 3 of this manual. The requirements for service testing aerial ladders and other aerial devices are contained in NFPA 1914. This standard specifies the frequency for testing, as well as the procedures to be used.

The frequency for service tests varies. Conduct visual inspections, operational tests, and load tests at least annually. These tests should also be conducted following the use of the aerial device under unusual conditions, exposure to excessive heat, stresses or loads, or any usage that exceeds the manufacturer's recommended operating procedures. Do not place the aerial device back into service until these tests have been successfully completed. Any sign of failure during these tests is cause to remove the aerial device from service. Repairs should be made by a technician who is approved by the manufacturer.

Complete inspection and testing of the aerial device, including nondestructive testing, must be conducted at least every five years. Nondestructive testing (NDT) must also be conducted when other service tests indicate there is a potential problem with the aerial device.

The following sections contain the information necessary to perform tests that are required on an annual basis. The procedures for the NDT that must be conducted every five years are not covered, as they would typically not be conducted by fire department personnel, especially the driver/operator. Departments that require driver/operators to participate in or perform annual service testing may find it helpful to develop a checklist they can follow while performing the tests. This will help ensure that no areas are missed.

The following service test procedures pertain only to aerial ladders. The aerial ladder portion of aerial ladder platforms must conform to these requirements and the tests outlined later in this chapter for elevating platforms.

Review Service Records

Begin any service testing of an aerial device with a review of the service records of that particular piece of apparatus. The importance of accurate service records cannot be overemphasized. When properly kept, service records provide the inspector(s) with a complete history of the particular apparatus being examined. By reviewing the records, the inspector(s) will be able to note any past problems that have been found with that piece of apparatus. Note on the service record any existing conditions that may indicate the potential for problems during testing. Tests may be conducted more accurately, completely, and safely when all existing factors are made known from the beginning.

Visual Inspection

Prior to any operational or load testing being conducted, the aerial device and stabilization systems should be subjected to a thorough visual examination (Figure 10.11). Check all components of these systems for any visible signs of damage or excessive wear. Also note any loose bolts or other parts, cracked welds, and leaking fluids. Correct any deficiencies that are noted at this point before any further tests are conducted.

Stabilizer Examination

It is important that the apparatus stabilization system be examined before inspection of the aerial device itself. Any problems with the stabilization system could result in disastrous problems should the aerial device be lifted from its resting position. The following list is a summary of items to check:

- Check all bolts, nuts, or other fasteners to ensure that bolts of a proper grade and fit, according to manufacturer's specifications, are used. Use a properly calibrated torque wrench to make sure that all bolts are tightened to manufacturer's specifications. Visually check bolts for cracks or flaws of any type.
- Inspect accessible weldments for defects (cracks, holes, etc.), and check welds for obvious fractures (Figure 10.12).

Figure 10.11 A thorough visual inspection should be conducted on all components of the stabilization and aerial device systems.

Figure 10.12 Visually inspect the welds for obvious flaws.

- Visually ensure that the stabilizer pads are present, appropriate for the aerial device being inspected, and in good condition.
- Inspect the stabilizer mounting to the frame or torque box to ensure that there are no loose bolts, cracked welds, or other defects that could affect the stability of the apparatus when the aerial device is deployed.
- Inspect all hydraulic lines and hoses for kinks, cuts, cracks, and abrasions. Check for hydraulic fluid leakage at all connections and fittings (Figure 10.13).
- Check the stabilizer interlock system and/or warning devices for proper operation.
- Inspect stabilizer extension cylinder pins for proper alignment, installation, lubrication, operation, and retention. Check stabilizer cylinder rods to make sure that they are not pitted, scored, or otherwise defective. Perform a drift test by properly setting the stabilizers and marking the cylinder position (Figure 10.14). Allow the apparatus to stand for one hour with the engine off and the hydraulic fluid at the ambient temperature. The results should not exceed the manufacturer's recommendations for allowable drift. Check the holding valves for external leakage.
- Check the stabilizer controls to ensure that they operate smoothly, return to the neutral position when released, and are free of any hydraulic fluid leakage.
- Check to ensure that the selector (diverter) valve is free of hydraulic fluid leakage.
- Check the extension stops to ensure that they will not allow the stabilizers to overextend.
- Make sure that the apparatus can be stabilized within the time limit established by the apparatus manufacturer.
- If the apparatus is equipped with manual spring locks or tractor spring locks, make sure that they are in good condition and proper working order.
- If the apparatus is equipped with manual stabilizers, make sure that they deploy and operate as they were designed to do.

Figure 10.13 Check hydraulic system hoses and tubes for damage or leaks.

Figure 10.14 The stabilizer system deployed for a drift test.

Turntable and Torque Box Examination

The turntable and torque box should be the first portion of the aerial device to be inspected. The following list is a summary of items to check:

- Check all bolts, nuts, or other fasteners to ensure that bolts of a proper grade and fit are used. Use a properly calibrated torque wrench to make sure that all bolts are tightened to manufacturer's specifications (Figure 10.15). Visually check bolts for cracks or flaws of any type.

- Check the torque box to frame mounting system to make sure that all bolts and/or welds are in good condition. Bolts should be tight and show no signs of cracks or other deformities. Welds should show no signs of cracking.
- Inspect all accessible weldments for defects (cracks, holes, etc.), and check welds for obvious fractures.
- Inspect the rotation gear and bearing for missing, worn, or damaged teeth, proper pinion-to-gear alignment, proper lubrication, and backlash. Also check the rotation gear reduction box to make sure that bolts are tight or welds are secure.
- Check all turntable structural elements for excessive wear, deformities, or cracked welds.
- Inspect all hydraulic lines and hoses for kinks, cuts, and abrasions. Check the hydraulic hoses, lines, connections, hydraulic pump, power take-off, and rotation swivel for external hydraulic fluid leakage (Figure 10.16).
- Check the extension, elevation, and rotation locks for external hydraulic fluid leakage. Engage each lock and then operate the aerial device to ensure that the locks will hold. No movement should be detected.
- Check the collector rings to make sure that there is no foreign material buildup.
- Examine thoroughly the elevation cylinder assemblies. Inspect the anchor ears and plates for defects and weld fractures. Inspect cylinder pins for proper alignment, installation, lubrication, operation, and retention. Check cylinder rods to make sure that they are not pitted, scored, or otherwise defective.

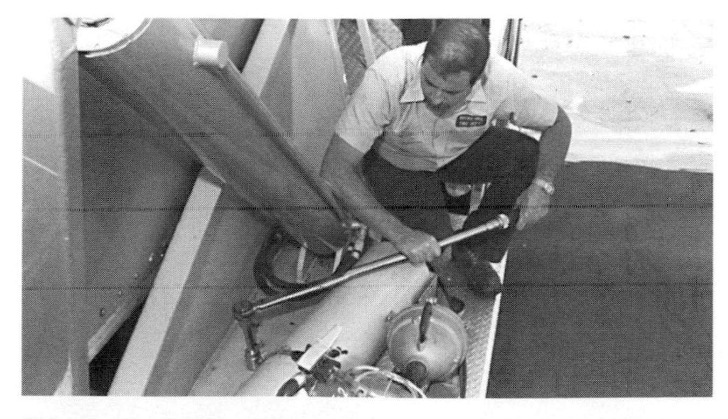

Figure 10.15 The only proper way to check for proper tightness on bolts is to use a torque wrench.

- Perform a drift test by placing the aerial device at a 60° elevation at full extension and marking the cylinder position (Figure 10.17). Manually close the lock valves and allow the aerial ladder to stand for one hour with the engine off with the hydraulic fluid at ambient temperature. The results should not exceed the manufacturer's recommendations for allowable drift. Check the holding valves for external leakage.

Figure 10.16 Check the turntable rotation swivel for signs of hydraulic fluid leakage.

Figure 10.17 Special equipment is used to measure the amount of cylinder drift during the aerial device drift test. *Courtesy of Underwriters Laboratories, Inc.*

- Check the following devices for proper operation:
  - —Load limit indicator (Figure 10.18)
  - —Emergency hand crank control
  - —Inclimeter
  - —Auxiliary hydraulic power
  - —Throttle control
  - —Communications system
  - —Hydraulic pressure relief system
  - —Transmission/aerial device interlocks
  - —Engine speed interlocks
  - —Breathing air system (Figure 10.19)
- Ensure that a turntable alignment indicator is present (Figure 10.20).

Aerial Ladder Examination

The aerial ladder itself should be thoroughly examined to ensure its structural and mechanical soundness. If, at any point during the testing, serious flaws are found or failure of the aerial device appears imminent, further testing should be halted until the problem can be corrected. The aerial ladder should then be subjected to the entire battery of tests again. The following list is a summary of items to check:

- Check all bolts, nuts, or other fasteners to ensure that bolts of a proper grade and fit are used. Use a properly calibrated torque wrench to make sure that all bolts are tightened to manufacturer's specifications. Visually check bolts for cracks or flaws of any type.
- Inspect all accessible weldments for defects (cracks, holes, etc.), and check welds for obvious fractures (Figure 10.21).
- Measure the amount of ladder section twisting or bowing in the aerial ladder. This amount must not exceed the manufacturer's recommended amount for twisting and bowing (Figure 10.22). The manufacturer may also have a preferred method for how this particular test is conducted.
- Inspect all hydraulic lines and hoses for kinks, cuts, and abrasions. Check for hydraulic fluid leakage at all connections and fittings.
- Check any pneumatic or electric lines located in the ladder sections for proper wear, mounting, or signs of defects.

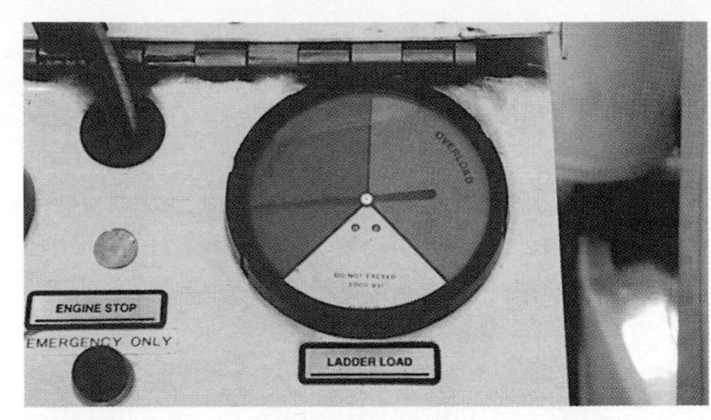

Figure 10.18 Check the load limit indicator for proper operation.

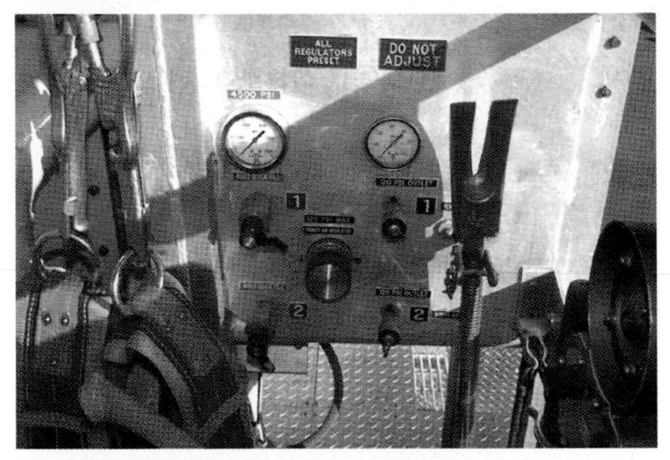

Figure 10.19 The breathing air system should be inspected for proper operation.

Figure 10.20 Ensure that a turntable alignment indicator is present.

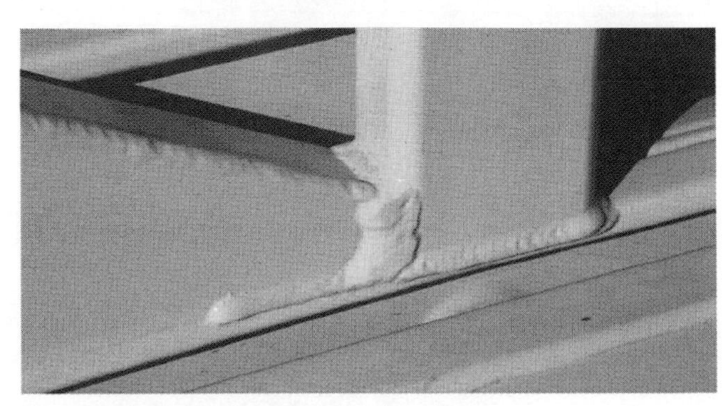
Figure 10.21 Check all welds for defects.

Figure 10.22 Excessive bowing or twisting is cause for removing an aerial device from service. *Courtesy of Harvey Eisner.*

- Check the aerial ladder for repairs or modifications that were not authorized by the manufacturer of the aerial ladder.
- Inspect the top rails and base rails for proper alignment. Look for signs of ironing, dents, corrosion, or other signs of serious wear on base rails.
- Inspect all rungs and folding steps (if present) for damage or defects. Rung covers should be present and firmly attached (Figure 10.23).
- Inspect all rollers for proper lubrication, operation, and signs of wear or excessive dirt.
- Inspect all guides, babbitts, wear strips, pads, and slide blocks for wear, defects, and proper mounting or alignment.
- Inspect extension sheaves and sheave mounting brackets for wear and defects. Inspect welds for fractures.
- Inspect extension cables for excessive wear. Inspect the cable separation guide to ensure free travel and proper alignment.

Figure 10.23 Make sure the rung covers are in place.

Figure 10.24 The ladder should be fully elevated and extended about 10 feet (3 m).

- Check the extension and retraction motor for signs of hydraulic fluid leakage, brake wear, and brake alignment with the shaft.
- Check the winch and brake holding abilities individually by fully elevating the aerial ladder and extending the ladder 10 feet (3 m) for at least 5 minutes (Figure 10.24). Any slippage that occurs must remain within the manufacturer's recommendations.

- Check the extension and elevation indicators for legibility, clarity, and accuracy.
- Inspect the fly locks for proper mounting, alignment, lubrication, and operation.
- Check the ladder cradle and bed locks for proper alignment. Check the bed locks for hydraulic fluid leaks.
- Check for the presence and proper operation of overload and maximum extension warning devices. Make sure that all ladder stops work as designed to prevent overextension or over-retraction of the ladder.
- Check all lights mounted on the aerial ladder to ensure proper working condition.
- Inspect extension cylinder anchor ears and plates for weld fractures and defects (Figure 10.25). Check the extension cylinders, cylinder rods, and cylinder pins for scoring, pitting, and other defects.
- Conduct a drift test by placing the aerial ladder at full elevation and 10 feet (3 m) of extension. After marking the piston or the second ladder section, allow the aerial ladder to stand for one hour with the engine turned off and the hydraulic fluid at the ambient temperature. The amount of drift should not exceed the manufacturer's specifications.
- Check the extension cylinder holding valves to make sure that there is no internal or external hydraulic fluid leakage.
- Inspect the secondary operating control panel at the tip of the aerial device (if aerial ladder is so equipped) for proper marking of the controls, control instructions, and necessary warning placards. Verify that the controls operate properly and that the controls at the turntable control station will override the tip controls when both are operated simultaneously. The controls at the tip of the device should not allow the aerial device to be operated at a speed faster than the manufacturer's allowable speed.

Operational Testing

Operational testing is performed to determine if the apparatus is still capable of operating within its full range of movement and in a reasonable amount of time. These differ from load tests that determine whether the aerial ladder is capable of safely holding its rated load capacity.

Figure 10.25 Check the extension cylinder anchor ears to make sure that they are free of defects.

The operational test of an aerial ladder should cover the complete range of actions required to place the aerial ladder in service. Start the apparatus engine, engage the power take-off, and set the stabilizers. Next, raise the ladder from its bed to full elevation, rotate it 90°, then extend it fully. This must be performed in the time allowed by the standard in effect when the aerial device was manufactured. While the ladder is extended, check to see whether there is any twist to the ladder. If there is, stop the test, and notify the manufacturer for recommended corrections.

If the test is proceeding satisfactorily, retract the ladder, complete the 360° rotation, and lower the ladder back to its bed. After this operational test, thoroughly inspect all moving parts, especially the ladder cables. Note whether the controls operated the ladder smoothly and without excessive vibration.

After completion of the operational testing, an oil sample is taken from the aerial device hydraulic reservoir tank and submitted to a laboratory for spectrochemical analysis. Include a copy of the report from the laboratory performing the test with the final report.

Load Testing

The purpose of load testing is to determine the ladder's ability to handle its rated load. It is important to do the following when performing any load testing:

- Conduct testing when wind velocity is less than 10 mph (16 km/h).
- Maintain a close watch during all load tests. Permit only those personnel essential to conduct the test near the apparatus during the test. If the ladder shows any twist or deformation at any time, discontinue the test immediately, and place the aerial ladder out of service (Figure 10.26). The aerial ladder must be repaired in accordance with the manufacturer's written recommendations and fully tested before it is placed back into service.

Figure 10.26 On occasion, an aerial ladder will fail dramatically during use or testing. *Courtesy of Harvey Eisner.*

- Conduct all load tests with the aerial apparatus on a firm, level surface or road. All stabilizers must be properly deployed and stabilizer pads in place.
- Use the manufacturer's load chart or operator's manual to determine the maximum rated live load in the horizontal or elevated positions. If full extension is not permitted, use the maximum permissible extension with a specified live load for the purpose of the horizontal load test.
- Attach and properly center a test cable hanger to the top rung of the top ladder section (Figure 10.27).
- For single-chassis, rear-mounted aerial apparatus, rotate the ladder, if necessary, until the ladder is positioned over the rear and parallel to the vehicle centerline (Figure 10.28). A midship-mounted aerial ladder will have to be rotated slightly off-center so that it will not contact the

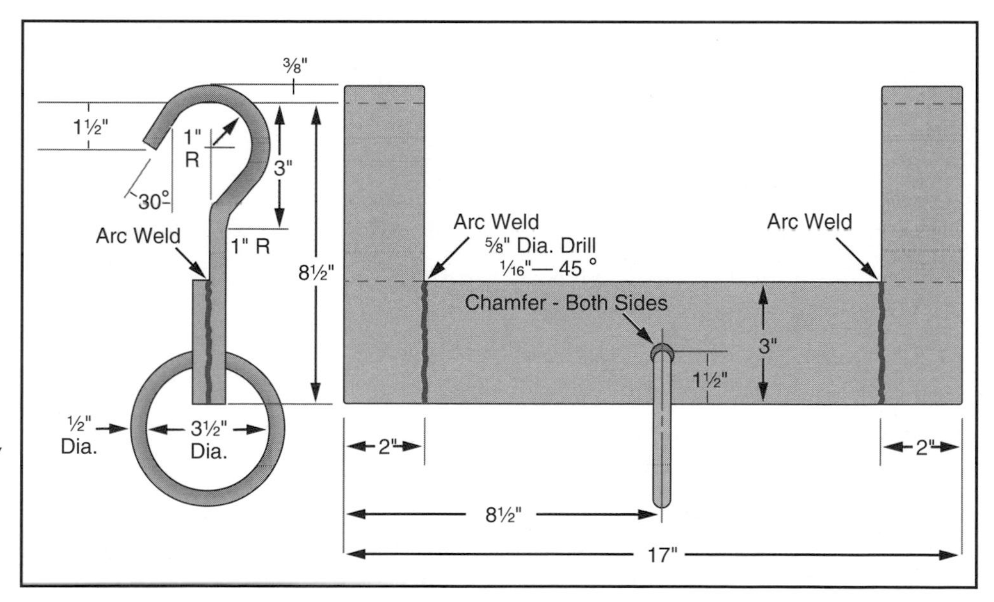

Figure 10.27 The design of a proper test cable hanger. *Reprinted with permission from NFPA 1914-1997,* Testing Fire Department Aerial Devices., *copyright© 1997 National Fire Protection Association, Quincy, MA 02269. This is not the complete and official position of the NFPA on the referenced subject which is represented only by the standard in its entirety.*

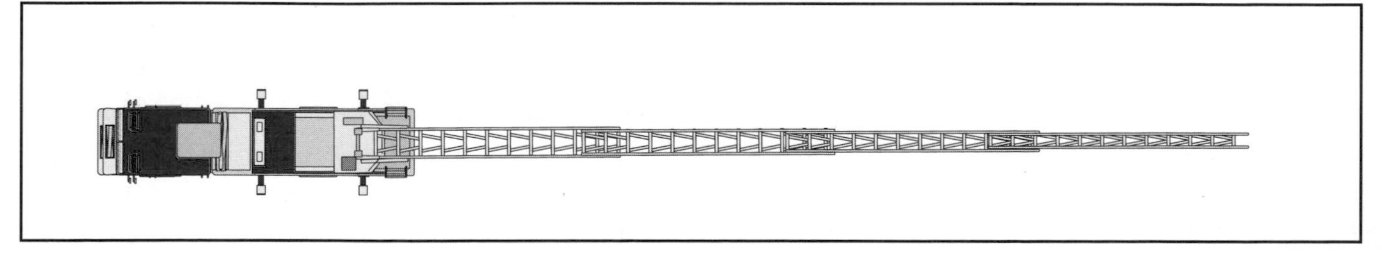

Figure 10.28 On a rear-mount apparatus, the aerial device is extended directly over the rear of the apparatus.

ladder storage cradle during testing (Figure 10.29). Position tractor-drawn apparatus in the most stable position as recommended by the manufacturer.

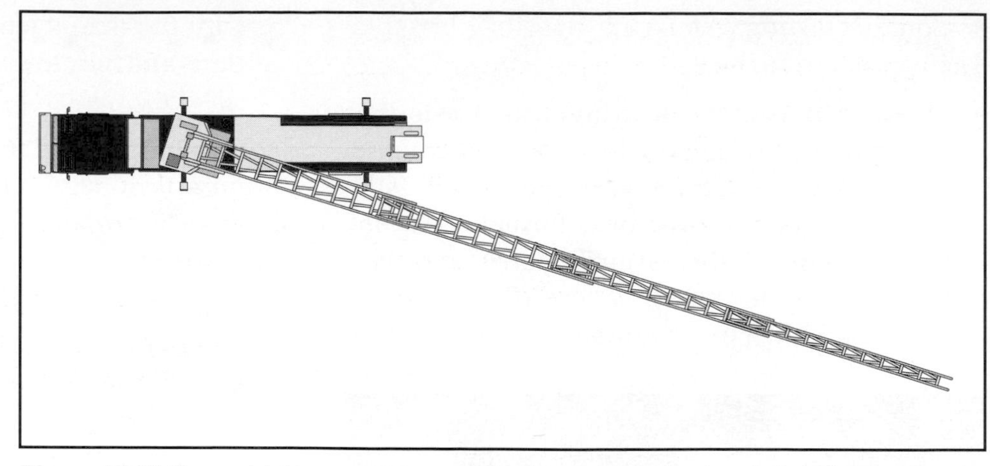

Figure 10.29 On a midship-mount apparatus, the aerial device is rotated slightly off center and then extended off the rear of the apparatus.

The following sections highlight, in accordance with NFPA 1914, various types of load tests for aerial ladders.

Horizontal Load Testing

The procedure for horizontal load testing of aerial ladders is as follows:

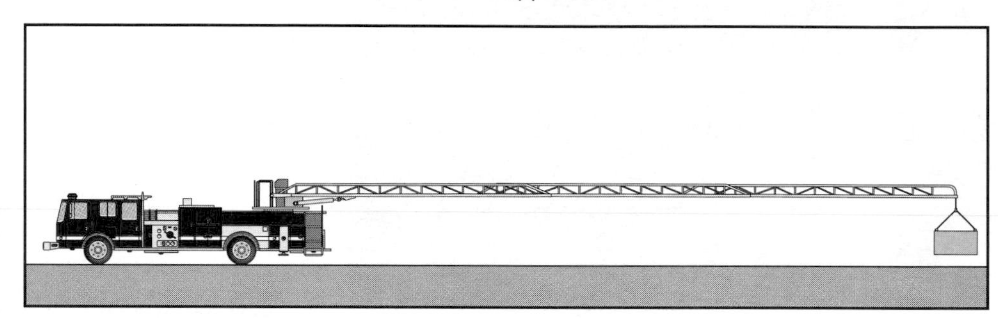

Figure 10.30 The horizontal load test position.

Step 1: Place the ladder in the horizontal position and extend to full or maximum permitted extension as determined by the manufacturer's requirements. Do not allow the base section to rest in the bed.

Step 2: Properly close or apply the ladder section locks, either manual pawls or hydraulic holding valves, if this is in accordance with the manufacturer's guidelines.

Step 3: Properly close or apply the elevation cylinder integral holding valve or shutoff safety valves. The twist of the ladder section rung rails or beams must not exceed the manufacturer's tolerance for twist.

Step 4: Gradually apply a weight equal to the manufacturer's rated live load to the top rung of the aerial ladder (Figure 10.30). Do this by using the test weight container or another suitable means of applying the weight. The unsupported aerial ladder must sustain the test weight for 5 minutes.

(**CAUTION:** The total weight of the supporting hangers, containers, and test weight are taken as a whole and shall not exceed the rated live load. Dropping the weights and shock loading the ladder must not be permitted.)

Step 5: The test weight shall hang freely from the tip of the aerial ladder. If the test weight hanger and ladder deflection are such that the test weight comes to rest on the ground, raise the ladder elevation slightly above the horizontal position.

> **WARNING!**
>
> **DO NOT move the ladder with the test weight applied at any time during the load test. Moving the ladder with the test weight applied can result in failure of the aerial ladder and injury to personnel conducting or observing the test. If the ladder must be raised, first remove the test weight, then raise the ladder, and finally reapply the test weight.**

Any visually detectable signs of damage or permanent deformation constitutes noncompliance with the load test requirements. The aerial device must also function properly after the load test.

Maximum Elevation Load Testing

The procedure for elevation load testing for aerial ladders is as follows:

Step 1: Elevate the ladder to maximum height and extend fully with the test weight hanger, cable, and empty load bucket attached. Suspend the bucket by the cable so that when the weight is added the bucket will be no more than 3 feet (1 m) above the ground (Figure 10.31).

Step 2: Apply the ladder section locks, either manual pawls or hydraulic holding valves, if this is in accordance with the manufacturer's guidelines.

Step 3: Close or apply the elevation cylinder integral holding valve or shutoff safety valve. The ladder section rung rails or beams must be parallel to each other and without twist before adding any test weight.

Step 4: Gradually apply weight equal to the manufacturer's rated live load. Do this by filling the load bucket with a predetermined amount of water, sand bags, or other weights. Do not throw the weight into the bucket. The unsupported aerial ladder must sustain the test weight for five minutes. The test weight shall hang freely from the tip of the aerial ladder.

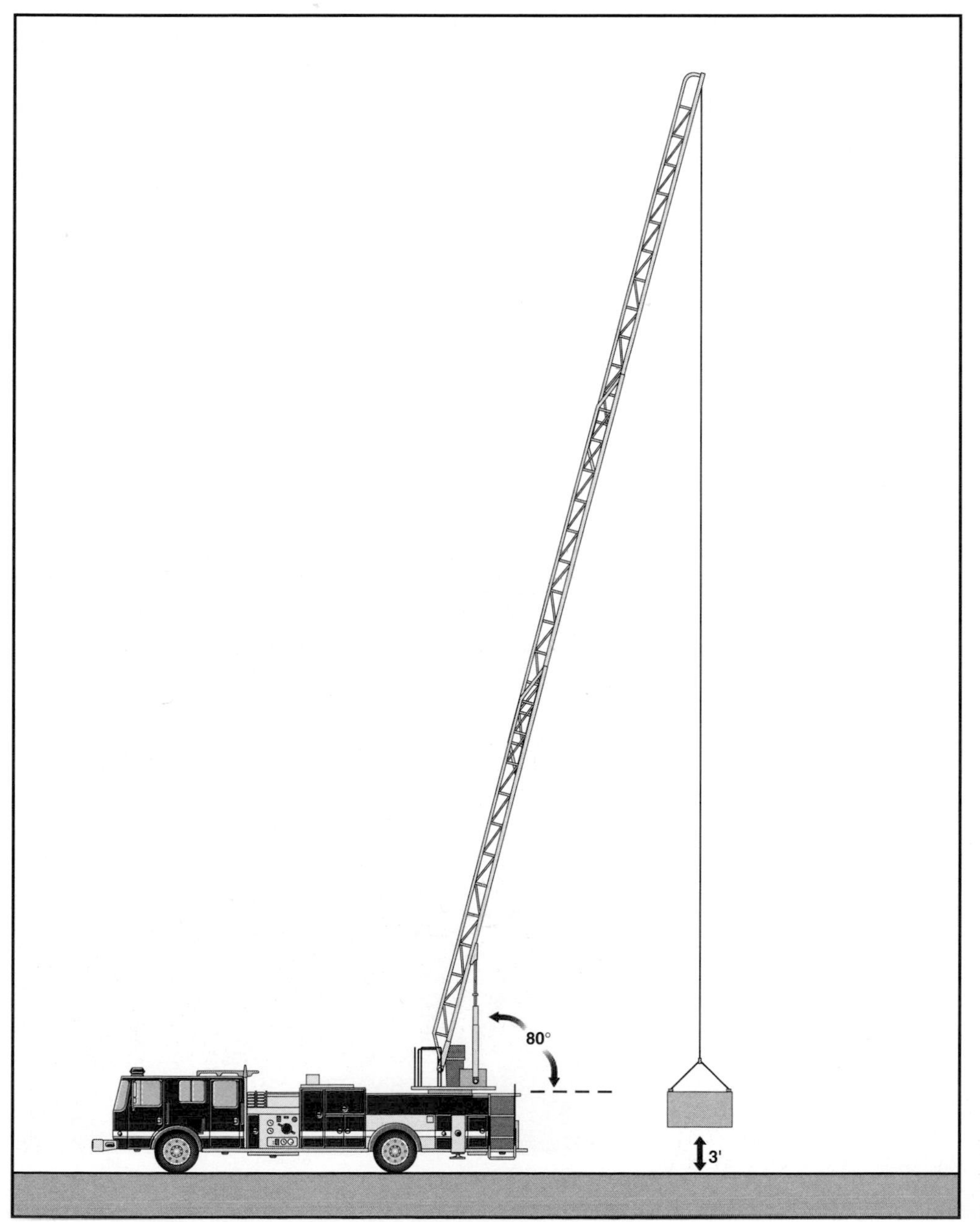

Figure 10.31 Test position for the maximum elevation load test.

(**CAUTION:** The total weight of the supporting hangers, containers, and test weight are taken as a whole and shall not exceed the rated live load. Dropping the weights and shock loading the ladder must not be permitted.)

WARNING!

DO NOT move the ladder with the test weight applied at any time during the load test. Moving the ladder with the test weight applied can result in failure of the aerial ladder and injury to personnel conducting or observing the test. If the ladder must be raised, first remove the test weight, then raise the ladder, and finally reapply the test weight.

Any visually detectable signs of damage or permanent deformation shall constitute noncompliance with the load test requirements. The aerial device must also function properly after the load test.

Water System Testing

Water system testing is performed on all aerial ladders that have permanently piped waterway systems. Before any pressurization of the system, check the entire system for proper operation of all components. Inspectors should ensure that there is no serious corrosion of components or blockage within the system. The brackets that attach the system to the aerial ladder should be checked for loose, missing, or otherwise defective bolts; cracked welds; or any other defects. Once these steps have been taken, pressurization testing may be conducted using the following tests. Remember that when the tests are complete, the water system must be completely drained and at least one end of it should be open while the aerial device is being retracted.

Retracted Aerial Ladder Pressure Test

Step 1: Equip the discharge end of the water system with some type of valve.

Step 2: Elevate the ladder between 0° and 10°, but keep the ladder fully retracted (Figure 10.32). Fill the system with water, and bleed all air from the system. Close the valve.

Step 3: Pressurize the waterway system statically to the pressure recommended by the manufacturer of the device. Be careful not to overheat the pump that is supplying pressure for this test.

Step 4: Raise the aerial ladder to its maximum permissible elevation and rotate 360° (Figure 10.33). Check for the presence of leaks in the swivel area of the water system while it is under pressure.

Figure 10.32 The aerial device is elevated 10°, but not extended.

Figure 10.33 The aerial device is fully elevated, but not extended, and is rotated 360°.

Extended Aerial Ladder Pressure Test

Following the pressure test where the aerial ladder was kept fully retracted, use the following procedure to complete the pressure testing:

Step 1: Elevate the ladder between 0° and 10° and fully extend the aerial ladder (Figure 10.34). Fill the system with water, and bleed all air from the system. Close the valve.

Step 2: Pressurize the waterway system statically to the pressure recommended by the manufacturer of the device. Be careful not to overheat the pump that is supplying pressure for this test (Figure 10.35). Check for the presence of leaks in the water system while it is under pressure.

Figure 10.34 The aerial device is elevated 10° and fully extended.

Figure 10.35 The aerial device is fully extended and elevated.

Flowmeter Accuracy Test

If the aerial device is equipped with a flowmeter to monitor the amount of water flowing through the waterway system, check the flowmeter to make sure that it provides an accurate reading. Conduct the test at the maximum flow for which the waterway system is rated. The flow from the discharge end of the aerial device may be measured using a properly calibrated portable flowmeter that is attached to the discharge or by taking a pitot reading from a smoothbore nozzle attached to the tip (Figure 10.36). The pressure from the pitot reading can then be used to calculate the actual flow being discharged. The figures from the flowmeter being tested and those calculated from the actual discharge must be within ± 10 percent. If they vary more than that, the flowmeter must be recalibrated or repaired.

Pressure Gauge Test

If the waterway system is equipped with a pressure gauge, its accuracy should be checked at a minimum of three pressure points, including:

- 150 psi (1 034 kPa)
- 200 psi (1 379 kPa)
- 250 psi (1 723 kPa)

The exception to this rule would be if the waterway system is designed for a maximum pressure of 200 psi (1 379 kPa). In that case, the 250 psi (1 723 kPa) test is omitted. If the gauge readings are off by more than 10 psi (70 kPa), repair or replace the gauge.

Figure 10.36 A pitot tube is attached to the nozzle to measure the flow through the system. *Courtesy of Underwriters Laboratories, Inc.*

Waterway Relief Valve Test

If the waterway system is equipped with a pressure relief valve, test it to ensure that it will activate and relieve pressure that is in excess of the manufacturer's maximum recommended operating pressure. Simply make sure that the relief is set to the maximum recommended operating pressure and charge the system slightly above the set pressure to see if the valve activates. If it does not, repair the valve and repeat the test.

Service Testing Elevating Platform Aerial Apparatus

The following sections apply to all types of elevating platform aerial apparatus, including aerial ladder platforms, telescoping platforms, and articulating platforms. Some of the procedures will vary slightly between the different types of elevating platforms, and these differences are noted throughout these sections.

Review Service Records

Any service testing of an aerial device should begin with a review of the service records of that particular piece of apparatus. The importance of accurate service records cannot be overemphasized. When properly kept, service records provide the inspector(s) with a complete history of the particular apparatus being examined. By reviewing the records, the inspector(s) will be able to note any past problems that have been found with that piece of apparatus. Any existing conditions that may indicate the potential for problems during testing should also be noted on the service record. Tests may be conducted more accurately, completely, and safely when all existing factors are made known from the beginning.

Visual Inspection

Prior to any operational or load testing being conducted, the aerial device and stabilization systems should be subjected to a thorough visual examination (Figure 10.37). Check all components of these systems for any visible signs of damage or excessive wear. Loose bolts or other parts and cracked welds should also be noted. Correct any deficiencies that are noted at this point before any further tests are conducted.

Stabilizer Examination

The procedure used to examine the stabilization system of elevating platform apparatus is the same as that described in the section for aerial ladders.

Turntable and Torque Box Examination

The procedure used to examine the turntable and torque box of elevating platform apparatus is the same as that described in the section for aerial ladders.

Elevating Platform Examination

Examination of the elevating platform should include the following:

- Visually inspect all platform mounting brackets for cracks, dents, or bends.
- Inspect the platform for cracks, dents, bends, or poor welds (Figure 10.38).
- Check all hydraulic, pneumatic, and electrical lines for proper installation, kinks, excessive wear, and abrasions.

Figure 10.37 Perform a visual inspection of the elevating platform.

Figure 10.38 Check the welds on the platform for obvious defects.

Figure 10.39 If the platform is equipped with an auxiliary winch, check it for proper operation.

- Check auxiliary winches mounted on the platform for proper mounting bolts (Figure 10.39). Use a torque wrench to inspect the bolts for manufacturer's recommended torque. Examine the winch mounting for weld fractures. Check winch controls for proper control identification and operation.
- Make sure that the platform load capacity identification chart is legible and in a plainly visible location (Figure 10.40).
- Inspect platform gate hinges and latches for proper operation and condition (Figure 10.41).
- Check platform hinge pins visually for flaws or defects (Figure 10.42).
- Check the controls on the platform control station for proper identification and operation. Verify that the lower control station will override the platform controls.
- Check to ensure that the ground-to-platform control station is in good working order.

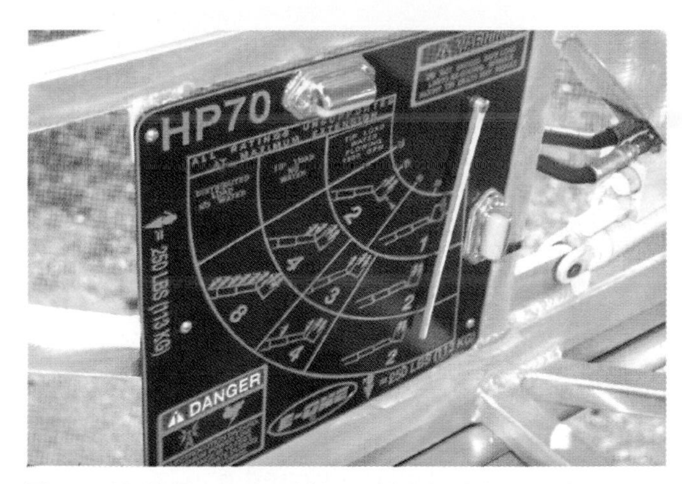

Figure 10.40 There should be a plainly visible load capacity chart somewhere on the aerial device.

Figure 10.41 Make sure that the platform gate operates properly.

Figure 10.42 Check the platform hinge pins for signs of excessive wear.

- Ensure that no modifications or alterations have been made to the platform without the manufacturer's approval. Pay particular attention to the addition of equipment that adds weight to the platform. This reduces the weight of people that the platform may carry and also affects the amount of weight that must be placed on the platform during load testing.
- Check the platform monitor nozzle(s) for a solid mounting and for proper operation.
- Check all lights on the platform to ensure that they are in good working condition.

Aerial Boom Examination

The procedure for examining the booms of the aerial device will vary slightly depending on the particular type of aerial device being tested. The following sections explain the procedure for each type.

Articulating Aerial Device — Lower Boom Examination

The examination of the lower boom on an articulating device should include the following:

- Inspect all bolts, nuts, hinge pins, or other fasteners to ensure that bolts of a proper grade and fit are used. Use a properly calibrated torque wrench to ensure that all bolts are tightened to manufacturer's specifications. Visually check bolts for cracks or flaws.
- Check all accessible weldments for defects (cracks, holes, etc.), and check all welds for obvious fractures (Figure 10.43).
- Inspect all hydraulic lines and hoses for kinks, cuts, cracks, and abrasions. Check for hydraulic fluid leakage at all connections and fittings. Pay particular attention to lines that run through the knuckle between the lower and upper booms.
- Examine lower boom elevation cylinder anchor ears and plates for weld fractures and defects. Check the lower boom elevation cylinders and cylinder rods for scoring, pitting, and other defects. Inspect the cylinder rod-to-barrel seat to make certain there is no external leakage.
- Conduct a drift test to ensure that the amount of drift does not exceed the manufacturer's specifications. The manufacturer of the aerial device should provide information on how exactly the drift test is to be conducted (Figure 10.44). Check the boom elevation cylinder holding valves to ensure that there is no internal or external hydraulic fluid leakage. Visually inspect the cylinder link pins for signs of defect.
- Inspect the boom assembly itself for signs of wear, damage, or bends.
- Check platform level linkages for physical damage and weld defects.

Figure 10.43 Check the welds on the lower boom.

Figure 10.44 The aerial device must be elevated for the drift test.

- Inspect the following components for wear, defects, proper lubrication (if applicable), and proper operation:
  —Cables
  —Chains
  —Rods
  —Sprockets
  —Pulleys
  —Hooks
- Check the boom support for damage to the welds or main structure (Figure 10.45).
- Confirm the proper operation of the lower boom angle indicator light.
- Check any pneumatic or electric lines that are located in the boom section for proper wear, mounting, or signs of defects. Also include the knuckle area between the two boom sections in this check (Figure 10.46).

Articulating Aerial Device — Upper Boom Examination

The examination of the upper boom on an articulating aerial device should include the following:

- Inspect all bolts, nuts, or other fasteners to ensure that bolts of a proper grade and fit are used. Use a properly calibrated torque wrench to make sure that all bolts are tightened to manufacturer's specifications. Visually check bolts for cracks or flaws.
- Check all accessible weldments for defects (cracks, holes, etc.), and check all welds for obvious fractures.
- Inspect all hydraulic lines and hoses for kinks, cuts, cracks, and abrasions. Check for hydraulic fluid leakage at all connections and fittings. Pay close attention to lines that run through the knuckle between the lower and upper booms.
- Examine platform level linkages for physical damage and weld defects (Figure 10.47).
- Check the upper boom for proper alignment with the lower boom.
- Check boom boost cylinder brackets for damage or other defects. Examine the boom boost cylinders for signs of external leakage. Visually inspect the cylinder link pins for signs of defect.

Figure 10.45 While the aerial device is raised, check the boom support cradle for signs of damage.

Figure 10.46 Air and electric lines in the boom knuckle should be examined for signs of wear.

Figure 10.47 Check the linkages in the platform leveling system for signs of problems.

- Inspect the following components for wear, defects, proper lubrication (if applicable), and proper operation:
  —Cables
  —Chains
  —Rods
  —Sprockets
  —Pulleys
  —Hooks

- Confirm the presence and proper operation of safety stop devices.
- Examine upper boom elevation cylinder anchor ears and plates for weld fractures and defects. Check the upper boom elevation cylinders and cylinder rods for scoring, pitting, and other defects. Inspect the cylinder rod-to-barrel seat to make certain there is no external leakage. Check the holding valves on these cylinders for signs of external hydraulic fluid leakage.
- Check the upper boom hold-down devices for proper operation and structural integrity (Figure 10.48).
- Check any pneumatic or electric lines that are located in the boom section for proper wear, mounting, or signs of defects.
- Conduct a drift test to ensure that the amount of drift does not exceed the manufacturer's specifications. The manufacturer of the aerial device should provide information on how exactly the drift test is to be conducted. Check the boom elevation cylinder holding valves to ensure that there is no internal or external hydraulic fluid leakage. Visually inspect the cylinder link pins for signs of defect.

Figure 10.48 Make sure that the upper boom hold-down device works properly.

Telescoping Elevating Platforms — Boom Examination

The examination of the boom on a telescoping elevating platform should include the following:

- Check all bolts, nuts, or other fasteners to ensure that bolts of a proper grade and fit are used. Use a properly calibrated torque wrench to make sure that all bolts are tightened to manufacturer's specifications. Visually check bolts for cracks or flaws.
- Inspect all accessible weldments for defects (cracks, holes, etc.), and check all welds for obvious fractures (Figure 10.49).
- Inspect the boom assembly itself for signs of wear, damage, or bends.
- Check the following components for wear, defects, proper lubrication (if applicable), and proper operation:
  - —Cables
  - —Chains
  - —Rods
  - —Sprockets
  - —Pulleys
  - —Hooks
- Inspect the boom support for damage to the welds or main structure.

Figure 10.49 Inspect the welds on the telescoping boom.

- Examine thoroughly the elevation cylinder assemblies. Inspect the anchor ears and plates for defects and weld fractures (Figure 10.50). Check cylinder pins for proper alignment, installation, lubrication, operation, and retention. Check cylinder rods to make sure that they are not pitted, scored, or otherwise defective.
- Confirm the proper operation of the audible maximum extension warning device.
- Inspect platform leveling cylinders and their mountings for defects. Perform NDT on all welds and bolts.
- Check the ancillary boom ladder for any defects (Figure 10.51).
- Inspect all hydraulic lines and hoses for kinks, cuts, cracks, and abrasions. Check for hydraulic fluid leakage at all connections and fittings.
- Check all guides, babbitts, wear strips, pads, and slide blocks for wear, defects, and proper mounting or alignment.

Figure 10.50 Check the elevating cylinders for signs of wear or fluid leaks.

- Inspect extension sheaves and sheave mounting brackets for wear and defects. Check welds for fractures.
- Check extension cables for excessive wear. Inspect the cable separation guide to ensure free travel and proper alignment.
- Make sure that the elevation indicator is legible and plainly visible (Figure 10.52).

Figure 10.51 Inspect the condition of the ancillary ladder on the telescoping aerial platform.

Figure 10.52 There should be an accurate elevation indicator on the aerial device.

- Inspect extension cylinder anchor ears and plates for weld fractures and defects. Check the extension cylinders and cylinder rods for scoring, pitting, and other defects. Conduct a drift test by placing the aerial device at full elevation and 10 feet (3 m) of extension (Figure 10.53). After marking the piston or the second boom section, allow the aerial device to stand for one hour with the engine turned off. The amount of drift should not exceed the manufacturer's specifications. Also check the extension cylinder holding valves to make sure that there is no internal or external hydraulic fluid leakage.
- Check any pneumatic or electric lines that are located in the boom sections for proper wear, mounting, or signs of defects.

Operational Testing

Operational testing is performed to determine if the apparatus is still capable of operating within its full range of movement and in a reasonable amount of time. The test procedure for elevating platforms covers operation from both the lower (ground) and platform control stations. The following sections highlight the tests from each station.

Operational Tests from Lower Control Station

The elevating platform should be put through the following exercises from the lower control station:

Step 1: Start the apparatus engine, engage the power take-off, and set the stabilizers. Set the engine speed at the maximum speed permitted by the manufacturer.

Step 2: Operate the aerial device through its entire range of motion. This operation should include but not be limited to the following movements:

- Movement of the platform from the ground to maximum elevation.
- Revolve the aerial device 360° to the left and right with the platform at its maximum horizontal reach.

The boom shall operate without any unusual noises or vibrations.

Step 3: Release controls. They should automatically return to the neutral position. All safety devices must be in proper working condition.

Figure 10.53 The aerial device should be fully elevated and extended about 10 feet (3 m).

Step 4: Check leveling indicators for accuracy and legibility.

Step 5: Check proper alignment and operation on telescoping elevating platforms, rollers, sheaves, etc.

Step 6: Return the aerial device to its fully stowed position to prepare for the time portion of the testing procedure.

Step 7: Raise the elevating platform from its bed to full elevation and rotate it 90°. This must be accomplished in the amount of time specified by the manufacturer of the apparatus (Figure 10.54).

Step 8: Fully retract the aerial device, and complete the rotation through 360°.

Step 9: Stow the aerial device, and perform a visual inspection to determine if any problems have occurred.

Figure 10.54 The aerial device should be fully elevated and retracted and then rotated perpendicular to the chassis.

Operational Tests from Platform Control Station

The elevating platform should be put through the following exercises from the platform control station:

Step 1: Set the engine speed at its maximum allowable rpm according to the manufacturer.

Step 2: Raise the elevating platform to its maximum height. Place it in its position of greatest horizontal reach and rotate it 360° in both directions. No unusual noises, bouncing, or movement should occur during these tests.

Step 3: Release controls. They should automatically return to the neutral position. All safety devices must be in proper working condition.

Step 4: The platform control deactivation switch, located on the lower control station, should be shown to work properly (Figure 10.55). With that control in the "off" position, the platform controls should be deactivated.

Figure 10.55 Make sure that the platform activation switch operates properly.

Keep the platform level through all phases of the test. The mechanical override on the hydraulic leveling system (if present) should keep the platform level when the boom is being lowered without hydraulic power under emergency conditions.

Load Testing

The following load test is prescribed by NFPA 1914 for elevating platforms. The purpose of the test is to determine the platform's ability to handle its rated load.

Step 1: Conduct all load tests with the aerial apparatus on a firm, level surface or road. All stabilizers must be properly deployed and stabilizer pads in place.

Step 2: With the platform basket placed near the ground, carefully load the platform with sandbags or other suitable weights to the manufacturer's rated payload capacity.

Make sure to take into account any equipment that was placed on the platform after delivery from the factory. Make certain the load is properly secured.

Step 3: Operate the aerial device through its full range of motion from the lower control station. Do not exceed the manufacturer's operational limits. All boom movements must not exhibit any abnormal noise, vibration, or deflection. If the stabilizers show any signs of instability, stop testing and reposition the apparatus. Notify the manufacturer if the stabilizers cannot be made stable after repositioning. The platform basket must level properly as the booms are moved through all allowable positions.

Step 4: At the conclusion of the load test, inspect the weld joints at the following locations for signs of deterioration:

- Stabilizer to frame point
- On the stabilizers themselves
- All frame components
- Turntable
- Cylinder anchors
- Boom joints
- Leveling system
- Platform basket
- Pivot pin bosses

Water System Testing

The procedure for testing the water system on elevating platform apparatus is the same as previously described for aerial ladders with permanently piped waterways.

Service Testing Water Tower Apparatus

The procedures for service testing water tower apparatus are basically the same as those for testing articulating and telescoping elevating platform apparatus, depending on whether the water tower articulates or telescopes. The only particular tests that are different are operational tests and water system tests. These are covered in the following sections. It should be noted that there are no load tests required of water towers.

Operational Testing

The procedure for the operational testing of water tower apparatus is as follows:

Step 1: Conduct all tests with the aerial apparatus on a firm, level surface or road.

Step 2: Start the apparatus engine, engage the power take-off, and set the stabilizers.

Step 3: Raise the water tower from its bed to full elevation and rotate it 90° (Figure 10.56). On telescoping water towers, extend the aerial device fully. This must be accomplished within the time parameter established by the aerial device manufacturer.

Step 4: Retract the water tower, rotate the turntable 360°, then bed into the travel position, and follow with a thorough examination of all parts.

At some point in the procedure, every control on the control pedestal must be operated and shown to be in good working condition.

Figure 10.56 The water tower should be fully elevated and extended and rotated perpendicular to the chassis.

Water System Testing

The procedures for testing the water system on water tower apparatus are the same as those described for aerial ladders with permanently mounted waterways.

 Post-Testing Activities

Keep complete records of all phases of the testing process on each piece of apparatus. This will be an important part of the vehicle's service record file. If a third-party testing organization is certifying the aerial device, store the certification paperwork in the service record file.

If any problems are found during the testing, stop testing and correct the problem. All maintenance that is performed must have the approval of the apparatus manufacturer. Only after the proper corrections have been made should the testing continue, and then it should be restarted from the beginning.

Aerial Apparatus Manufacturers Contact Information

American LaFrance Corporation
11710 Statesville Boulevard
P.O. Box 39
Cleveland, NC 27013
(704) 278-6200
800-932-3523
FAX: (704) 278-6221

Crash Rescue Equipment Service, Inc.
SNOZZLE,
11122 Morrison Lane
P.O. Box 29044
Dallas, TX 75229
(972) 243-3307
FAX: (972) 243-6504
www.crashrescue.com

Emergency One, Inc.
1601 S.W. 37th Avenue
P.O. Box 2710
Ocala, FL 34474
(352) 237-1122
FAX: (352) 237-4369

Ferrara Fire Apparatus, Inc.
27855 James Chapel Road North
P.O. Box 249
Holden, LA 70744
(504) 567-7100
800-443-9006
FAX: (504) 567-5260

KME Fire Apparatus
World Headquarters
1 Industrial Complex
Nesquehoning, PA 18240
(717) 669-5132
800-235-3928
FAX: (717) 645-7007

Ladder Towers, Inc.
A subsidiary of American LaFrance
64 Cocalico Creek Road
Ephrata, PA 17522-9403
(717) 859-1176
FAX: (717) 859-2774
www.laddertowers.com

Pierce Manufacturing, Inc.
2600 American Drive
P.O. Box 2017
Appleton, WI 54913-2017
(414) 832-3231

Schwing America, Inc.
Fire Apparatus Division
5900 Centerville Road
St. Paul, MN 55127
(612) 429-0999

Seagrave Fire Apparatus
105 East 12th Street
Clintonville, WI 54929
(715) 823-2141
FAX: (715) 823-5768

Smeal Fire Apparatus Company
Highway 91 West
P.O. Box 8
Snyder, NE 68664
(402) 568-2224
FAX: (402) 568-2346

Snorkel
Lake Avenue & Arizona Street
P.O. Box 1160
St. Joseph, MO 64502
(816) 238-4503
800-338-0889
FAX: (816) 238-8744

Sutphen Corporation
7000 Col-Mrsvil Road
P.O. Box 0158
Amlin, OH 43002-0158
(614) 889-1005
800-848-5860
FAX: (614) 889-0874

Providence Fire Department

Daily Apparatus / Equipment Inspection

Company- ______________ Group- ______________

Date - __________ Chauffeur- __________________

1st Day Check	O.K.	Needs Service
S.C.B.A. - Tank Level, Operation, Harness, P.A.S.S.		
ENGINE - Oil, Coolant, Trans. Power Steering, Brake Fluids / Leaks. Belts / Hoses, Fuel Level Batteries Condition		
TIRES - Tread, Condition Check for Flats (w/ hammer)		
LIGHTS - Headlights, Brake/Tail Signal, Warning Backup/Alarm		
CAB - Clean Interior, Books, Keys Note Paper, Clean Windows Wipers, Seats/Belts		
RADIO/SIREN/HORN - Operation		
E.M.S. EQUIPMENT - Oxygen, Latex Gloves, Infectious Control, BLS Supplies		
OTHER - Handlights, Igloo Cooler		
WASH - Apparatus / Equipment as needed		
PUMPERS - Water Level, Primer, Relief Valve, Transfer Valve, Clean Relief Valve Screen, Hydrant Fittings/Wrenches		
AERIALS / TOWERS - Hydraulic Oil, Level & Operation, PTO, Jacks/Outriggers/Interlocks, Pedestal, Bucket		
RESCUE - Main Oxygen, Stair Chair, Stretcher, Back Boards, Lifepak, ALS Equipment		
OTHER -		

Date - __________ Chauffeur- __________________

2nd Day Check	O.K.	Needs Service
S.C.B.A. - Tank Level, Operation, Harness, P.A.S.S.		
HAND TOOLS - Clean And Inspect All Hand Tools		
GENERATORS & INVERTERS - Oil level, Fuel Level, Condition Circuit Breakers Check & Operate Electrical Equipment (Fans, Etc.)		
NOZZLES - Appliances and Fittings Clean and Operate		
EXTINGUISHERS - Pressure, Condition		
PORTABLE PUMPS - Fuel Oil Level		
LIGHTING - Scene Lights, Cords, Reels, Outlet Boxes, Adapters		
HOSE - Preconnects, Booster Hose & Reel Operations, Supply Lines, High Rise Paks, Fittings		
ROPES - Utility, Lifelines, Bags, and Other Equipment		
CHOCKS - Wheel Chocks, Blocks Planks, Shoring, Etc.		
PUMPERS - Water Level, Foam Level & Equipment, Operate Gates & Drains		
AERIALS / TOWERS - Inspect Waterways, Ladder Pipes, Salvage & Overhaul Equip. Lifebelts & Harnesses, Saw - Oil & Fuel Levels, Operate (Per Manuf. Recom.) Spare Fuel, Oil & Blades		
RESCUE - Splints, Hare Traction, KED, MAST, Linen, Extrication & Hand Tools Restraints, Supplies		
OTHER -		

Courtesy of Providence (RI) Fire Department

Providence Fire Department – Apparatus Defect Report

Date- ______ Co.- ______ Officer- ______________ Chauffeur- ______________

MAJOR DEFECTS ONLY
Immediate Repair or Change Over Indicated

Apparatus Satisfactory Except as Noted Below Initial-		Defective (Driver)	Repaired (Mechanic)
Brakes–	Soft/Low Pedal		
	Pulls/Grabs		
	Excess Noise/Heat		
	Emergency/Parking Brake		
	Other–		
Electrical–	Total Failure		
	Headlights/Brake Lights		
	All Warning Lights		
	Radio/Siren		
	Charging System		
	Other–		
Engine–	No Power		
	Over Heats		
	Cuts out/Stalls		
	Throttle Controls		
	Rough/Noisy		
	Hard/Doesn't Start		
	Belts/Hoses		
	Other–		
Transm./ Drivetrain–	Does Not Shift		
	Slips		
	Noise/Vibration		
	Other–		
Air System–	No Pressure		
	Slow Pressure Build-up		
	Rapid Loss of Air Pressure		
	Other–		
Suspension–	Spring/Spring Hanger		
	Vehicle Leans (Excessively)		
	Loss/Lack of Control		
	Other–		

Reported To (name) – ______________
Time – ______ hrs.
Corrective Action Taken On All Items
(Mechanics Signature) – ______________
Corrective Action Taken – Except On Items Listed Below
(Mechanics Signature) – ______________
Items Not Repaired – ______________

Apparatus Satisfactory Except as Noted Below Initial-		Defective (Driver)	Repaired (Mechanic)
Tires–	Condition		
	Tread		
	Other–		
	Which Tire? (Circle)	Right / Left	Front / Rear — Inside / Outside
Steering–	Hard		
	Pulls		
	Shimmies/Vibrates		
	Loss of/No Control		
	Other–		
Cab Body–	Doors Don't Close/Latch		
	Windshield Wipers		
	Seats/Seat Belts		
	Major Body Damage		
	Other–		
Exhaust–	Filter System		
	Excessive Smoke		
	Fumes		
	Other–		
Aerials– Towers	PTO Won't Engage		
	Jacks/Outriggers		
	Controls		
	Hydraulics		
	Cable/Guides		
	Tiller/Trailer Chassis		
	Other–		
Pump–	Won't Engage		
	Relief Valve		
	No Pressure		
	Other–		

Remarks/Explanations–

Courtesy of Providence (RI) Fire Department

Weekly Emergency Vehicle Report

Page No. ________

Name of Company ____________________

Address ____________________

Emergency Vehicle MFG. ____________________

Year ____________ Serial No. ____________ Type ____________

Required Tire Pressure: ____________

Date Inspection Completed	Inspector	Battery Check	Braking System	Electrical System, Lights & Sirens	Tires & Wheel Lugs	Fuel Level	Oil Level Eng & Hyd	Hydraulic System	Pump Check	Cooling System	Lubrication Pump & Ladder	Engine Check	Booster Tank Level	Doors - Compartment & Cab	Portable Equipment	Special Remarks On Road Test Inspection Use Other Side

Courtesy of Volunteer Fireman's Insurance Services.

REMARKS: (Please itemize procedure taken on unsatisfactory inspection items noted on opposite side)

Inspection Date:	Repair Date:	Comments:	Repairs Completed -	
			By:	Date:

Printed in U.S.A.

Item No. C10:007

Courtesy of Volunteer Fireman's Insurance Services.

VFIS

...a subsidiary of the Glatfelter Insurance Group

Emergency Vehicle Maintenance Record Card

Vehicle Description ______________ Manufacturer's Serial No. ______________

Model Year ______________ Plate No. ______________

Tire Record

Make	Warranty (Life)	Date Installed	Odometer

Battery Record

Motor Oil & Oil Filter Record

Date	Months Or Miles	Quarts Of Oil	Filter	Remarks

Lubrication Record

Date	Remarks	Date	Remarks

Printed in U.S.A. C10:005

Courtesy of Volunteer Fireman's Insurance Services.Courtesy of Volunteer Fireman's Insurance

Maintenance And Repair Record

Date	Nature Of Repairs & / Or Maintenance Service	Repaired By	Cost

Courtesy of Volunteer Fireman's Insurance Services.

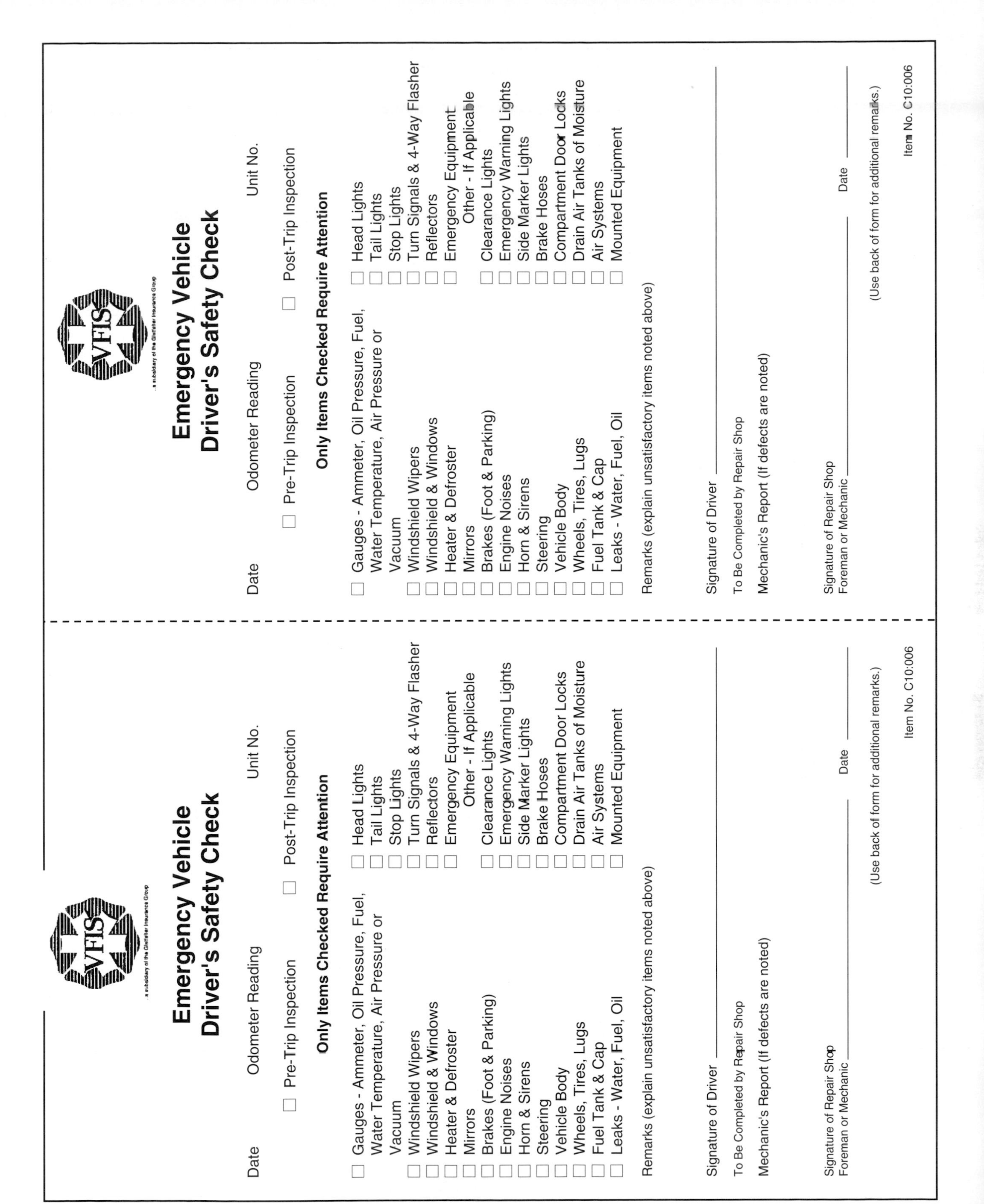

VFIS
...a subsidiary of the Glatfelter Insurance Group

Emergency Vehicle
Driver's Safety Check

Date Odometer Reading Unit No.

☐ Pre-Trip Inspection ☐ Post-Trip Inspection

Only Items Checked Require Attention

- ☐ Gauges - Ammeter, Oil Pressure, Fuel, Water Temperature, Air Pressure or Vacuum
- ☐ Windshield Wipers
- ☐ Windshield & Windows
- ☐ Heater & Defroster
- ☐ Mirrors
- ☐ Brakes (Foot & Parking)
- ☐ Engine Noises
- ☐ Horn & Sirens
- ☐ Steering
- ☐ Vehicle Body
- ☐ Wheels, Tires, Lugs
- ☐ Fuel Tank & Cap
- ☐ Leaks - Water, Fuel, Oil
- ☐ Head Lights
- ☐ Tail Lights
- ☐ Stop Lights
- ☐ Turn Signals & 4-Way Flasher
- ☐ Reflectors
- ☐ Emergency Equipment Other - If Applicable
- ☐ Clearance Lights
- ☐ Emergency Warning Lights
- ☐ Side Marker Lights
- ☐ Brake Hoses
- ☐ Compartment Door Locks
- ☐ Drain Air Tanks of Moisture
- ☐ Air Systems
- ☐ Mounted Equipment

Remarks (explain unsatisfactory items noted above)

Signature of Driver ________

To Be Completed by Repair Shop

Mechanic's Report (If defects are noted)

Signature of Repair Shop Foreman or Mechanic ________ Date ________

(Use back of form for additional remarks.)

Item No. C10:006

VFIS
...a subsidiary of the Glatfelter Insurance Group

Emergency Vehicle
Driver's Safety Check

Date Odometer Reading Unit No.

☐ Pre-Trip Inspection ☐ Post-Trip Inspection

Only Items Checked Require Attention

- ☐ Gauges - Ammeter, Oil Pressure, Fuel, Water Temperature, Air Pressure or Vacuum
- ☐ Windshield Wipers
- ☐ Windshield & Windows
- ☐ Heater & Defroster
- ☐ Mirrors
- ☐ Brakes (Foot & Parking)
- ☐ Engine Noises
- ☐ Horn & Sirens
- ☐ Steering
- ☐ Vehicle Body
- ☐ Wheels, Tires, Lugs
- ☐ Fuel Tank & Cap
- ☐ Leaks - Water, Fuel, Oil
- ☐ Head Lights
- ☐ Tail Lights
- ☐ Stop Lights
- ☐ Turn Signals & 4-Way Flasher
- ☐ Reflectors
- ☐ Emergency Equipment Other - If Applicable
- ☐ Clearance Lights
- ☐ Emergency Warning Lights
- ☐ Side Marker Lights
- ☐ Brake Hoses
- ☐ Compartment Door Locks
- ☐ Drain Air Tanks of Moisture
- ☐ Air Systems
- ☐ Mounted Equipment

Remarks (explain unsatisfactory items noted above)

Signature of Driver ________

To Be Completed by Repair Shop

Mechanic's Report (If defects are noted)

Signature of Repair Shop Foreman or Mechanic ________ Date ________

(Use back of form for additional remarks.)

Item No. C10:006

Courtesy of Volunteer Fireman's Insurance Services.

FORM NO. VBFD 76 REV. 7/91

DAILY APPARATUS AND EQUIPMENT CHECK

UNIT #   I.D.#   TYPE: MONTH:

DATE

EQUIPMENT	1	2	3	4	5	6	7	8	9	10	11	12	13	14	15	16	17	18	19	20	21	22	23	24	25	26	27	28	29	30	31
A. VEHICLE																															
1. Engine Crankcase Oil																															
2. Radiator																															
3. Hoses, Radiator Heater																															
4. Fan Belts																															
5. Primer Oil																															
6. Chassis Grease Reservoir																															
7. Seat Belts																															
8. Steering																															
9. Mirrors																															
10. Brakes																															
11. Lights and Warning Signals																															
12. Dashboard Gauges																															
13. Fuel																															
14. Windshield Wipers and Washers																															
15. Windshield Defroster																															
16. Radios																															
17. Sirens																															
18. Horns Air Electric																															
19. Tires and Wheels																															
20. Pump Panel																															
21. Water Tank																															
22. Hose Reel																															
23. Portable Generator																															
24. All Vehicles																															
B. INVENTORY																															
1. Paperwork and Binoculars																															
2. S.C.B.A./S.A.B.A.																															
3. Spare S.C.B.A./S.A.B.A.																															
4. P.A.S.S. Devices																															
5. Oxygen Bottles																															
6. Portable Fire Extinguishers																															
7. High Rise Pack																															
8. Ropes																															
9. Forcible Entry Tools																															
10. Hose Tools																															
11. Hose Appliances and Nozzles																															
12. Foam and Application Equipment																															
13. Traffic Flares and Cones																															
14. Salvage Equipment																															
15. Portable Lights																															
16. Extra Gasoline																															
17. Electric Cords, Adapters, & Fans																															
18. Hose Loads																															
19. Power Saws																															
20. Basic Life Support Equipment																															
21. Ladders																															
22. Life Jackets																															
23. Haz-Maz Bag																															
24. Tool Box																															
25. Hearing Protection																															
C. SPECIAL DEVICES & EQUIP																															
1. Aerial Devices																															
2. Brush Trucks																															
3. Special Operation Vehicles																															
INSPECTED BY:																															

(✔) Accepted   (X) Discrepency   (O) Unable To Check   (M) Missing

Courtesy of Virginia Beach (VA) Fire Department.

DATE	PROBLEM	ACTION TAKEN	DATE CORRECTED

Courtesy of Virginia Beach (VA) Fire Department.

FORM NO. VBFD 28 REV. 7/91

WEEKLY APPARATUS AND EQUIPMENT CHECK

UNIT #	TYPE:															PERIOD ENDING:
EQUIPMENT	JAN APR					FEB MAY					MAR JUN					REMARKS
	JUL OCT					AUG NOV					SEP DEC					
	1	2	3	4	5	1	2	3	4	5	1	2	3	4	5	
1. Parking Brakes																
2. Air Brakes																
3. Air Tank																
4. Radiator																
5. Power Steering																
6. Portable Generator																
7. Power Saws																
8. Batteries																
9. Tires																
10. Springs																
11. Drive Train																
12. Exhaust System																
14. Seagrave Governor																
15. Slack Adjusters RF"																
Slack Adjusters LF"																
Slack Adjusters RR"																
Slack Adjusters LR"																
Slack Adjusters RRR"																
Slack Adjusters LRR"																
16. Road Test																
17. Pump Operation																
Engage Pump																
Operate Valves (if necessary)																
Primer																
Governor/Relief Valve																
Auxiliary Cooler																
Change Over Valve																
Pump Panel																
18. Refill Tank & Primer																
19. Clean Apparatus																
20. Ladder Truck/Telesquirt																

(✔) Accepted (X) Discrepancy (O) Unable To Check (M) Missing

Courtesy of Virginia Beach (VA) Fire Department.

DATE	PROBLEM	ACTION TAKEN	DATE CORRECTED

Courtesy of Virginia Beach (VA) Fire Department.

FORM NO. VBFD 92

MONTHLY APPARATUS AND EQUIPMENT CHECK

UNIT 3	TYPE						YEAR					
	JAN	FEB	MAR	APR	MAY	JUNE	JULY	AUG	SEPT	OCT	NOV	DEC
1. Doors and Windows												
2. Compartment & Fenders												
3. Emergency Stop												
4. Air Filter												
5. Ladder Maintenance												
6. Clean Equipment												
7. Hard/Soft Suction												
8. Pump Transmission Oil												
9. Hydrostatic Check												
10. Aerial Devices												
11. Seagrave Ladder												
12. Portable Water Pumps												
13. Pump Bearings												
14. Seagrave/Waterous Pump												
15. Change Over Valve												
16. Engine Transmission												
17. Primer Motor (Elec.)												
18. Hose Reel												
19 Steering Box												
20. Rear End												
21. Elec. Pump Shift Oil												
22. Back Flush Pump												
23. Rope Check												
24. Rope Card Filled Out												
(✔) Accepted	(X) Discrepancy			(O) Unable To Check				(M) Missing				

Courtesy of Virginia Beach (VA) Fire Department.

DATE	PROBLEM	ACTION TAKEN	DATE CORRECTED

Courtesy of Virginia Beach (VA) Fire Department.

FORM NO. VBFD 77, Rev. 9/89

SUBSEQUENT APPARATUS AND EQUIPMENT CHECK

UNIT #	TYPE						YEAR					
	JAN	FEB	MAR	APR	MAY	JUNE	JULY	AUG	SEPT	OCT	NOV	DEC
QUARTERLY:												
1. Lubricate Grease Fittings												

SEMI-ANNUAL:	MILEAGE/DATE OF COMPLETION	REMARKS
1. Engine Oil		
2. Engine Oil Filters		
3. Auxiliary Generator Oil		
4. Fuel Filters		
5. Ladder Trucks		
6. Air Box Drain Tube		
7. Clean and Polish		
8. Air Filters		
9. Auxiliary Generator		
10. Extra Gasoline		
11. Power Saws		
12. Brush Truck Pump Oil Change		

ANNUAL:	MILEAGE/DATE OF COMPLETION	REMARKS
1. Transmission Oil & Filter		
2. Ladder Trucks		
3. Miscellaneous		
4. Pump Transmission		

(✔) Accepted (X) Discrepancy (O) Unable To Check (M) Missing

Courtesy of Virginia Beach (VA) Fire Department.

DATE	PROBLEM	ACTION TAKEN	DATE CORRECTED

Courtesy of Virginia Beach (VA) Fire Department.

EAST HAVEN FIRE DEPARTMENT AERIAL 1

DATE _______ SHIFT _______ DRIVER _______

MOTOR

CRANK CASE OIL LEVEL _____
RADIATOR _____
BRAKE OPERATION/EFFECTIVENESS _____
PARKING BRAKE OPERATION _____
BLEED AIR TANKS _____
BATTERY FLUID LEVEL _____
STARTING BANK 1&2 _____
TIRE DAMAGE/WEAR/PRESSURE _____
RADIO OPERATION _____
RADIO/EXTERIOR OPERATION _____

LIGHTS

EMER. LIGHTING _____
HEAD LIGHTS _____
TAIL LIGHTS _____
BRAKE LIGHTS _____
BACK UP LIGHTS _____
BACK UP ALARM _____
SPOT LIGHTS _____
ICC LIGHTS _____
AIR HORN _____
COMP. LIGHTS _____

CAB

CAIRNS IRIS HELMET & CASE _____
1 INTERSPRIRO SCBA _____
2 PORTABLE RADIOS _____
2 KOHLER 100S _____
1 DOOR OPENER _____

GLOVE COMPARTMENT

1 CO DETECTOR _____
OPERATORS GUIDE _____
2 SPRINKLER TONGS _____
DOT GUIDE BOOK 1993 _____

RIGHT SIDE JUMP SEAT

HOSE CLAMP _____
2 WHEEL CHOCKS _____
1 SCBA _____
2 KOHLER 100S _____
4 PORTABLE FLOODLIGHTS _____

RIGHT SIDE PUMP PANEL

ROLL FIRE LINE TAPE _____
1 PR. SAFETY GLASSES _____
1 BOX GLOVES _____
ABC DRY CHEM. EXT. _____
PRESSURIZED WATER EXT. _____
2 STORZ SPANNERS _____
2 STD. SPANNERS _____
1 HYDRANT WRENCH _____

RIGHT SIDE ABOVE FRONT OUTRIGGER

1 EMS BAG _____
BURN KIT _____

RIGHT SIDE BEHIND OUTRIGGER

K-12 SAW _____
ECHO SAW _____
GAS CAN _____
4 45 MIN. BOTTLES _____
3 30 MIN. BOTTLES _____
CHAIN SAW _____

RIGHT SIDE UPPER FRONT MIDDLE

4 30 MIN. SCBA _____

RIGHT SIDE ABOVE REAR TIRES

3 RED SALVAGE _____
1 YELLOW RUNNER _____
1 BLUE _____
10 LIGHTWEIGHT BLUE _____

RIGHT SIDE FRONT OR REAR OUTRIGGER

2 45 MIN. BOTTLES _____

RIGHT SIDE REAR NEAR STEP

THREE-WAY SIAMESE _____
2 LIFE BELTS _____
2 SPANNERS/HYDRANT WRENCH _____
HYDRANT PUMP _____

LEFT SIDE JUMP SEAT

1 SCBA _____
2 WHEEL CHOCKS _____

Courtesy of East Haven (CT) Fire Department.

EAST HAVEN FIRE DEPARTMENT AERIAL 1

DATE _______ SHIFT _______ DRIVER _______

PUMP OPERATORS COMP.

RUBBER MALLET ____
1 25' 3" SUCTION HOSE ____
2 HYDRANT WRENCHES ____
WATER CO. HYDRANT WRENCH ____
2 SPANNERS ____
2 STORZ SPANNERS ____
GATE VALVE ____
3-1 1/2 DBL. MALE ____
3-1 1/2 DBL. FEMALE ____
3-2 1/2 DBL. FEMALE ____
3-2 1/2 DBL. MALE ____
1 STORZ 5" TO 3" ____
1 STORZ 5" TO 4" ____

LEFT SIDE ABOVE FRONT OUTRIGGER

4 HOUSE CURRENT CORDS ____

LEFT FRONT LOWER

1 SUPER VAC. FAN ____
1 PPV FAN ____
25' EXT. CORD ____

LEFT FRONT UPPER

5-3 WIRE CORD ____
1 JUNCTION BOX ____

RIGHT SIDE ABOVE REAR TIRES

3 RED SALVAGE ____
1 YELLOW RUNNER ____
1 BLUE ____
10 LIGHTWEIGHT BLUE ____

LEFT OVER REAR TIRES

HALLIGAN ____
BOLT CUTTER 36" ____
BOLT CUTTER 24" ____
HUX BAR ____
PICK AXE ____
2 FLAT HEAD AXES ____
ZIAMATIC FAN BRACKET ____
SLEDGE HAMMER ____
PRY BAR ____
CLOSET POLE ____

LEFT LOWER FRONT OR OUTRIGGER

25' LDH ____
SMALL ASH CAN ____

LEFT SIDE NEXT TO STEP

TOOL BOX ____
2 LIFE BELTS ____
2 HOSE ROLLERS ____

SECOND REAR COMP. ABOVE OUTRIGGER

2 ROPES (ONE IN EACH) ____

LADDER COMP.

28' LADDER ____
10' FOLDING ____
20' ROOF ____
6' PIKE POLE ____
2-35' LADDERS ____
12' FOLDING ____
16' ROOF ____
12' PIKE POLE ____
8' PIKE POLE ____
SHEETROCK PULLER ____

TOP OF VEHICLE

5 SHOVELS ____
14' EXT. LADDER ____
1 BOX CRIBBING ____
2 CHIMNEY CHAINS ____
1 BACK BOARD ____
4 SQUEEGEES ____
2-3 WIRE CORD/REEL ____
2 SCBA TANKS 2200 PSI ____
4 CONES ____
2 GAS CANS ____
BUNDLE FOR STRIPS ____
1 DOLLY ____

BUCKET

SM 100-1000 GPM NOZZLE ____
SMOOTH TIPS 1½; 1¾; 2" ____
PICK AXE ____
6' PIKE POLE ____
12' PIKE POLE ____
2 SPANNER WRENCHES ____
2 INTERSPIRO HOSELINES ____
SCBA REG. (SET AT 85-90 PSI) ____
1-25' EXT.CORD ____

Revised 3/17/99

Glossary

A

Accelerator — Device, usually in the form of a foot pedal, used to control the speed of a vehicle by regulating the fuel supply.

Acceptance Testing (Proof Test) — Preservice tests on fire apparatus or equipment performed at the factory or after delivery to assure the purchaser that the apparatus or equipment meets bid specifications.

Accessibility — Ability of fire apparatus to get close enough to a building to conduct emergency operations.

Accident — Unplanned, uncontrolled event that results from unsafe acts of people and/or unsafe occupational conditions, either of which can result in injury.

Aerial Apparatus — Fire fighting vehicle equipped with a hydraulically operated ladder or elevating platform for the purpose of placing personnel and/ or water streams in elevated positions.

Aerial Device — General term used to describe the hydraulically operated ladder or elevating platform attached to a specially designed fire apparatus.

Aerial Ladder — Power-operated (usually hydraulically) ladder mounted on a special truck chassis.

Aerial Ladder Platform — Power-operated (usually hydraulically) ladder with a passenger-carrying device attached to the end of the ladder.

Aerial Ladder Truss — Assembly of bracing bars or rods in triangular shapes to form a rigid framework for the aerial device.

A-Frame Stabilizer — Stabilizing device that extends at an angle down and away from the chassis of an aerial fire apparatus.

AFFF — Abbreviation for Aqueous Film Forming Foam.

Ammeter — Gauge that indicates both the amount of electrical current being drawn from and provided to the vehicle's battery.

Angle of Approach — Angle formed by level ground and a line from the point where the front tires of a vehicle touch the ground to the lowest projection at the front of the apparatus.

Angle of Departure — Angle formed by level ground and a line from the point where the rear tires of a vehicle touch the ground to the lowest projection at the rear of the apparatus.

Anti-Electrocution Platform — Slide-out platform mounted beneath the side running board or rear step of an apparatus equipped with an aerial device. This platform is designed to minimize the chance of the driver/operator being electrocuted should the aerial device come in contact with energized electrical wires or equipment.

Apparatus Bay (Apparatus Room) — Area of the fire station where apparatus are parked.

Apparatus Engine — Diesel or gasoline engine that powers the apparatus drive chain and associated fire equipment. Also called Power Plant.

Aqueous Film Forming Foam (AFFF) — Synthetic foam concentrate that, when combined with water, is a highly effective extinguishing and blanketing agent on hydrocarbon fuels.

ARFF — Acronym for Aircraft Rescue and Fire Fighting.

Articulating Aerial Platform — Aerial device that consists of two or more booms that are attached with hinges and operate in a folding manner. A passenger-carrying platform is attached to the working end of the device.

Articulating Boom — Arm portion of the articulating aerial platform.

Auxiliary Hydraulic Pump — Electrically operated, positive displacement pump used to supply hydraulic oil through the hydraulic system of an aerial device in the event that the main hydraulic pump fails.

B

Bangor Ladder — *See* Pole Ladder.

Base Section — *See* Bed Section.

Basket Stabilizer — Device used to support the platform (basket) portion of an elevating platform device in the stowed position during road travel.

Bed Ladder — Lowest section of a multisection ladder.

Bed Ladder Pipe — Nontelescoping section of pipe, usually 3 or 3½ inches (77 mm or 90 mm) in diameter, attached to the underside of the bed section of the aerial ladder for the purpose of deploying an elevated master stream.

Bed Section — Bottom section of an extension ladder. Also called Base Section.

Booster Tank — *See* Water Tank.

Bourdon Tube — Part of a pressure gauge that has a curved, flat tube that changes its curvature as pressure changes. This movement is then transferred mechanically to a pointer on the dial.

Box Stabilizer — Two-piece aerial apparatus stabilization device consisting of an arm that extends directly out from the vehicle and a lifting jack that extends from the end of the extension arm to the ground. Also called H-Jack.

Brake Limiting Valve — Valve that allows the vehicle's brakes to be adjusted for the current road conditions.

Braking Distance — Distance the vehicle travels from the time the brakes are applied until it comes to a complete stop.

Bumper — Structure designed to provide front or rear end protection of a vehicle.

Bumper Line — Preconnected hoseline located on the apparatus bumper.

C

Cable Hanger — Device used to test the structural strength of aerial ladders.

Cantilever Operation — *See* Unsupported Tip.

Capacity — Maximum ability of a pump or water distribution system to deliver water.

Cascade Air Cylinders — Large air cylinders that are used to refill smaller SCBA cylinders.

Certification Tests — Preservice tests for the aerial device, ladder, pump, and other equipment which is conducted by an independent testing laboratory prior to delivery of an apparatus. These tests ensure that the apparatus or equipment will perform as expected after being placed into service.

Chamois — Soft pliant leather used for drying furniture and contents or for removing small amounts of water.

Chassis — Frame upon which the body of the fire apparatus rests.

Chauffeur — *See* Fire Apparatus Driver/Operator.

Chocks — Wooden, plastic, or metal blocks constructed to fit the curvature of a tire; placed against the tire to prevent the apparatus from rolling. Also called Wheel Blocks.

Class A Foam — Foam specially designed for use on Class A combustibles. Class A foams are essentially wetting agents that reduce the surface tension of water and allow it to soak into combustible materials easier than plain water.

Commercial Chassis — Truck chassis produced by a commercial truck manufacturer. The chassis is in turn outfitted with a rescue or fire fighting body.

Company — Basic fire fighting organizational unit consisting of firefighters and apparatus; headed by a company officer.

Company Officer — Individual responsible for command of a company. This designation is not specific to any particular fire department rank (may be a firefighter, lieutenant, captain, or chief officer if responsible for command of a single company).

Control Pedestal — Central location for most or all of the aerial device controls. Depending on the type and manufacturer of the apparatus, the control pedestal may be located on the turntable, on the rear or side of the apparatus, or in the elevating platform. Also called Pedestal.

D

Deck Gun — *See* Turret Pipe.

Deck Pipe — *See* Turret Pipe.

Defensive Attack — Exterior fire attack with emphasis on exposure protection.

Defensive Mode — Commitment of a fire department's resources to protect exposures when the fire has progressed to a point where an offensive attack is not effective.

Dike — Temporary or permanent barriers that prevent liquids from flowing into certain areas or that direct the flow as desired.

Diverter Valve — *See* Selector Valve.

Dogs — *See* Pawls.

Double-Acting Hydraulic Cylinder — Hydraulic cylinder capable of transmitting force in only two directions.

Driver/Operator — *See* Fire Apparatus Driver/Operator.

Driver Reaction Distance — Distance a vehicle travels while a driver is transferring the foot from the accelerator to the brake pedal after perceiving the need for stopping.

E

Elevated Master Stream — Fire stream in excess of 350 gpm (1 400 L/min) that is deployed from the tip of an aerial device.

Elevating Master Stream Device — *See* Water Tower.

Elevating Platform — Work platform attached to the end of an articulating or telescoping aerial device.

Elevating Water Device — Articulating or telescoping aerial device added to a fire department pumper to enable the unit to deploy elevated master stream devices. These devices range from 30 to 75 feet (9 m to 23 m) in height.

Elevation Cylinder — Hydraulic cylinder used to lift the aerial device from its bed to a working position. Also called Hoisting Cylinder.

Elevation Loss — *See* Elevation Pressure.

Elevation Pressure — Gain or loss of pressure in a hoseline due to a change in elevation. Also called Elevation Loss.

Engine — Fire department pumper.

Engineer — *See* Fire Apparatus Driver/Operator.

Extend — To increase the reach of an extension ladder or aerial device by raising the fly section.

Extension Cylinders — Hydraulic cylinders that control the extension and retraction of the fly sections of an aerial device.

Extension Fly Locks — Devices that prevent the fly sections of a ground or aerial ladder from retracting unexpectedly.

External Water Supply — (1) Any water supply to a fire pump from a source other than the vehicle's own water tank. (2) Any water supply to an aerial device from a source other than the vehicle's own fire pump.

F

FDC — Abbreviation for Fire Department Connection.

Fender — Exterior body portion of a vehicle adjacent to the front or rear wheels.

Finished Foam — Completed product after the foam solution reaches the nozzle and air is introduced into the solution (aeration). Also called Foam.

Fire Apparatus — Any fire department emergency vehicle used in fire suppression or other emergency situation.

Fire Apparatus Driver/Operator — Firefighter charged with the responsibility of operating fire apparatus to, during, and from the scene of a fire operation or any other time the apparatus is in use. The driver/operator is also responsible for routine maintenance of the apparatus and any equipment carried on the apparatus. This is typically the first step in the fire department promotional chain. Also called Chauffeur or Engineer.

Fire Department Connection (FDC) — Point at which the fire department can connect into a sprinkler or standpipe system to boost the water flow in the system. This connection consists of a clappered siamese with two or more 2½-inch (65 mm) intakes or one large-diameter (4-inch [100 mm] or larger) intake. Also called Fire Department Sprinkler Connection.

Fire Department Pumper — Piece of fire apparatus having a permanently mounted fire pump with a rated discharge capacity of 750 gpm (3 000 L/min) or greater. This apparatus may also carry water, hose, and other portable equipment.

Fire Department Sprinkler Connection — *See* Fire Department Connection.

Fire Pump — Water pump on a piece of fire apparatus.

Fire Stream — Stream of water or other water-based extinguishing agent after it leaves the fire hose and nozzle until it reaches the desired point.

Flowmeter — Mechanical device installed in a discharge line that senses the amount of water flowing and provides a readout in units of gallons per minute (liters per minute).

Foam — Extinguishing agent formed by mixing a foam concentrate with water and aerating the solution for expansion; for use on Class A and Class B fires. Foam may be protein, synthetic, aqueous film forming, high expansion, or alcohol type. Also called Finished Foam.

Foam Blanket — Covering of foam applied over a burning surface to produce a smothering effect; can be used on nonburning surfaces to prevent ignition.

Foam Concentrate — Raw chemical compound solution that is mixed with water and air to produce foam.

Foam Proportioner — Device that injects the correct amount of foam concentrate into the water stream to make the foam solution.

Foam Solution — Mixture of foam concentrate and water after it leaves the proportioner but before it is discharged from the nozzle and air is added to it.

Fog Stream — Water stream of finely divided particles used for fire control.

Friction Loss — Loss of pressure created by the turbulence of water moving against the interior walls of the hose or pipe.

Front Bumper Well — Hose or tool compartment built into the front bumper of a fire apparatus.

G

Gallon — Unit of liquid measure. One U.S. gallon (3.785 L) has the volume of 231 cubic inches (3 785 cubic centimeters). One imperial gallon equals 1.201 U.S. gallons (4.546 L).

Gallons Per Minute (gpm) — Unit of volume measurement used in the U.S. fire service for water movement.

Gaskets — Rubber seals used in fire hose couplings and pump intakes to prevent the leakage of water at connections.

Gauge — Instrument used to show the operating conditions of an appliance or piece of equipment.

Generator — Auxiliary electrical power generating device. Portable generators are powered by small gasoline or diesel engines and generally have 110- and/or 220-volt capacities.

Governor — Built-in pressure-regulating device to control pump discharge pressure by limiting engine rpm.

GPM — Abbreviation for gallons per minute.

Gradability — Ability of a piece of apparatus to traverse various terrain configurations.

Grade — Natural, unaltered ground level.

Ground Ladder — Ladders specifically designed for fire service use that are not mechanically or physically attached permanently to fire apparatus and do not require mechanical power from the apparatus for ladder use and operation.

Guy Ropes — Ropes attached between the tip of a raised aerial device and an object on the ground to stabilize the device during high wind conditions; should be used only if approved by the manufacturer of the aerial device.

H

Handline — Small hoselines (2½-inch [65 mm] or less) that can be handled and maneuvered without mechanical assistance.

Hazardous Material — Any material that possesses an unreasonable risk to the health and safety of persons and/or the environment if it is not properly controlled during handling, storage, manufacture, processing, packaging, use, disposal, or transportation.

Head — Water pressure due to elevation. For every 1-foot increase in elevation, 0.434 psi is gained (for every 1-meter increase in elevation, 10 kPa is gained). Also called Head Pressure.

Head Pressure — *See* Head.

Heavy Stream — *See* Master Stream.

H-Jack — *See* Box Stabilizer.

Hoisting Cylinder — *See* Elevation Cylinder.

Hold-Down Locks — Locks that secure the aerial device in its cradle during road travel.

Hose Bed — Main hose-carrying area of a pumper or other piece of apparatus designed for carrying hose. Also called Hose Body.

Hose Body — *See* Hose Bed.

Hose Clamp — Mechanical or hydraulic device used to compress fire hose to stop the flow of water.

I

Intake — Inlet for water into the fire pump.

Intake Pressure — Pressure coming into the fire pump.

Inverter — Auxiliary electrical power generating device. The inverter is a step-up transformer that converts the vehicle's 12- or 24-volt DC current into 110- or 220-volt AC current.

J

Jack Pad — *See* Stabilizer Pad.

Jack Plate — *See* Stabilizer Pad.

Jump Seat — Seats on a fire apparatus that are behind the front seats.

K

Kilopascal (kPa) — Metric unit of measure for pressure; 1 psi = 6.895 kPa, 1 kPa = 0.1450 psi.

Kink — Severe bend in a hoseline that increases friction loss and reduces the flow of water through the hose.

L

Ladder Company — Group of firefighters assigned to a fire department aerial apparatus who are primarily responsible for search and rescue, ventilation, salvage and overhaul, forcible entry, and other fireground support functions. Also called Truck Company.

Ladder Locks — *See* Pawls.

Ladder Pipe — Master stream nozzle mounted on the fly of an aerial ladder.

Large Diameter Hose (LDH) — Relay-supply hose of 3½ to 6 inches (90 mm to 150 mm)in diameter; used to move large volumes of water quickly with a minimum number of pumpers and personnel.

Level I Staging — Used on all multiple-company emergency responses. The first-arriving vehicles of each type proceed directly to the scene, and the others stand by a block or two from the scene and await orders.

Level II Staging — Used on large-scale incidents where greater alarm companies are responding. These companies are sent to a specified location to await assignment.

Life Safety Harness — Any harness that meets the requirements of NFPA 1983, *Standard on Fire Service Life Safety Rope, and System Components.*

Load Testing — Aerial device test intended to determine whether or not the device is capable of safely carrying its rated weight capacity.

Lugging — Condition that occurs when the throttle application is greater than necessary for a given set of conditions. It may result in an excessive amount of carbon particles issuing from the exhaust, oil dilution, and additional fuel consumption. Lugging can be eliminated by using a lower gear and proper shifting techniques.

M

Maintenance — Keeping equipment or apparatus in a state of usefulness or readiness.

Manifold — (1) Hose appliance that divides one larger hoseline into three or more small hoselines. Also called Portable Hydrant. (2) Hose appliance that combines three or more smaller hoselines into one large hoseline. (3) Top portion of the pump casing.

Manual Stabilizer — Manually deployed stabilizing device for aerial apparatus that consists of an extension arm with a jack attached to the end of it.

Manufacturer's Tests — Fire pump or aerial device tests performed by the manufacturer prior to delivery of the apparatus.

Mars Light — Single-beam, oscillating warning light.

Master Stream — Any of a variety of heavy, large-caliber water streams; usually supplied by siamesing two or more hoselines into a manifold device delivering 350 gpm (1 400 L/min) or more. Also called Heavy Stream.

Master Stream Nozzle — Nozzle capable of flowing in excess of 350 gpm (1 400 L/min).

Mattydale Hose Bed — *See* Transverse Hose Bed.

MDH — *See* Medium Diameter Hose.

Medium Diameter Hose (MDH) — 2½- or 3-inch (65 mm or 77 mm) hose that is used for both fire fighting attack and for relay-supply purposes.

N

National Fire Protection Association (NFPA) — Nonprofit educational and technical association located in Quincy, Massachusetts devoted to protecting life and property from fire by developing fire protection standards and educating the public.

Nonconforming Apparatus — Apparatus that does not conform to the standards set forth by NFPA standards.

Nondestructive Testing — Method of testing metal objects that does not subject them to stress-related damage.

Nozzle — Appliance on the discharge end of a hoseline that forms a fire stream of definite shape, volume, and direction.

Nozzle Pressure — Velocity pressure at which water is discharged from the nozzle.

Nozzle Reaction — Counterforce directed against a person holding a nozzle or a device holding a nozzle by the velocity of water being discharged.

O

Offensive Fire Attack —ggressive fire attack that is intended to stop the fire at its current location. Also called Offensive Mode Attack.

Offensive Mode Attack — *See* Offensive Fire Attack.

Outrigger — *See* Stabilizer.

Overthrottling — Process of injecting or supplying the diesel engine with more fuel than can be burned.

P

Parapet — (1) Extension of the exterior walls above the roof. (2) Any required fire walls surrounding or dividing a roof or surrounding roof openings such as light/ventilation shafts.

Pattern — Shape of the water stream as it is discharged from a fog nozzle.

Pawls — Devices attached to the inside of the beams on fly sections used to hold the fly section in place after it has been extended. Also called Ladder Locks.

PDP — Abbreviation for Pump Discharge Pressure.

Pedestal — *See* Control Pedestal.

Piercing Nozzle — Nozzle with an angled, case-hardened steel tip that can be driven through a wall, roof, or ceiling to extinguish hidden fire. Also called Puncture Nozzle.

Pitot Tube — Instrument containing a Bourdon tube that is inserted into a stream of water to measure the velocity pressure of the stream. The gauge reads in units of pounds per square inch (psi) or kilopascals (kPa).

Playpipe — Base part of a three-part nozzle that extends from the hose coupling to the shutoff.

Pole Ladder — Large extension ladder that requires tormentor poles to steady the ladder as it is raised and lowered. Also called Bangor Ladder.

Portable Equipment — Those items carried on the fire apparatus that are not permanently attached to or a part of the apparatus.

Portable Hydrant — *See* Manifold.

Portable Ladder Pipe — Portable, elevated master stream device clamped to the top two rungs of the aerial ladder when needed and supplied by a 3- or 3½-inch (77 mm or 90 mm) fire hose.

Positive Displacement Pumps — Self-priming pump that moves a given amount of water or hydraulic oil through the pump chamber with each stroke or rotation. These pumps are used for hydraulic pumps on aerial device hydraulic systems and for priming pumps on centrifugal fire pumps.

Pounds Per Square Inch (psi) — U.S. unit for measuring pressure. Its metric equivalent is kilopascals.

Power Plant — *See* Apparatus Engine.

Power Take-Off (PTO) — Rotating shaft that transfers power from the engine to auxiliary equipment.

Preconnect — (1) Attack hose connected to a discharge when the hose is loaded; this shortens the time it takes to deploy the hose for fire fighting. (2) Soft intake hose that is carried connected to the pump intake.

Prefire Inspection — *See* Pre-Incident Planning.

Prefire Planning — *See* Pre-Incident Planning.

Prefire Inspection — *See* Pre-Incident Planning.

Pre-Incident Planning—Act of preparing to handle an incident at a particular location or a particular type of incident before an incident occurs. Also called Prefire Planning, Preplanning, Prefire Inspection, or Pre-Incident Inspection.

Preservice Tests — Tests performed on fire pumps or aerial devices before they are placed into service. These tests are broken down into manufacturer's tests, certification tests, and acceptance tests.

Preplanning — *See* Pre-Incident Planning.

Pressure— Force per unit area measured in pounds per square inch (psi) or kilopascals (kPa).

Pressure Governor — Pressure control device that controls engine speed and therefore eliminates hazardous conditions that result from excessive pressures.

Proportioner — Device used to introduce the correct amount of foam concentrate into a stream of water.

PSI — Abbreviation for Pounds Per Square Inch.

PSIG — Abbreviation for Pounds Per Square Inch Gauge.

PTO — Abbreviation for Power Take-Off.

Pump Can—Water-filled pump-type extinguisher. Also called Pump Tank.

Pump Discharge Pressure (PDP) — Actual velocity pressure (measured in pounds per square inch) of the water as it leaves the pump and enters the hoseline.

Pumping Apparatus— Fire department apparatus that has the primary responsibility to pump water.

Pump Operator— Firefighter charged with operating the pump and determining the pressures required to operate it efficiently.

Pump Panel — Instrument and control panel located on the pumper.

Pump Tank — *See* Pump Can.

Puncture Nozzle — *See* Piercing Nozzle.

Q

Quad — Four-way combination fire apparatus; sometimes referred to as quadruple combination. A quad combines the water tank, pump, and hose of a pumper with the ground ladder complement of a truck company.

Quint— Fire apparatus equipped with a fire pump, water tank, ground ladders, and hose bed in addition to the aerial device.

R

Radiated Heat — *See* Radiation.

Radiation — Transfer of heat energy through light by electromagnetic waves. Also called Radiated Heat.

Relay— Use of two or more pumpers to move water distances that would require excessive pressures if only one pumper was employed.

Relay Operation — Using two or more pumpers to move water over a long distance by operating them in series. Water discharged from one pumper flows through hoses to the inlet of the next pumper, and so on. Also called Relay Pumping.

Relay Pumping — *See* Relay Operation.

Relief Valve— Pressure control device designed to eliminate hazardous conditions resulting from excessive pressures by allowing this pressure to bypass to the intake side of the pump.

Road Tests — Preservice apparatus maneuverability tests designed to determine the road worthiness of a new vehicle.

Roof Ladder — Straight ladder with folding hooks at the top end. The hooks anchor the ladder over the roof ridge.

Rotary Gear Positive Displacement Pump — Type of positive displacement pump commonly used in hydraulic systems. The pump imparts pressure on the hydraulic fluid by having two intermeshing rotary gears that force the supply of hydraulic oil into the pump casing chamber.

Rotary Vane Pump — Type of positive displacement pump used commonly in hydraulic systems. A rotor with attached vanes is mounted off-center inside the pump housing. Pressure is imparted on the water as the space between the rotor and the pump housing wall decreases.

Rung — Step portion of a ladder running from beam to beam.

S

Safety Bar — Hinged bar designed to protect firefighters from falling out of the open jump seat area of a fire apparatus.

Safety Gates — Protective guards that are placed over the apparatus jump seat opening to prevent firefighters from falling off the apparatus.

Selector Valve — Three-way valve on a fire department aerial apparatus that directs oil to either stabilizer control valves or the aerial device control valves. Also called Diverter Valve.

Service Records — Detailed description of maintenance and repair work for a particular apparatus or piece of equipment.

Service Test — Series of tests performed on apparatus and equipment in order to ensure operational readiness of the unit. These tests should be performed at least yearly or whenever a piece of apparatus or equipment has undergone extensive repair.

Siamese — Hose appliance used to combine two or more hoselines into one. The siamese generally has female inlets and a male outlet and is commonly used to supply the hose leading to a ladder pipe.

Single-Acting Hydraulic Cylinder — Hydraulic cylinder capable of transmitting force in only one direction.

Small Diameter Hose (SDH) — Hose of ¾ to 2 inches (20 mm to 50 mm) in diameter; used for fire fighting purposes.

Solid Stream — Hose stream that stays together as a solid mass as opposed to a fog or spray stream.

SOP — *See* Standard Operating Procedure.

Speedometer — Dashboard gauge that measures the speed at which the vehicle is traveling.

Spotter — Firefighter who walks behind a backing apparatus to provide guidance for the driver/operator.

Spotting — Positioning the apparatus in a location that provides the utmost efficiency for operating on the fireground.

Spray Curtain Nozzle — Fog nozzle mounted to the underside of an elevating platform to provide a protective shield against convected heat for firefighters operating in the platform.

Stabilizer — Devices that transfer the center of gravity of the apparatus and prevent it from tipping as the aerial device is extended away from the center line of the chassis. Also called Outrigger.

Stabilizer Boot — Flat metal plate attached to the bottom of the aerial apparatus stabilizer to provide firm footing on the stabilizing surface.

Stabilizer Pad — Unattached, flat metal plate that is larger in area than the stabilizer boot. The stabilizer pad is placed on the ground beneath the intended resting point of the stabilizer boot to provide better weight distribution. Also called Jack Pad or Jack Plate.

Staging — Process by which noncommitted units responding to a fire or other emergency incident are stopped at a location away from the fire scene to await their assignment.

Staging Area — (1) Location away from the emergency scene where units assemble and wait until they are assigned a position on the emergency scene. (2) Location on the emergency scene where tools and personnel are assembled before being used or assigned.

Standard Apparatus — Apparatus that conforms to the standards set forth by the National Fire Protection Association standards on fire apparatus design.

Standard Operating Procedure (SOP) — Standard method in which a fire department carries out routine functions. Usually these procedures are written, and all firefighters should be well versed in their content.

Standpipe System — Wet or dry system of pipes in a large single-story or multistory building with fire hose outlets connected to them. The system is used to provide for quick deployment of hoselines during fire fighting operations.

T

Tachometer — Dashboard or pump panel gauge that measures the engine speed in revolutions per minute (rpm).

Tailboard — Back step of fire apparatus.

Tandem — Two-axle rear suspension.

Telescoping Aerial Platform Apparatus — Type of aerial apparatus equipped with an elevating platform; also equipped with piping systems and nozzles for elevated master stream operations. These apparatus are not meant to be climbed and are equipped with a small ladder that is to be used only for escape from the platform in emergency situations.

Telescoping Boom — Aerial device raised and extended via sections that slide within each other.

Third-Party Testing Agency — Independent agency hired to perform nonbiased testing on a specific piece of apparatus.

Throttle Control — Device that controls the engine speed.

Tiller — Rear steering mechanism on a tractor-trailer aerial ladder truck.

Tillerman — *See* Tiller Operator.

Tiller Operator — Driver/operator of the trailer section of a tractor-tiller aerial ladder apparatus. Also called Tillerman.

Torque Box — Structural housing that contains the rotational system for the aerial device between the apparatus chassis frame rails and the turntable.

Torque Wrench — Specially designed wrench that may be set to produce a particular amount of torque on a bolt.

Total Stopping Distance — Sum of the driver/operator reaction distance and the vehicle braking distance.

Tower Ladder — Term used to describe a telescoping aerial platform fire apparatus.

Traction — Act of exerting a pulling force.

Tractor-Tiller Aerial Ladder — Aerial ladder apparatus that consists of a tractor power unit and trailer (tiller) section that contains the aerial ladder, ground ladders, and equipment storage areas. The trailer section is steered independently of the tractor by a person called the tiller operator.

Traffic Control Device — Mechanical device that automatically changes traffic signal lights to favor the path of responding emergency apparatus.

Transverse Hose Bed — Hose bed that lies across the pumper body at a right angle to the main hose bed; designed to deploy preconnected attack hose to the sides of the pumper. Also called Mattydale Hose Bed.

Trash Line — Small diameter, preconnected hoseline intended to be used for trash or other small, exterior fires.

Triple-Combination Pumper — Fire department pumper that carries a fire pump, hose, and a water tank.

Truck — (1) Self-propelled vehicle carrying its load on its wheels; primarily designed for transportation of property rather than passengers. (2) Slang term for an aerial apparatus. (3) Ladder truck.

Truck Company — *See* Ladder Company.

Truss Construction Ladder — Aerial device boom or ladder sections that are constructed of trussed metal pieces.

Turntable — Rotational structural component of the aerial device. Its primary function is to provide continuous rotation on a horizontal plane.

Turret Pipe — Large master stream appliance mounted on a pumper or trailer and connected directly to a pump. Also called Deck Gun or Deck Pipe.

U

UL — Abbreviation for Underwriters Laboratories, Inc.

Ultrasonic Inspection — Nondestructive method of aerial device testing in which ultrasonic vibrations are injected into the aerial device. Deviance in the return of the waves is an indication that flaws exist.

Undercarriage — Portion of a vehicle's frame that is located beneath the vehicle.

Underwriters Laboratories, Inc. (UL) — Independent fire research and testing laboratory.

Unsupported Tip — Operation of an aerial device with the tip of the device or the platform, if so equipped, in the air and not resting on another object. Also called Cantilever Operation.

V

Valve — Mechanical device with a passageway that controls the flow of a liquid or gas.

Voltmeter — Device used for measuring the voltage existing in an electrical system.

W

Warning Devices — Any audible or visual devices, such as flashing lights, sirens, horns, or bells, added to an emergency vehicle to gain the attention of drivers of other vehicles.

Warning Lights — Lights on the apparatus designed to attract the attention of other motorists.

Water Curtain — Fan-shaped stream of water discharged from beneath an elevating platform to absorb radiant heat and protect the occupants of the platform.

Water Hammer — Force created by the rapid deceleration of water. It generally results from closing a valve or nozzle too quickly.

Water Supply — Any source of water available for use in fire fighting operations.

Water Tank — Water storage receptacle carried directly on the apparatus. NFPA 1901 specifies that Class A pumpers must carry at least 500 gallons (2 000 L). Also called Booster Tank.

Water Tower — Aerial device primarily intended for deploying an elevated master stream. Not generally intended for climbing operations. Also known as an Elevating Master Stream Device.

Waterway — Path through which water flows within a hose or pipe.

Wheel Blocks — *See* Chocks.

Winch — Pulling tool that consists of a length of steel chain or cable wrapped around a motor-driven drum. These are most commonly attached to the front or rear of a vehicle.

Index

COMMENT SHEET

DATE ____________________ NAME ____

ADDRESS ____

ORGANIZATION REPRESENTED ____

CHAPTER TITLE ___________________________________ NUMBER __________

SECTION/PARAGRAPH/FIGURE ________________________ PAGE ____________

1. Proposal (include proposed wording or identification of wording to be deleted), OR PROPOSED FIGURE:

2. Statement of Problem and Substantiation for Proposal:

RETURN TO: IFSTA Editor
Fire Protection Publications
Oklahoma State University
930 N. Willis
Stillwater, OK 74078-8045

SIGNATURE ____________________________

Use this sheet to make any suggestions, recommendations, or comments. We need your input to make the manuals as up to date as possible. Your help is appreciated. Use additional pages if necessary.

COMMENT SHEET

DATE ____________________ NAME ____

ADDRESS ____

ORGANIZATION REPRESENTED ____

CHAPTER TITLE ______________________________________ NUMBER ______________

SECTION/PARAGRAPH/FIGURE ____________________________ PAGE ________________

1. Proposal (include proposed wording or identification of wording to be deleted), OR PROPOSED FIGURE:

2. Statement of Problem and Substantiation for Proposal:

RETURN TO: IFSTA Editor
Fire Protection Publications
Oklahoma State University
930 N. Willis
Stillwater, OK 74078-8045

SIGNATURE ______________________________

Use this sheet to make any suggestions, recommendations, or comments. We need your input to make the manuals as up to date as possible. Your help is appreciated. Use additional pages if necessary.

Your Training Connection.....

The International Fire Service Training Association

We have a free catalog describing hundreds of fire and emergency service training materials available from a convenient single source: the International Fire Service Training Association (IFSTA).

Choose from products including IFSTA manuals, IFSTA study guides, IFSTA curriculum packages, Fire Protection Publications manuals, books from other publishers, software, videos, and NFPA standards.

Contact us by phone, fax, U.S. mail, e-mail, internet web page, or personal visit.

Phone
1-800-654-4055

Fax
405-744-8204

U.S. mail
IFSTA, Fire Protection Publications
Oklahoma State University
930 North Willis
Stillwater, OK 74078-8045

E-mail
editors@ifstafpp.okstate.edu

Internet web page
www.ifsta.org

Personal visit
Call if you need directions!